Edited by
Juan P. Torres and Lluis Torner

Twisted Photons

Related Titles

Clarke, D.

Stellar Polarimetry

2010

ISBN: 978-3-527-40895-5

Römer, H.

Theoretical Optics

An Introduction

2009

ISBN: 978-3-527-40776-7

Wang, L. V., Wu, H.-i

Biomedical Optics

Principles and Imaging

2007

ISBN: 978-0-471-74304-0

Meschede, D.

Optics, Light and Lasers

The Practical Approach to Modern Aspects of Photonics and Laser Physics

2007

ISBN: 978-3-527-40628-9

Busch, K., Lölkes, S., Wehrspohn, R. B., Föll, H. (eds.)

Photonic Crystals

Advances in Design, Fabrication, and Characterization

2004

ISBN: 978-3-527-40432-2

Edited by Juan P. Torres and Lluis Torner

Twisted Photons

Applications of Light with Orbital Angular Momentum

WILEY-VCH

WILEY-VCH Verlag GmbH & Co. KGaA

The Editors

Prof. Juan P. Torres
Universitat Politecnica de Catalunya
ICFO-The Institute of Photonic Sciences
Mediterranean Technology Park
Av Canal Olimpic s/n
08860 Castelldefels (Barcelona)
Spain
Juan.Perez@icfo.es

Prof. Lluis Torner
Universitat Politecnica de Catalunya
ICFO-The Institute of Photonic Sciences
Mediterranean Technology Park
Av Canal Olimpic s/n
08860 Castelldefels (Barcelona)
Spain
Lluis.Torner@icfo.es

Library of Congress Card No.: applied for

British Library Cataloguing-in-Publication Data
A catalogue record for this book is available from the British Library.

Bibliographic information published by the Deutsche Nationalbibliothek
The Deutsche Nationalbibliothek lists this publication in the Deutsche Nationalbibliografie; detailed bibliographic data are available on the Internet at <http://dnb.d-nb.de>.

Typesetting Laserwords Private Ltd., Chennai, India
Printing and Binding Fabulous Printers Pte Ltd., Singapore
Cover Design Grafik-Design Schulz, Fußgönheim

Printed in Singapore
Printed on acid-free paper

ISBN: 978-3-527-40907-5

Contents

Twisted Photons: Applications of Light with Orbital Angular Momentum.
Edited by Juan P. Torres and Lluis Torner

ISBN: 978-3-527-40907-5

Preface

The book *Twisted Photons: Applications of the Orbital Angular Momentum of Light* that we are honored to edit contains 12 salient contributions that focus on new applications that use one of the properties that characterizes electromagnetic waves in general, and light beams, in particular: the topology of their spatial shape. This is an important degree of freedom that adds up to the toolkit constituted by the other properties that characterize a light beam, namely, polarization, energy, and spectrum, thus putting forward a powerful enabling tool with widespread applications in several areas of science and technology where its use allows the exploration of unchartered territories, both in the realm of the very small and delicate (e.g., single atoms, *in vivo* cells, and micromachines) and in the realm of the very big (e.g., astronomy).

The topic has been extensively studied during the last two decades and many of the corresponding techniques are well understood, and conceptually and experimentally mastered. The goal of this book is to present the topic to a broad audience, and to illustrate its potential by examining examples of its use in different areas of application.

What is the Orbital Angular Momentum of Light?

Light carries energy and both, linear and angular momenta. The total angular momentum can contain a spin contribution associated with polarization, and an orbital contribution associated with the spatial profile of the light intensity and phase. By and large, a beam of light with a single intensity peak and smooth wave front, that is, a Gaussian-like shape that propagates in free space, shows no azimuthal phase variations, and the propagation of the energy flow follows a straight path along the direction of propagation of the beam.

Light with orbital angular momentum exhibits drastic differences, as illustrated in the images that appear in the cover of this book (see also *L. Allen and M. J. Padgett, The orbital angular momentum of light: an introduction*). The picture shows the simplest kind of light beams that carry orbital angular momentum. The intensity of the light beam, as depicted in the two figures on the left, presents a central dark area (the beam axis) with no intensity. Such light beams exhibit a corkscrew-like spiraling of the phase around the beam axis with no energy (top and

Twisted Photons: Applications of Light with Orbital Angular Momentum.
Edited by Juan P. Torres and Lluis Torner

ISBN: 978-3-527-40907-5

right), that is, an optical vortex. This spiraling, which represents a fundamentally new extra degree of freedom that researchers are exploring for a variety of novel natural phenomena, can be made visible with the help of an auxiliary plane wave, that is made to interfere with the optical vortex at a small angle, resulting in an interference pattern whose transverse shape depends on the concrete spiraling of the phase (bottom and right).

A beam carrying a single optical vortex represents one of the simplest cases of light beams carrying orbital angular momentum. However, one may engineer the properties of optical vortex beams to form a variety of complex transverse patterns (see *M. Padgett, Helically Phased Beams, and analogies with Polarization*), a property that might be a powerful asset in certain applications.

On the other hand, in a general situation, the polarization and spatial degrees of freedom are coupled by Maxwell equations. However, in beams with sizes much larger than the wavelength, which thus propagate in the paraxial regime, both properties may be controlled separately. Notwithstanding, different applications make use of the combination of the spatial shape of the beam and its polarization (see *A. Bekshaev and M. Vasnetsov, Vortex flow of light: "spin" and "orbital" flows in a circularly polarized paraxial beam*).

What can be done with the Orbital Angular Momentum of Light?

We present a list of applications that, although in no way aims at being extensive, presents nonetheless an overview at what can be done with twisted light. For instance, the orbital angular momentum of light can be transferred to trapped suitable material particles causing them to rotate (see *M. Mazilu and K. Dholakia, Trapping and rotation of particles in light fields with embedded optical vortices*), a property with important applications in micromanipulation (see *P. Galaja, L. Kelemen, L. Oroszi, P. Ormos, Rotational optical micromanipulation with specific shapes built by photopolymerization*) and in the design and operation of micromachines (see also *V. L. Y. Loke, T. Asavei, S. Parkin, N. R. Heckenberg, H. Rubinsztein Dunlop, and T. A. Nieminen, Driving optical micromachines with orbital angular momentum*).

PrefacePreface Light containing optical vortices might also be used in imaging and probing different sorts of physical and biological properties of matter (see *C. Maurer, S. Bernet, and M. Ritsch-Marte Spiral Phase Contrast Microscopy*), controlling technologically important materials (see *E. Santamato and B. Piccirillo Optical torques in liquid crystals*) and in astrophysics (*B. Thidé, N. M. Elias II, F. Tamburini, S. M. Mohammadi and J. T. Mendonca, Applications of Electromagnetic OAM in Astrophysics and Space Physics Studies*).

The concept also holds for single photons in the quantum world; thus, it can be used to encode quantum information that is carried by the corresponding photon states, to explore quantum features in higher-dimensional Hilbert spaces, as the observation of the violation of Bell inequalities in three–dimensional Hilbert spaces (see *G. Molina-Terriza and A. Zeilinger, Experimental control of the Orbital Angular Momentum of single and entangled photons*), to generate new quantum states (see *E. M. Wright, Optical Vortex Cat States and their utility for creating Macroscopic Superpositions of Persistent Flows*) or implement new tools to achieve full control of all degrees of freedom of atoms (see *K. Helmerson and W. D. Phillips, Rotating Atoms with Light*).

All these are illustrative examples of the wealth of possibilities afforded by the orbital momentum of light. Much more undoubtedly lay ahead. It is our intention that this book, contributed by some of the pioneers and world leading scientists in the different subareas and techniques, motivates further research into new ways by which "twisted light" is used to manipulate and to probe Nature.

We warmly thank Wiley for their timely vision to publish a book on this topic and all the authors for their generous time and efforts that were contributed to make it a reality. It is now the time for the readers to enjoy it and to multiply the uses of the orbital angular momentum of light for new applications.

ICFO, Barcelona, Spain

Juan P. Torres
Lluis Torner

List of Contributors

Les Allen
University of Strathclyde
Physics Department
Richmond Street
Glasgow G1 1XQ
Scotland
UK

and

University of Glasgow
Department of Physics and Astronomy
Kelvin Building
Glasgow G12 8QQ
Scotland
UK

Theodor Asavei
The University of Queensland
Quantum Science Laboratory
School of Mathematics and Physics
St. Lucia
Brisbane
Queensland 4072
Australia

Aleksandr Bekshaev
I.I. Mechnikov National University
Dvorianska 2
65082 Odessa
Ukraine

Stefan Bernet
Innsbruck Medical University
Division of Biomedical Physics
Müllerstr. 44
A-6020 Innsbruck
Austria

Kishan Dholakia
SUPA
University of St Andrews
School of Physics and Astronomy
North Haugh
Fife
Scotland KY16 9SS
UK

Nicholas M. Elias II
Zentrum für Astronomie der Universität Heidelberg
Königstuhl 12
D-69117 Heidelberg
Germany
EU

Twisted Photons: Applications of Light with Orbital Angular Momentum.
Edited by Juan P. Torres and Lluis Torner

ISBN: 978-3-527-40907-5

Péter Galaja
Institute of Biophysics
Biological Research Centre of the Hungarian Academy of Sciences
Temesvari krt. 62
6701 Szeged
Hungary

Norman R. Heckenberg
The University of Queensland
Quantum Science Laboratory
School of Mathematics and Physics
St. Lucia
Brisbane
Queensland 4072
Australia

Kristian Helmerson
Monash University
Faculty of Science
School of Physics
Wellington Road
Clayton
Victoria 3800
Australia

and

Joint Quantum Institute
National Institute of Standards and Technology
Gaithersburg
100 Bureau Drive
Maryland 20899-8424
USA

Lóránd Kelemen
Institute of Biophysics
Biological Research Centre of the Hungarian Academy of Sciences
Temesvari krt. 62
6701 Szeged
Hungary

Vincent L. Y. Loke
The University of Queensland
Quantum Science Laboratory
School of Mathematics and Physics
St. Lucia
Brisbane
Queensland 4072
Australia

Christian Maurer
Innsbruck Medical University
Division of Biomedical Physics
Müllerstr. 44
A-6020 Innsbruck
Austria

Michael Mazilu
SUPA
University of St Andrews
School of Physics and Astronomy
North Haugh
Fife
Scotland KY16 9SS
UK

José T. Mendonça
IPFN and CFIF
Instituto Superior Técnico
AV. Rovisco Pais 1
PT-1049-001 Lisbon
Portugal

Siavoush M. Mohammadi
Swedish Institute of
Space Physics
Ångström Laboratory
P.O. Box 537
SE-75121 Uppsala
Sweden
EU

Gabriel Molina-Terriza
Macquarie University
QSciTech and Department of
Physics & Astronomy
2109 New South Wales
Australia

Timo A. Nieminen
The University of Queensland
Quantum Science Laboratory
School of Mathematics and
Physics
St. Lucia
Brisbane
Queensland 4072
Australia

Pál Ormos
Institute of Biophysics
Biological Research Centre of
the Hungarian Academy of
Sciences
Temesvari krt. 62
6701 Szeged
Hungary

László Oroszi
Institute of Biophysics
Biological Research Centre of
the Hungarian Academy of
Sciences
Temesvari krt. 62
6701 Szeged
Hungary

Miles Padgett
University of Glasgow
Department of Physics and
Astronomy
Kelvin Building
Glasgow G12 8QQ
Scotland
UK

Simon Parkin
The University of Queensland
Quantum Science Laboratory
School of Mathematics and
Physics
St. Lucia
Brisbane
Queensland 4072
Australia

William D. Phillips
Joint Quantum Institute
National Institute of
Standards and Technology
100 Bureau Drive
Gaithersburg
Maryland 20899-8424
USA

Bruno Piccirillo
Università degli Studi di Napoli
"Federico II"
Dipartimento di Scienze Fisiche
Complesso Universitario di
Monte S. Angelo
via Cintia
80126 Naples
Italia

Monika Ritsch-Marte
Innsbruck Medical University
Division of Biomedical Physics
Müllerstr. 44
A-6020 Innsbruck
Austria

Halina Rubinsztein-Dunlop
The University of Queensland
Quantum Science Laboratory
School of Mathematics and
Physics
St. Lucia
Brisbane
Queensland 4072
Australia

Enrico Santamato
Università degli Studi di Napoli
"Federico II"
Dipartimento di Scienze Fisiche
Complesso Universitario di
Monte S. Angelo
via Cintia
80126 Naples
Italia

Fabrizio Tamburini
University of Padova
Department of Astronomy
vicolo dell'Osservatorio 2
IT-35122
Padova
Italy
EU

Bo Thidé
Swedish Institute of Space
Physics
Ångström Laboratory
P.O. Box 537
SE-75121 Uppsala
Sweden
EU

Mikhail Vasnetsov
Institute of Physics of the
National Academy of
Sciences of Ukraine
Department of Optical and
Quantum Electronics
Prospect Nauki av. 46
03028 Kiev
Ukraine

Ewan M. Wright
University of Arizona
College of Optics and
Department of Physics
Tucson
AZ 85721
USA

Anton Zeilinger
Institute of Quantum Optics and
Quantum Information
Austrian Academy of Sciences
Boltzmanngasse 3
1090 Wien
Austria

and

Universität Wien
Fakultät für Physik
Boltzmanngasse 5
A-1090 Wien
Austria

Color Plates

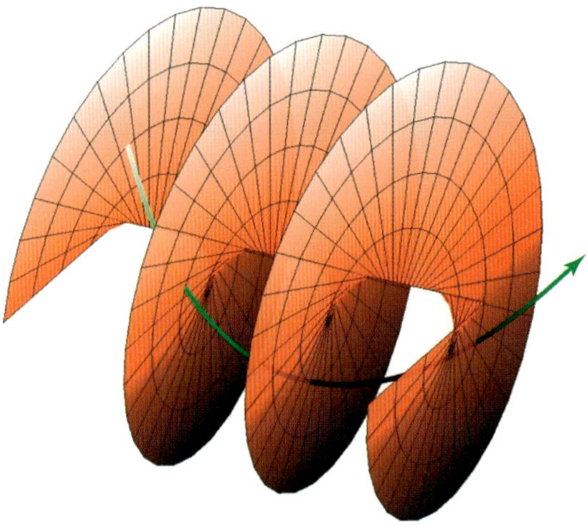

Figure 1.1 The helical wave fronts characterized by an azimuthal phase term ($l = 1$) and the associated Poynting vector, the azimuthal component of which gives rise to an orbital angular momentum. (This figure also appears on page 4.)

Twisted Photons: Applications of Light with Orbital Angular Momentum.
Edited by Juan P. Torres and Lluis Torner

ISBN: 978-3-527-40907-5

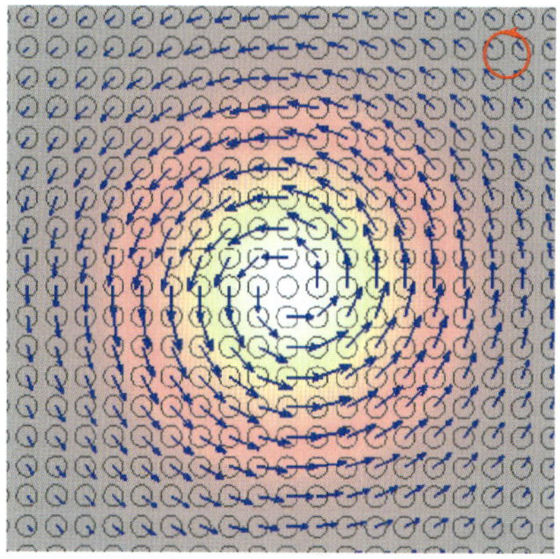

Figure 2.1 Map of the spin flow density of Eq. (2.16) for a left-polarized Gaussian beam ($\sigma = 1$, polarization handedness is shown in the upper right corner); lengths of arrows correspond to relative flow density, the intensity distribution and polarization ellipses (circles) are shown in the background, the beam is viewed against the propagation axis. (This figure also appears on page 18.)

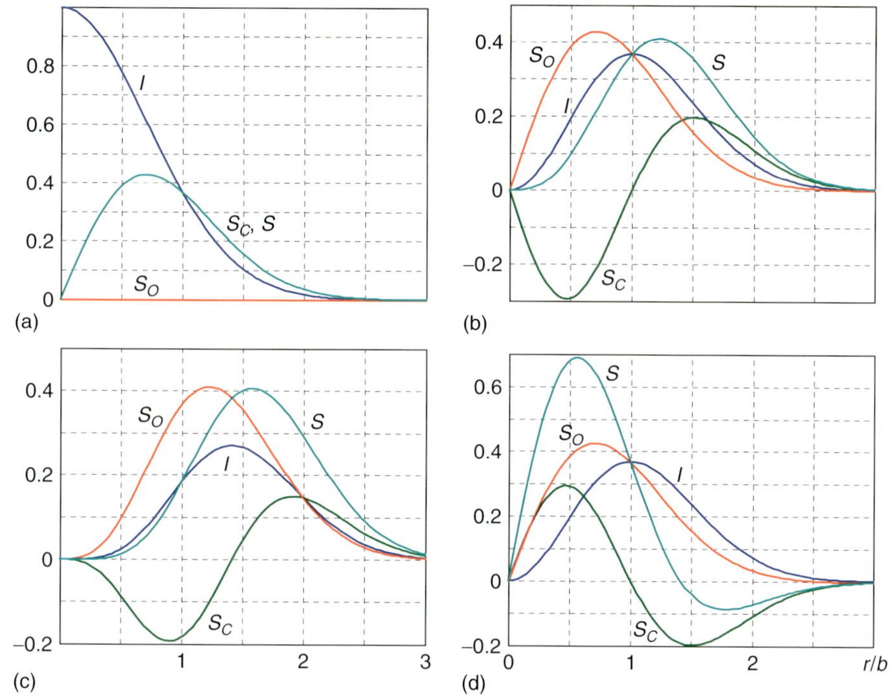

Figure 2.2 Radial profiles of (I) intensity (Eq. (2.17)) in units of $I_{\sigma 0}$, (S_C) spin flow density (Eq. (2.18)), (S_O) orbital flow density (Eq. (2.19)), and (S) total transverse flow density (Eq. (2.24)) (all in units of $I_{\sigma 0}/kb$), for the circularly polarized LG beams with zero radial index and the following sets of parameters: (a) $\sigma = 1, l = 0$ (Gaussian beam of Figure 2.1), (b) $\sigma = 1, l = 1$, (c) $\sigma = 1, l = 2$, (d) $\sigma = -1, l = 1$. (This figure also appears on page 19.)

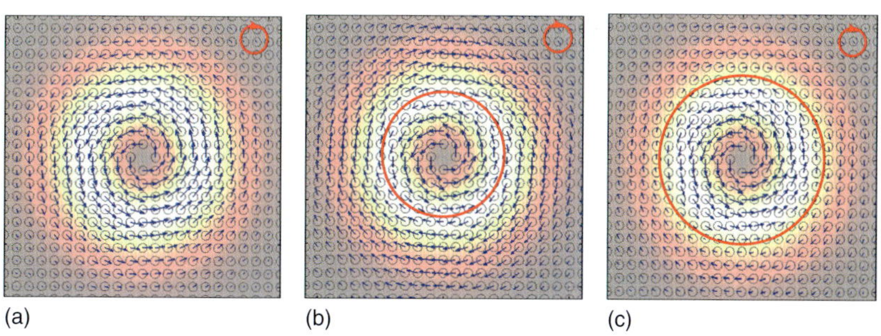

Figure 2.3 Maps of the (a) orbital $\mathbf{S}_O$, (b) spin $\mathbf{S}_C$, and (c) total $\mathbf{S}$ transverse energy flows in the cross-section of a right-polarized LG beam (Eq. (2.17)) with $l = 1, \sigma = -1$ (case of Figure 2.2d). At every point, polarization is the same as shown in the upper right corners; circular contours in panels (b) and (c) are contours where the corresponding flow component vanishes. (This figure also appears on page 20.)

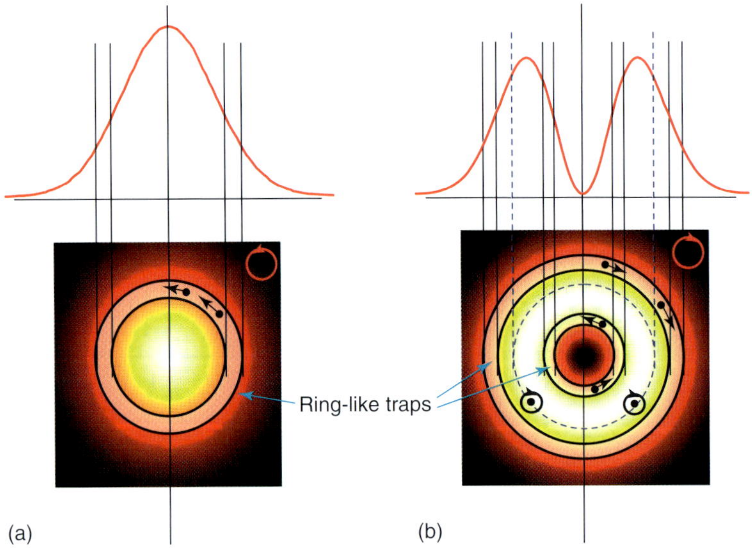

Figure 2.4 Possible schemes of orbital motion of the absorbing suspended particles confined in the ring-like traps within the circularly polarized field of (a) Gaussian beam of Figure 2.1 and (b) LG beam with $l = 1$, $\sigma = -1$ of Figure 2.3. Top row: diametric sections of the intensity profiles with boundaries of the ring-like traps, bottom row: views of the beam cross sections with the trap traces (polarization handedness is indicated in the upper right corners). Circles with arrows indicate the expected orbital motion of the trapped particles and dashed lines in panel (b) specify locations where orbital motion is not excited (see the circular contour in Figure 2.3c). Particles situated at this contour perform only the spinning motion (shown by the arrow loops); in all other positions the spinning motion is not shown but is also expected in addition to the orbital one. (This figure also appears on page 22.)

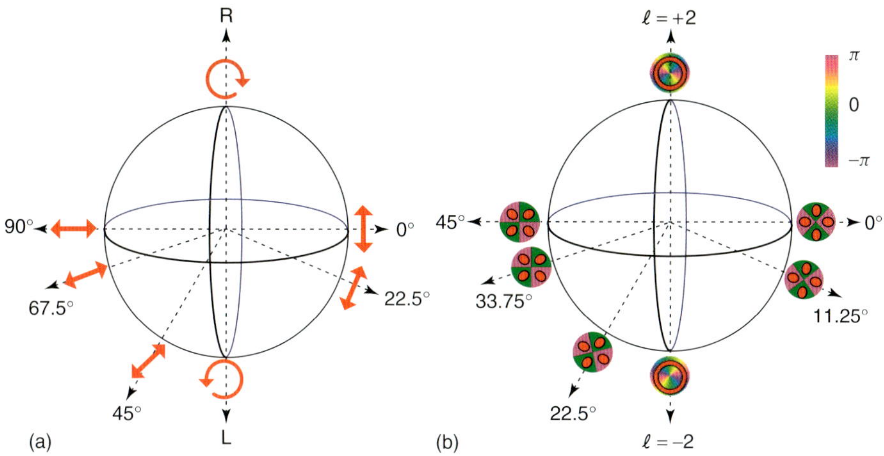

Figure 3.1 (a) The Poincaré sphere representation of polarization state. (b) An equivalent representation for beams formed from the superposition of Laguerre–Gaussian modes $\ell = 2, p = 0$ and $\ell = -2, p = 0$. (This figure also appears on page 27.)

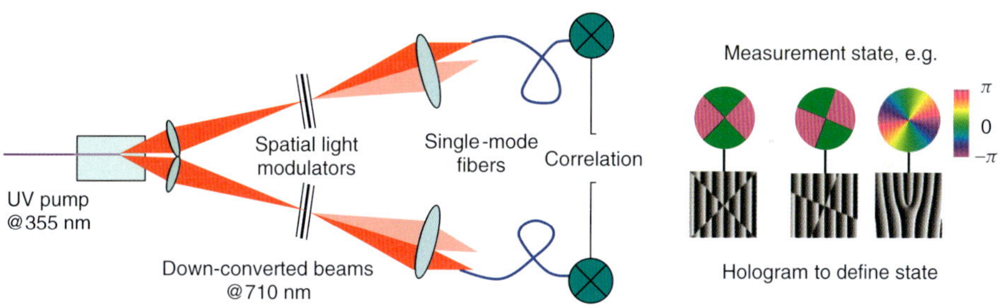

Figure 3.3 Observing correlations in the orbital angular momentum of down-converted beams. The use of spatial light modulators to define various holograms allows the measurement of an arbitrary spatial mode, including orbital angular momentum states analogous to those of polarization on the Poincaré sphere. (This figure also appears on page 31.)

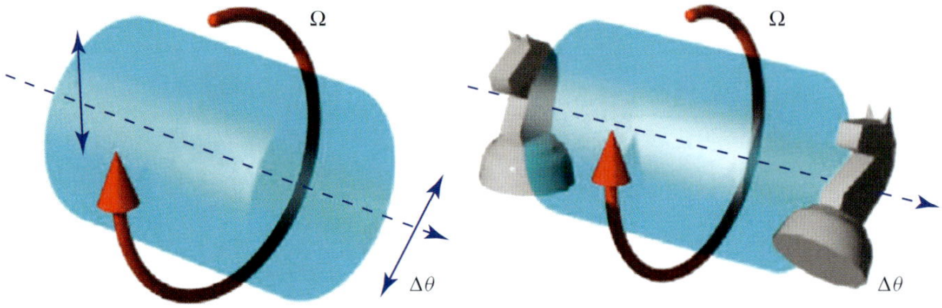

Figure 3.4 A spinning medium is predicted to slightly rotate both the polarization state and the image of the transmitted light. (This figure also appears on page 32.)

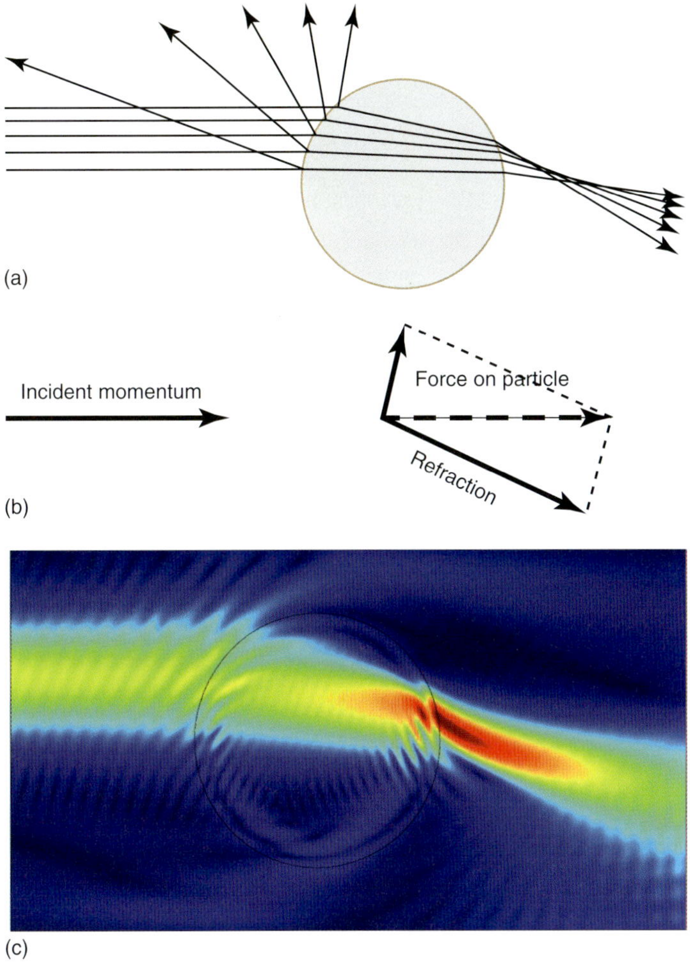

Figure 4.1 Optical forces arising from the conservation of momentum. (a) Bundle of rays being reflected and refracted by a transparent sphere. (b) Total momentum before (red) and after the optical interaction. (c) Refraction and reflection of a light beam by a dielectric sphere. (This figure also appears on page 42.)

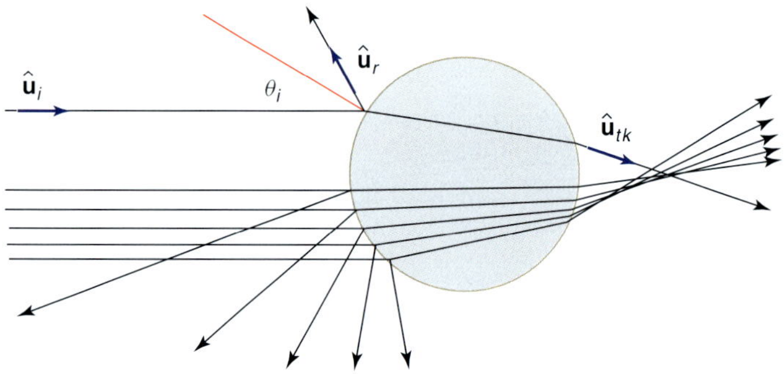

Figure 4.2 Rays reflecting and refracting from a transparent dielectric sphere. The various terms are defined in the text. (This figure also appears on page 43.)

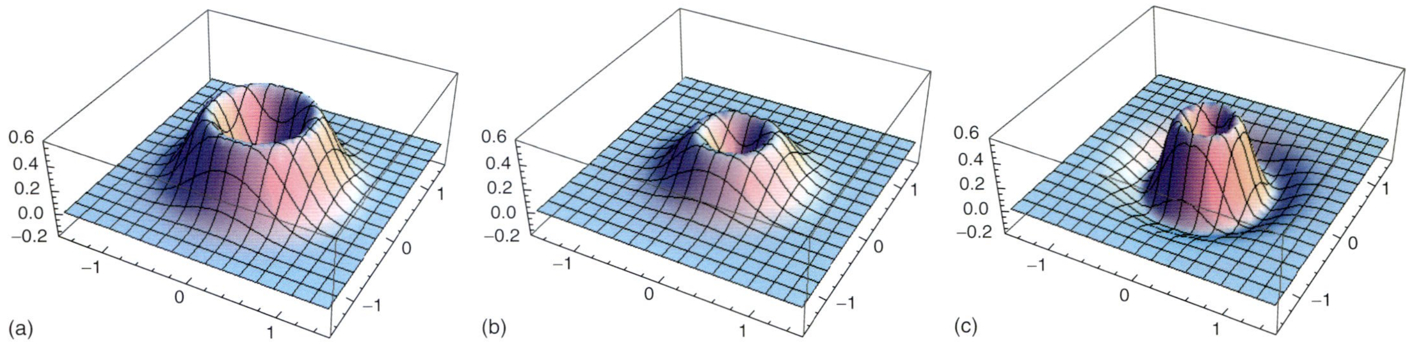

Figure 4.3 Optical torque arising in the transverse cross section of an LG beam ($l = 1$, $p = 0$) as defined by Eq. (4.9). (a) Spin and vortex charge are of the same sign. (b) Linearly polarized beam. (c) Spin and vortex charge are of opposite signs. The total integrated torque is double in the same sign case compared to the linearly polarized case and averages to zero in the opposite sign case. (This figure also appears on page 46.)

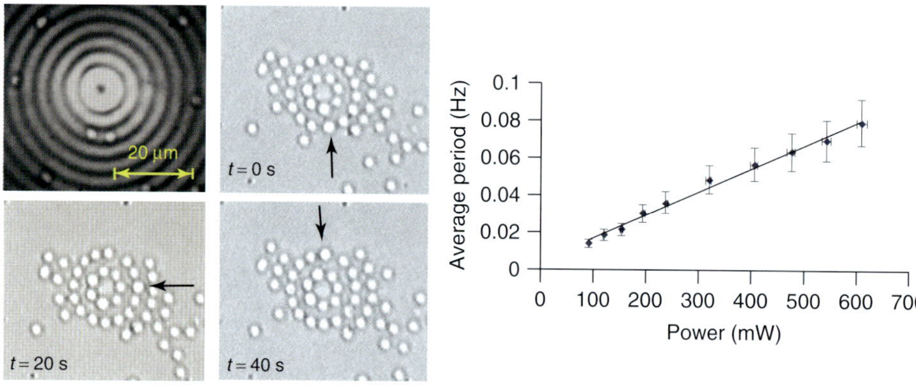

Figure 4.4 (a) The first frame shows the second-order Bessel beam used to trap 3 μm spheres in its rings. The rest of the frames show the trapped spheres rotating in the anticlockwise direction. To see the rotation one sphere in the second ring is highlighted by an arrow. (b) Average period of rotation in the inner ring as a function of the total power of the beam (Reprinted with permission from [28] © (2002) by IOP Publishing.) (This figure also appears on page 53.)

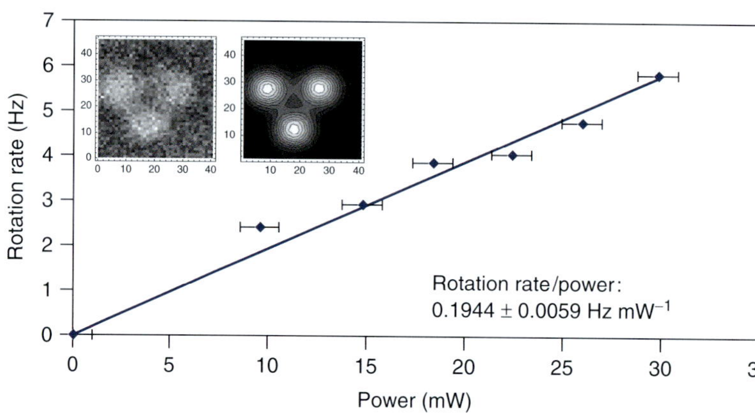

Figure 4.6 Three 1 μm spheres are optically trapped in the first bright annular ring of the focal spot of supercontinuum LG beam ($l = 3$, $p = 0$). The rotation rate is represented as a function of the incident power. The inset shows the trapped spheres and the tracking reconstruction (Reprinted with permission from [49].) (This figure also appears on page 57.)

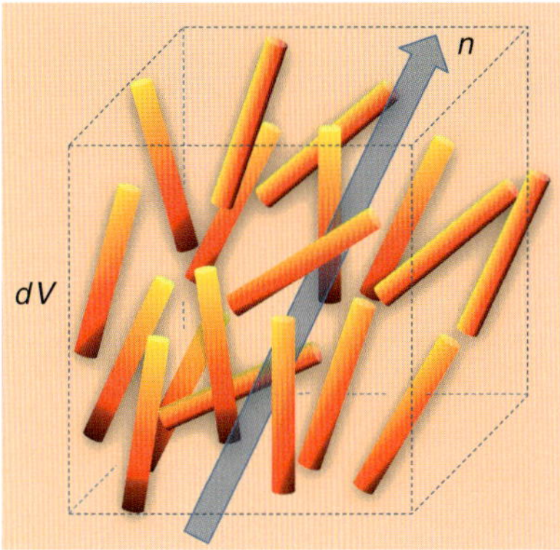

Figure 5.1 Schematic representation of the nematic phase. (This figure also appears on page 68.)

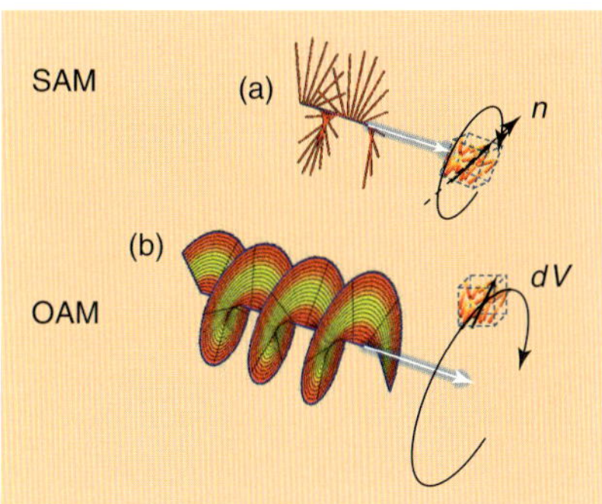

Figure 5.2 Pictorial representation of the separation of the orbital and spin parts of the total angular momentum of radiation on the grounds of the effects they produce inside matter. (a) Spin transfer induces a rotation of the director **n** (local optical axis). (b) Orbital transfer induces a rotational motion of the centers of mass of an elemental fluid volume. (This figure also appears on page 78.)

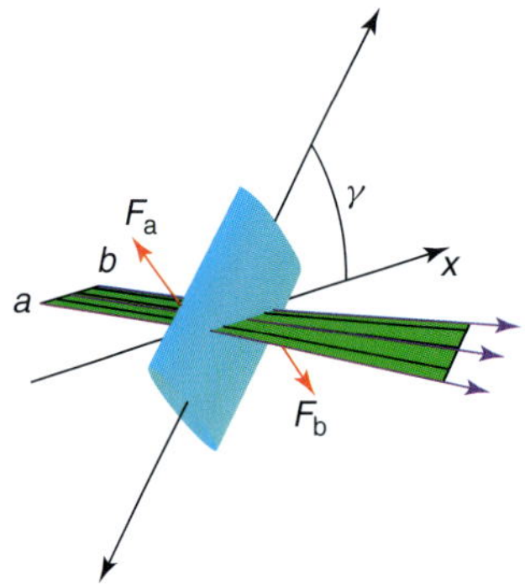

Figure 5.3 An NLC film optically distorted through an elliptically shaped laser beam behaves as a birefringent astigmatic microlens. Refraction of the incident beam by this lens results in a couple of forces $\mathbf{f}_a$ and $\mathbf{f}_b$ acting on the lens itself. (This figure also appears on page 81.)

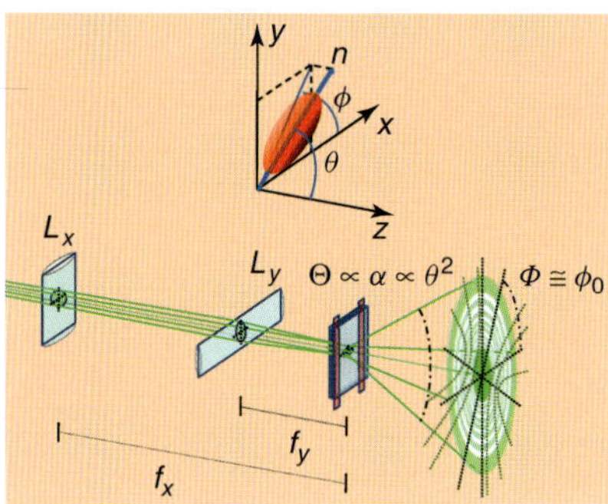

Figure 5.4 Schematic of the optical layout for photon OAM transfer in liquid crystals. Θ is the angular aperture of the far-field ring pattern and ϕ_0 is the inclination of its average polarization. The local direction of the optical field is tangent to the hyperbolic-like curves drown in the figure. (This figure also appears on page 82.)

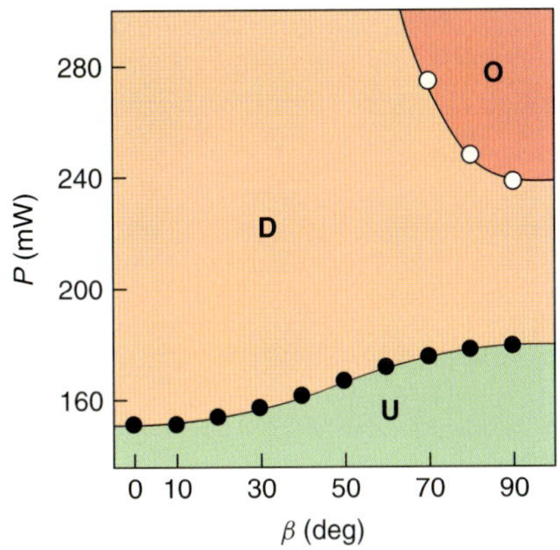

Figure 5.6 Map of the dynamical regimes in the parameter plane P, β. Three regions may be recognized: **U**, undistorted states; **D**, steady distorted states; **O**, oscillating states. The borderline between **U**- and **D**-regions represents the thresholds for the OFT and the borderline between **D**- and **O**-regions represents the thresholds for the oscillations start up as calculated from our model. Full circles on the first curve and open circle on the second are the experimental points. (This figure also appears on page 85.)

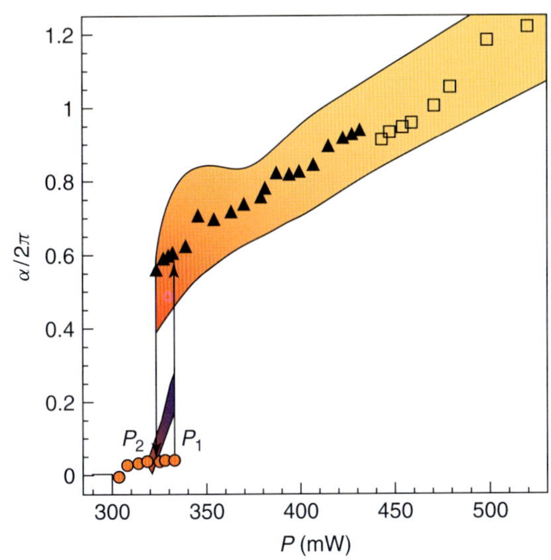

Figure 5.8 The optical phase difference α in unit of 2π as a function of the incident power P. (○) steady states; (▲) rotations; (□) intermittent states. The shadowed regions represent the oscillation amplitude of α and the experimental points mark the oscillation center. (This figure also appears on page 87.)

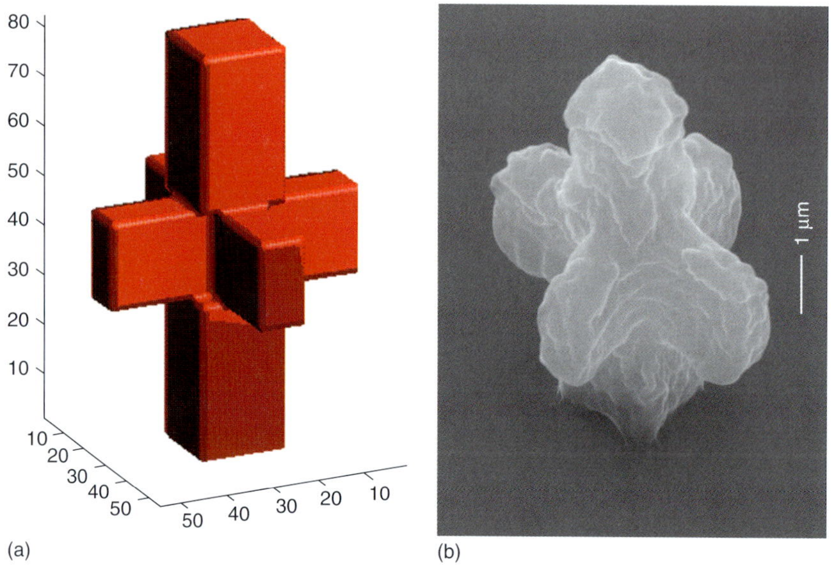

Figure 6.3 Microrotor: design and realization. (a) The design with a scale of 100 nm wide voxels and (b) a scanning electron microscope (SEM) image of the microfabricated structure attached to the cover slip. (This figure also appears on page 98.)

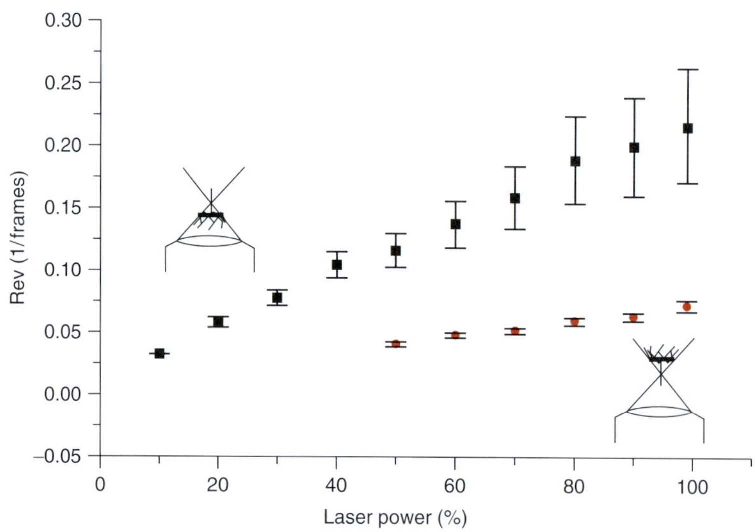

Figure 7.4 Dependence of the rate of revolution upon laser power. The two plots correspond to the two indicated positions the rotor can assume in the laser trap. (This figure also appears on page 123.)

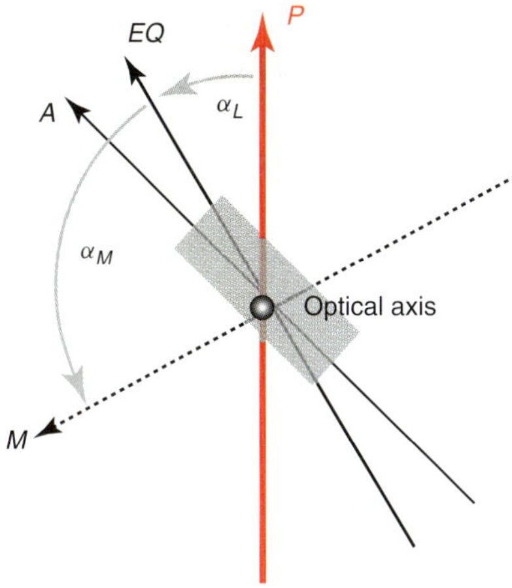

Figure 7.14 Explanation of the measurement of torque and torsional elasticity. *M* denotes the orientation where the molecular torsional strain vanishes, *P* indicates the plane of polarization of the trapping light, EQ is the equilibrium state where the two orienting torques (molecular and optical) cancel. The orientation of the trapped object fluctuates around EQ due to Brownian motion, *A* represents the actual orientation at a certain time. (This figure also appears on page 137.)

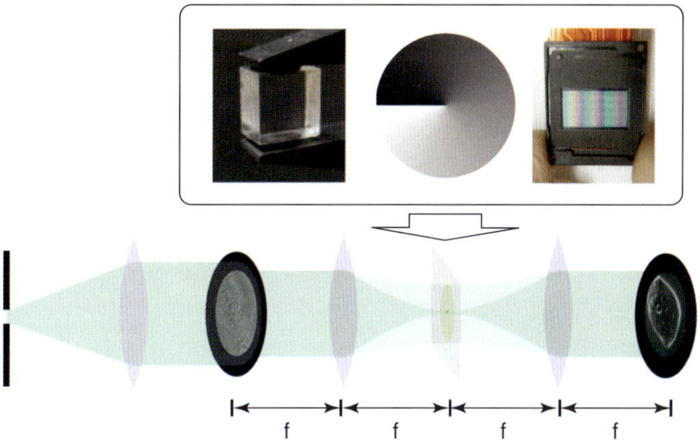

Figure 8.1 Generic setup for Fourier filtering in optical imaging: An object illuminated with a plane wave diffracts the light to outer regions in the Fourier plane, that is, the focal plane of the objective lens. One can access and specifically manipulate the various spatial frequencies there with a spatial light modulator or a phase plate. As an example, spiral phase filtering, which includes isotropic edge enhancement is shown in the upper part of the figure. (This figure also appears on page 145.)

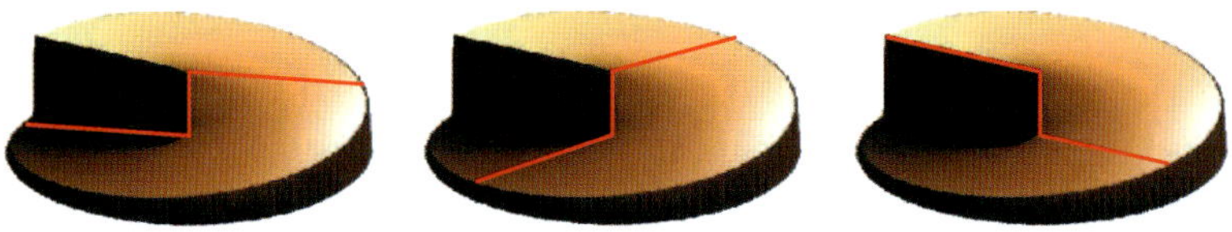

Figure 8.2 Spiral phase filter: The phase shift produced by an SLM or a phase plate corresponds to the helical phase profile typical for a Laguerre–Gauss beam. Note that opposite points are out of phase by π. (This figure also appears on page 146.)

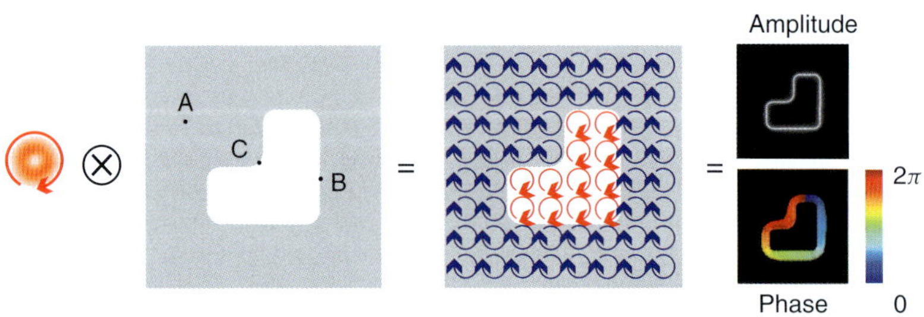

Figure 8.3 Graphical representation of the convolution of an object with a spiral phase filter: When the integration is carried out, the helical phase profile of the PSF of the spiral phase filter in unstructured regions leads to perfect cancellation of the signal by destructive interference, except at edges where either the phase or the amplitude of neighboring points differ. Note: convolution kernel not to scale. (This figure also appears on page 147.)

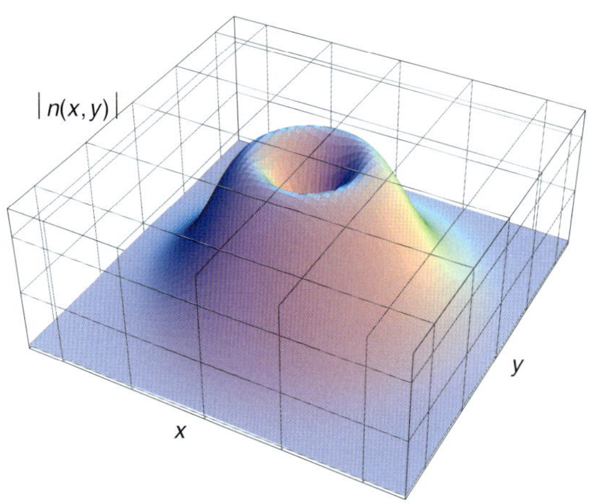

Figure 9.6 A phonon (ion-acoustic wave) in a plasma can carry OAM (but not spin angular momentum (SAM)). Source: From [40]. (This figure also appears on page 167.)

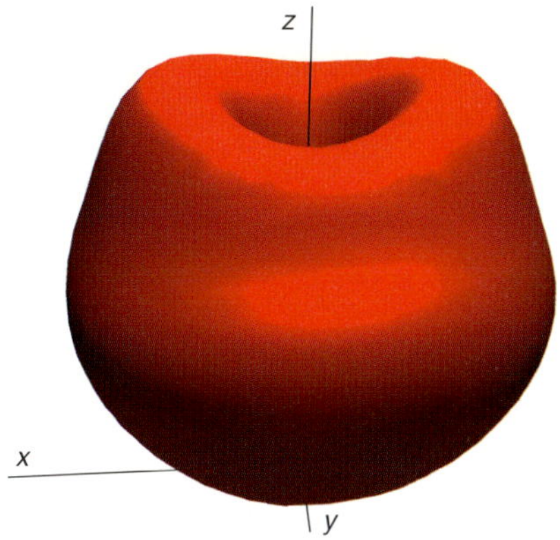

Figure 9.7 Plot of $|\mathbf{J}^{em}|$, that is, the intensity distribution of POAM, generated by a circular antenna array in the *xy* plane. Source: From [41]. (This figure also appears on page 168.)

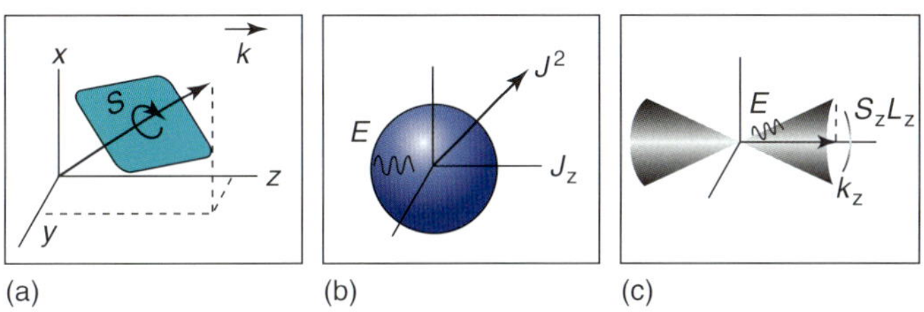

Figure 11.1 Different sets of modes that can be used for describing the electromagnetic field, and their respective set of parameters. (a) Plane wave, (b) multipolar mode, and (c) cylindrical mode. (This figure also appears on page 200.)

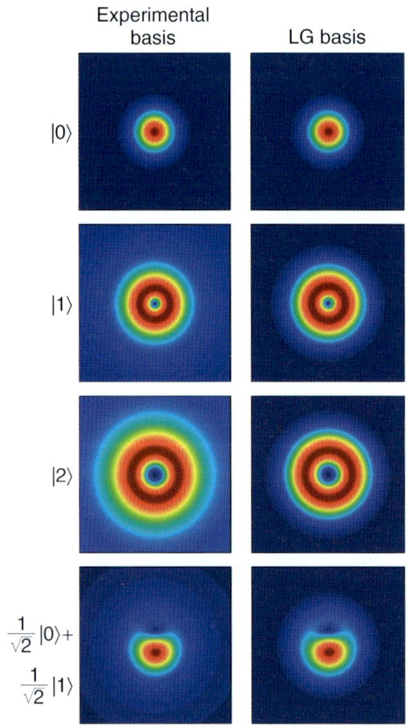

Figure 11.2 Orbital angular momentum modes: first column, experimentally achievable modes as explained; second column, Laguerre–Gaussian basis. The first three rows represent the amplitude distribution of the three lowest modes and last row shows the amplitude distribution of a superposition of two modes. Note that the three pure modes have an amplitude distribution that is rotationally invariant with respect to the propagation direction (perpendicular to the page). This is a trademark of the OAM eigenmodes. Also note the singularity in the center of the two higher-order pure modes. This singularity moves off-center in the superposition. (This figure also appears on page 204.)

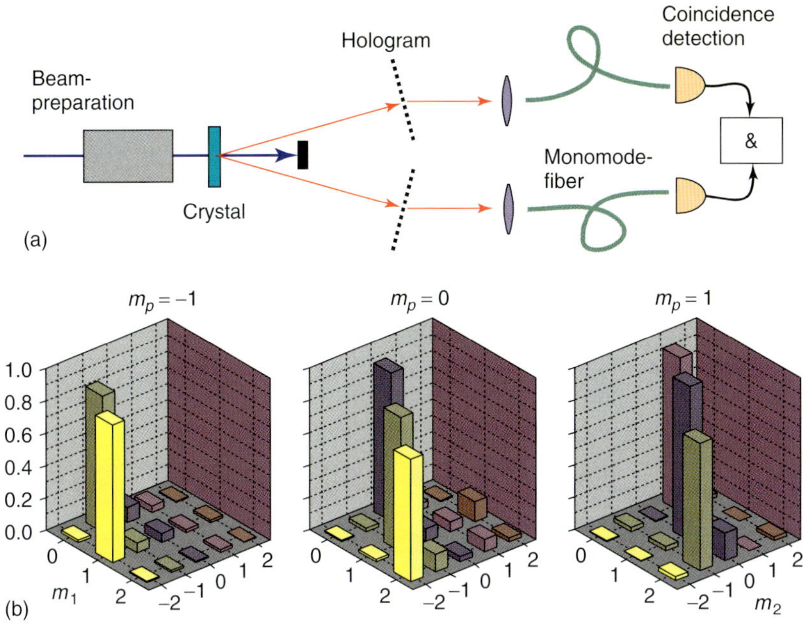

Figure 11.4 Conservation of OAM in the two-photon generation process : (a) Sketch of the experimental setup. (b) Results of the experiment, showing the relative number of coincident photons for different combinations of signal and idler modes. Every figure shows different pumping conditions. It can be noted how the OAM of the pump is transferred to the signal and idler and one only observes correlations in the outcomes when there is a conservation of the OAM between the three interacting modes. (Source: Reproduced from Mair *et al.* [12].) (This figure also appears on page 208.)

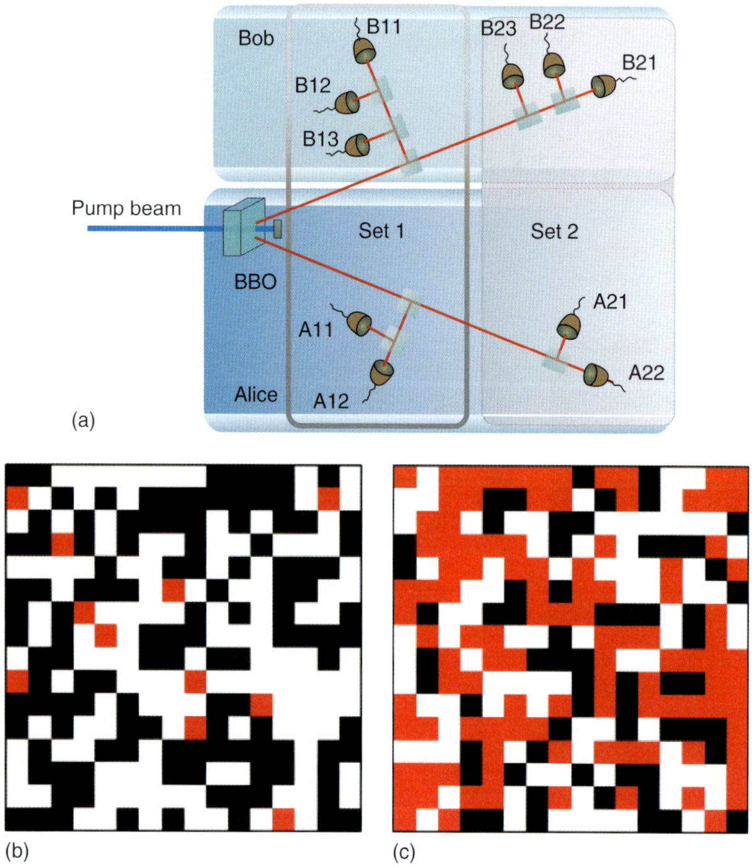

Figure 11.5 Experimental demonstration of the quantum coin tossing. (a) Sketch of the experimental setup: the different labels from *A*11 to *A*22 indicate the different projective measurements that Alice carries out. Each projective measurement consists of a hologram which transforms the state into a superposition of two modes, and a single mode fiber. The combination of the two pairs of projective measurements is effectively preparing the photon sent to Bob in two different mixed states. At the final step of the protocol, Bob can measure the resulting state with a set of projective measurements, then implementing the measurement on two different bases. (b) Result of an honest protocol: each small square represents one photon successfully received and measured by Bob. The outcome of the protocol can be both parties agreeing on a "heads" result (white square), a "tails" one (black) or a "failure" (red) that is, not agreeing on the result. "Failures" can happen because of dishonest parties or due to experimental errors. (c) Result of a protocol where Alice was cheating: in this case, it can be seen how the number of "failures" critically increases, showing the presence of a dishonest party. (This figure also appears on page 210.)

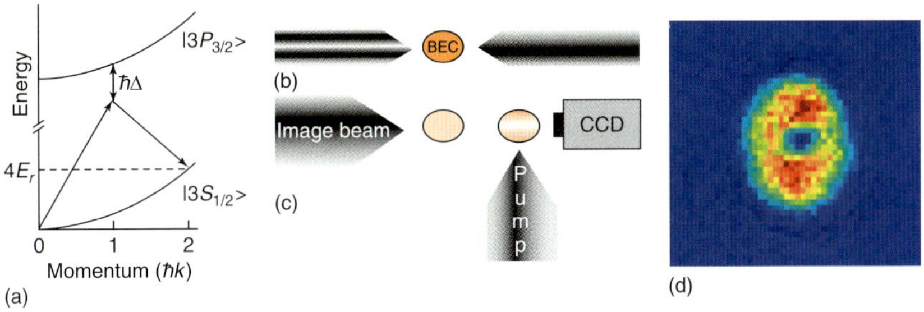

Figure 12.1 (a) Two-photon rotational Bragg scheme used to couple atoms between momentum states 0 and $2\hbar k$, transferring OAM in the process. (b) Schematic representation of the experiment. Counter propagating LG_0^1 and Gaussian laser beams, with the same linear polarization and a variable frequency difference of $\delta\omega/2\pi$, are applied to a BEC. (c) The atoms that have undergone the Raman transitions (right cloud) have been separated from those that did not (left cloud). A spatially localized "pump" beam enables independent imaging of each cloud by absorption of a probe beam propagating along the direction of linear momentum transfer. (d) Absorption image of a cloud that has undergone the Raman transition, taken along the axis of the LG_0^1 beam. The vortex core is seen as a hole in the cloud. (This figure also appears on page 217.)

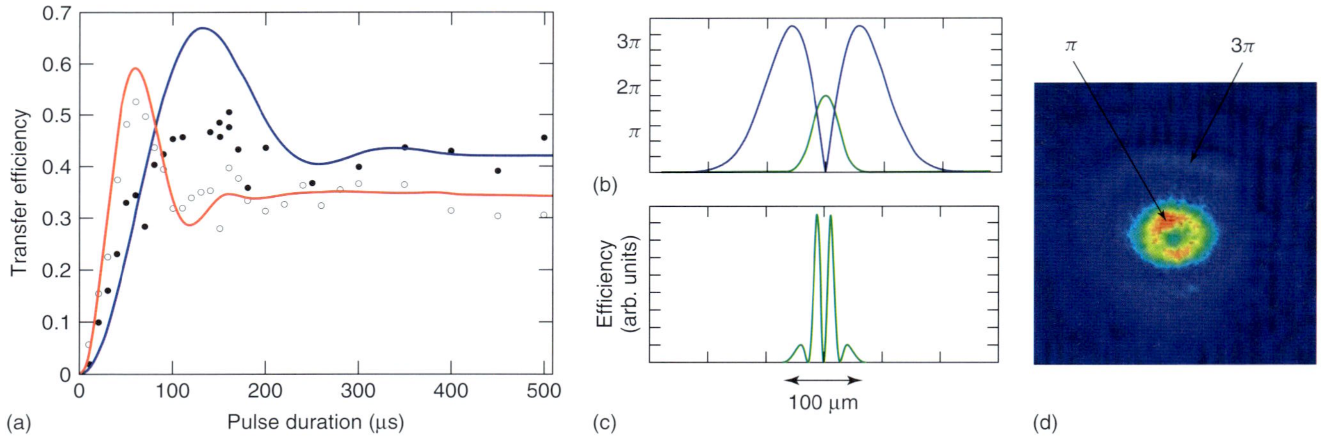

Figure 12.2 (a) Measured (dots) and calculated (lines) transfer efficiencies due to the mismatch of the spatial overlap of the Laguerre–Gaussian mode with the BEC. The experimental data points were obtained for Bragg detunings of 97.5 kHz (open dots) and 100 kHz (solid dots). The calculated curves are based on a full three-dimensional, time-dependent simulation of the BEC wavefunction [35], based on the spatial overlap depicted in (b). (b) Estimated spatial overlap of the LG beam with the BEC, approximated as a Gaussian. (c) Spatial dependence of the transfer efficiency (for a pulse duration of 130 μs) due to the spatial dependence of the Rabi frequency on the intensity profile of the LG mode shown in (b). (d) Observed spatial dependence of the transfer process of the BEC to the rotational state. The inner, intense ring results from an approximate π-pulse where the initial atom density is high, while the outer, weak ring corresponds to an approximate 3π pulse where the density is small. (This figure also appears on page 219.)

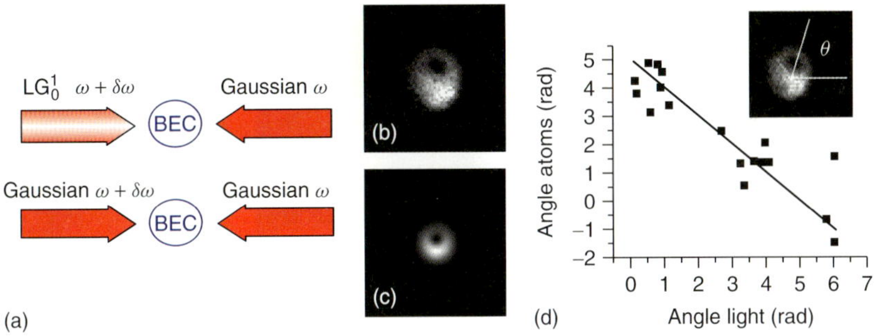

Figure 12.3 (a) Schematic representation of the laser pulse sequence used to generate and interfere the vortex state with the initial BEC. (b) Interference of a vortex state with $\hbar$ of angular momentum with a stationary state. The 2π phase winding associated with the rotating state results in a displacement of the hole, due to destructive interference. (c) Calculated interference pattern between a rotating and stationary state based on harmonic oscillator states. (d) Angle of the hole in the interference pattern between rotating and nonrotating atomic states as a function of the rotation angle of the optical interference pattern between the LG_0^1 and copropagating Gaussian beams. The straight line (to guide the eye) has slope −1. Inset: Image of the atomic interference between a rotating and nonrotating cloud. The hole is displaced from the center and its angular position θ depends on the relative phase between the interfering states. (This figure also appears on page 221.)

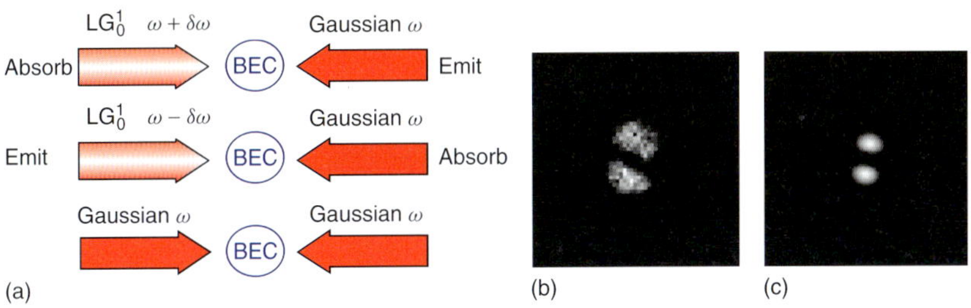

Figure 12.4 (a) Schematic representation of the laser pulse sequence used to generate and interfere the vortex state with an antivortex state. (b) Interference of a vortex state with $\hbar$ of angular momentum with a vortex state with $-\hbar$ of angular momentum. The interference of the rotating state with the counterrotating state results in a circular standing-wave. (c) Calculated interference pattern based on harmonic oscillator states between a rotating state (angular momentum $\hbar$) with a counterrotating state (angular momentum $-\hbar$). (This figure also appears on page 222.)

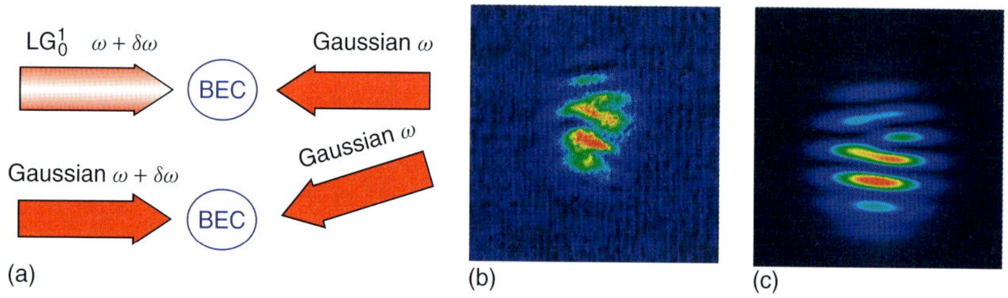

Figure 12.5 (a) Schematic representation of the laser pulse sequence used to generate and interfere the vortex state with the initial BEC. The second set of laser beams are not counterpropagating (in contrast to the first set), which results in the interfering clouds having a relative average velocity. (b) Interference of a vortex state with $\hbar$ of angular momentum with a nonrotating state. The fringes are a result of the relative velocity between the interfering clouds, while the singularity associated with the rotating state results in the fork-like structure. (c) Calculated interference pattern based on harmonic oscillator states. (This figure also appears on page 223.)

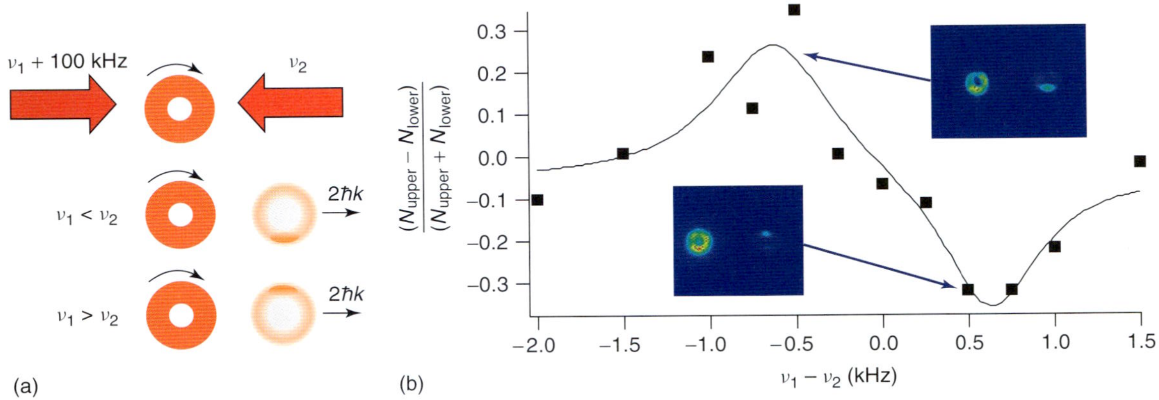

Figure 12.6 (a) Scheme for measuring rotational velocities using velocity-dependent Bragg diffraction. (b) The number of Bragg diffracted atoms in the upper half of the image minus the number of Bragg diffracted atoms in the lower half of an image, normalized to the total number of Bragg diffracted atoms, as a function of the relative detuning of the Bragg diffraction lasers. (This figure also appears on page 225.)

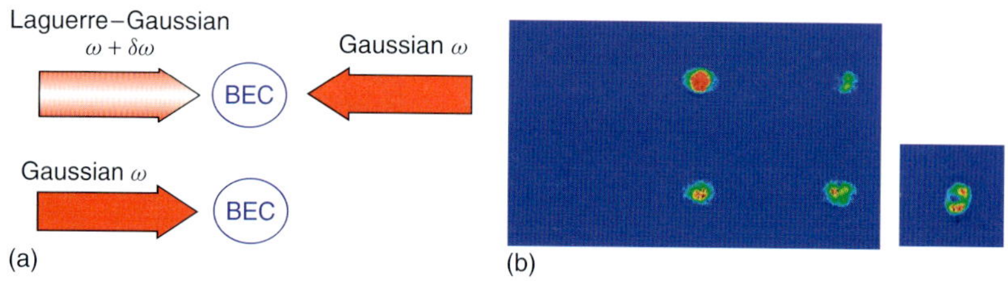

Figure 12.8 (a) Schematic representation of the laser pulse sequence used to amplify a vortex state generated in the initial BEC. (b) The upper image corresponds the generation of the "seed" vortex state from the initial BEC using only the LG and Gaussian beam pulse shown in the upper image of (a). The lower image shows the resulting amplification of the "seed" vortex from the two-pulse sequence of (a). The image on the far right is of the amplified vortex state taken along the direction of propagation. (This figure also appears on page 229.)

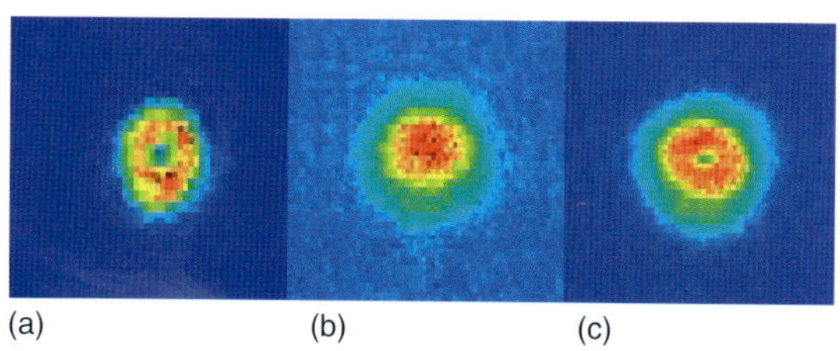

Figure 12.9 (a) Image of the condensate atoms in the hybrid magnetic TOP and blue-detuned laser trap. The absence of atoms in the central region is due to the laser. (b) TOF image of the condensate atoms released from the hybrid trap after being confined for 2 s. The interaction between the atoms causes the cloud to spread out and fill in the central region. (c) TOF image of the condensate atoms, which were in a vortex state about the blue-detuned laser beam. In this case the atoms in the vortex state were held in the hybrid trap for 2 s before being released. The presence of the core in the cloud of atoms after TOF indicates that the atoms were still in the vortex state when they were released from the hybrid trap. (This figure also appears on page 231.)

1
The Orbital Angular Momentum of Light: An Introduction

Les Allen and Miles Padgett

1.1
Introduction

Most physicists know that polarized light is associated with the spin angular momentum of the photon. It is almost certainly true that the idea of orbital angular momentum is a good deal less understood. Perhaps the simplest and most obvious display of both the spin and orbital angular momentum of light beams comes from an examination of the ratio of their angular momentum to their energy.

For an idealized, circularly-polarized plane wave, the spin angular momentum is given by $J_z = N\hbar$ and the energy by $W = N\hbar\omega$, where N is the number of photons. The angular momentum to energy ratio is thus,

$$\frac{J_z}{W} = \frac{\hbar}{\hbar\omega} = \frac{1}{\omega} \tag{1.1}$$

In fact the ratio in Eq. (1.1) is derivable from classical electromagnetism without any need to invoke the concept of a photon or any other quantum phenomenon [1].

A slightly more general result for elliptically polarized light, characterized by $-1 \leqslant \sigma \leqslant +1$, (with $\sigma = \pm 1$ for left- and right-handed circularly polarized light respectively and $\sigma = 0$ for linearly polarized light) is given by

$$\frac{J_z}{W} = \frac{\sigma}{\omega} \tag{1.2}$$

We can show for a light beam which has an l-dependent azimuthal phase angle such that the field amplitude is given by $u\left(x, y, z, \phi\right) = u_0\left(x, y, z\right) e^{-ikz} e^{+il\phi}$, that Eq. (1.2) becomes [2]

$$\frac{J_z'}{W} = \frac{l \pm \sigma}{\omega} \tag{1.3}$$

Here $\hbar\sigma$ describes the spin angular momentum per photon, while $l\hbar$ describes the orbital angular momentum per photon. In the absence of the phase term $\exp\left(il\phi\right)$, Eq. (1.3) would be the usual plane wave ratio of spin angular momentum divided by energy, namely, $\hbar\sigma/\hbar\omega$ or $\hbar\sigma$ per photon.

It transpires that this simple result is true both in the limit of the paraxial approximation and for fields described by a rigorous and unapproximated application of

Twisted Photons: Applications of Light with Orbital Angular Momentum.
Edited by Juan P. Torres and Lluis Torner

ISBN: 978-3-527-40907-5

Maxwell's equations [3]. In the paraxial approximation, other than assuming that $u(x, y, z)$ is normalizable and leads to a finite energy in the beam, no assumption has been made about the form of the distribution. In other words even for $\sigma = 0$, when the light is linearly polarized, there remains an angular momentum related to the spatial properties of the beam and dependent on l.

The fact that the simple paraxial result, Eq. (1.3), is fully justified by rigorous theory [4] enables a number of essentially simple conclusions to be drawn. The paraxial fields appropriate for linearly polarized light are

$$\mathbf{B} = \mu_0 \mathbf{H} = ik\left[u\hat{\mathbf{y}} + \frac{i}{k}\frac{\partial u}{\partial y}\hat{\mathbf{z}}\right]e^{ikz} \tag{1.4}$$

and

$$\mathbf{E} = ik\left[u\hat{\mathbf{x}} + \frac{i}{k}\frac{\partial u}{\partial x}\hat{\mathbf{z}}\right]e^{+ikz} \tag{1.5}$$

These allow evaluation of the time-averaged Poynting vector, $\varepsilon_0\, \mathbf{E} \times \mathbf{B}$, namely,

$$\begin{aligned}\varepsilon_0 \langle \mathbf{E} \times \mathbf{B} \rangle &= \frac{\varepsilon_0}{2}\left[\left\langle \mathbf{E}^* \times \mathbf{B}\right\rangle + \left\langle \mathbf{E} \times \mathbf{B}^*\right\rangle\right] \\ &= i\omega\frac{\varepsilon_0}{2}\left(u\nabla u^* - u^*\nabla u\right) + \omega k\varepsilon_0 |u|^2\,\hat{\mathbf{z}}\end{aligned} \tag{1.6}$$

For a field such as $u(r, \phi, z) = u_0(r, z)\, e^{+il\phi}$ the ϕ-component of linear momentum density is

$$\varepsilon_0 \langle \mathbf{E} \times \mathbf{B} \rangle_\phi = \varepsilon_0 \omega l\, |u|^2 / r \tag{1.7}$$

while its cross product with r gives an angular momentum density of magnitude $j_z = \varepsilon_0 \omega \ell\, |u|^2$. The energy density of such a beam is

$$w = c\varepsilon_0 \langle \mathbf{E} \times \mathbf{B} \rangle_z = c\varepsilon_0 \omega k\, |u|^2 = \varepsilon_0 \omega^2\, |u|^2 \tag{1.8}$$

Thus,

$$\frac{j_z}{w} = \frac{l}{\omega}$$

When the angular momentum density is integrated over the x–y plane, the ratio of angular momentum to energy per unit length of the beam is simply,

$$\frac{J_z}{W} = \frac{\iint r dr d\varphi\, (r \times \langle \mathbf{E} \times \mathbf{B} \rangle)_z}{c \iint r dr d\phi\, \langle \mathbf{E} \times \mathbf{B} \rangle_z} = \frac{l}{\omega} \tag{1.9}$$

The same straightforward calculation for fields that include polarization, again produces Eq. (1.3), but it is now for physically realizable fields and not just plane wave fields of infinite extent.

The earliest work on the orbital angular momentum of light beams took an LG (Laguerre–Gaussian) mode as the most easily available source of light possessing an azimuthal phase. This amplitude distribution, $u_{p,l}$, has the requisite $\exp(il\phi)$

term and is now well known. It readily follows for such a distribution that the linear momentum density is [2]

$$p = \varepsilon_0 \left(\frac{\omega k r z}{z_R^2 + z^2} \hat{\mathbf{r}} + \frac{\omega l}{r} \hat{\phi} + \omega k \hat{\mathbf{z}} \right) \left| u_{p,l} \right|^2 \tag{1.10}$$

and the cross product with r gives the angular momentum density,

$$\mathbf{j} = \mathbf{r} \times \mathbf{p} = \varepsilon_0 \left(\frac{-\omega l z}{r} \hat{\mathbf{r}} - \omega k r \left(\frac{z_R^2}{z_R^2 + z^2} \right) \hat{\phi} + \omega l \hat{\mathbf{z}} \right) \left| u_{p,l} \right|^2 \tag{1.11}$$

The expression for linear momentum p, (Eq. (1.10)), shows that at a constant radius, r, the Poynting vector maps out a spiral path of well-defined pitch,

$$z_p = \frac{2\pi k r^2}{l} \tag{1.12}$$

However, such a picture is misleading as it ignores the radial component of the Poynting vector and, hence, the spreading of the beam upon propagation [5]. For constant $r(z)/w(z)$, the angle of rotation, θ, of the Poynting vector from the beam waist at $z = 0$ is

$$\theta = \frac{l}{2} \left(\frac{w(z)}{r(z)} \right)^2 \arctan \left(\frac{z}{z_R} \right) \tag{1.13}$$

For a $p = 0$ mode, for which the intensity distribution is a single ring, the radius of the maximum amplitude in the mode is given by

$$r(z)_{\text{Max Int.}} = \sqrt{\frac{w(z)\, l}{2}} \tag{1.14}$$

and so for $p = 0, \ell \neq 0$, it follows that $\theta = \arctan\left(\frac{z}{z_R}\right)$ which, surprisingly, is independent of ℓ. Rather than describing a multiturn spiral as one might have presumed, the Poynting vector rotates only by $\pi/2$ either side of the beam waist as the light propagates to the far field. Perhaps even more surprisingly, the locus of the vector is simply a straight line at an angle to the axis of the beam [6, 7]. Note that the arctan term is simply proportional to the Gouy phase of the Gaussian beam and that, in free space, the Poynting vector is at all points parallel to the wavevector.

Simple though these results are, in hindsight, they were not known until the early 1990s. Their application to a number of conceptually straightforward experiments enables simple comparisons to be made, at least in the paraxial regime, between the behavior of spin and orbital angular momenta and enables the observation of a number of phenomena to be elucidated. This phenomenology provides much of the basis for the exploration and exploitation of the current understanding of the subject outlined in later chapters of this book. Although everything may be justified formally using a quantum approach, there is, outside of entanglement, little need to leave this classical formulation. In the nonparaxial case, the separation of spin and orbital angular momentum is more complicated [4, 8–10].

The use of the flow of angular momentum flux across a surface, rather than angular momentum density, allows the separation of the spin and orbital angular momentum parts in a gauge invariant way. This holds beyond the paraxial approach but confirms the simple values obtained for the ratio of angular momentum to energy [11].

1.2
The Phenomenology of Orbital Angular Momentum

Simple comparisons of the behavior of spin and orbital angular momenta in different situations prove to be a fruitful way to demonstrate their properties. First, however, we need to distinguish the general structures of light emitted by a laser and also its properties when converted to, for instance, an LG beam. Laser beams usually have spherical wave fronts while the azimuthal phase leads to beams with l intertwined helical wave fronts (Figure 1.1). The LG beam is not the only example of a helical wave front; Bessel beams [12], Mathieu beams [13], and Ince–Gaussian beams [14] can also carry orbital angular momentum. In all cases, the interference of these helical wave fronts with a plane wave gives rise to characteristic spiral interference fringes [15–17].

The production of a pure, high-order LG mode from a laser beam was first achieved using a mode convertor based on cylindrical lenses [18]. Although the details are interesting, they need not concern us here, as an approach based on simple holograms achieves a similar beam much more easily. Prior to the generation of LG beams with lenses, similar beams containing the same azimuthal phase term had also been produced using diffractive optical elements [19]. These components are simple diffraction gratings, that contain an edge dislocation,

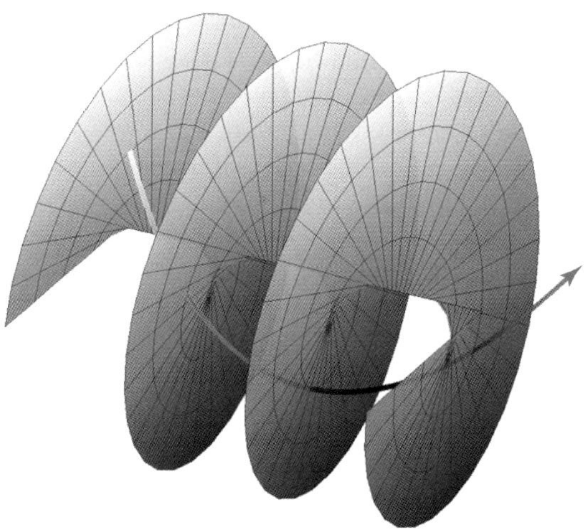

Figure 1.1 The helical wave fronts characterized by an azimuthal phase term ($l = 1$) and the associated Poynting vector, the azimuthal component of which gives rise to an orbital angular momentum. (Please find a color version of this figure on the color plates.)

coincident with the axis of the illuminating beam. Such "forked-gratings" give rise to a first order diffracted spot with an annular intensity cross section, which is a natural consequence of the $\exp(il\phi)$ phase structure. Indeed, similar beams have been widely studied as examples of optical phase singularities [20], also called *optical vortices* [21, 22]. However, in none of the earlier works had their angular momentum properties been recognized. These diffractive optical components can be readily designed, and are frequently referred to as *computer generated holograms.* Although easy to implement and producing perfect helical wave fronts, the resulting intensity distribution only approximates to that of a pure LG mode. Most recently, these "forked diffraction gratings" have been employed within the image train of a microscope to impose a point-spread function corresponding to a helical mode, giving an edge enhancement of the image [23, 24].

Rather than using the diffractive optical component, it is possible to form a refractive optical equivalent. A spiral phaseplate has an optical thickness, t, given by $t = \lambda l\phi/2\pi$, where ϕ is the azimuthal angle [25]. Upon transmission, a plane wave input beam is transformed into a helically phased beam again characterized by an azimuthal phase structure of $\exp(il\phi)$. Such spiral phase plates are not easy to manufacture but offer very high conversion efficiency. Interestingly, the azimuthal refraction of the ramped surface gives a skew angle of l/kr to each transmitted ray. For the linear momentum of the photon of $\hbar k$, this gives an azimuthal component $l\hbar/r$ and hence an angular momentum of $l\hbar$ per photon [26]. Thus we see that for a ray optical model, the orbital angular momentum of the photon is describable by skew rays [27].

Despite the various approaches that have been developed to generate helically phased beams, they are not a feature unique to advanced optical experiment. Interference between two plane waves yields sinusoidal fringes. Interference between three [28] or more [29] plane waves leads to points within the field cross section of perfect destructive interference around which the phase advances or retards by 2π. Nowhere is this more apparent than when examining the optical speckle resulting from laser light being scattered from a rough surface, where each black speck is a perfect phase singularity. Of course, the specks are dark and hence carry neither energy nor momentum. However, the light in the immediate vicinity of each is characterized by a helical phase front and does carry both energy and orbital angular momentum. Over the extent of the speckle pattern, there are an equal number of clockwise and anticlockwise singularities; and hence the overall orbital angular momentum tends to zero. These phase singularities map out lines of complete darkness in space, with both fractal [30] and topological [31] properties.

In order to generate pure LG modes, the cylindrical lens mode converter remains a convenient approach. The fidelity of the mode transformation means that when light with orbital angular momentum is passed through a cylindrical lens mode convertor it behaves in a mathematically analogous way to polarized (spin) light through a quarter waveplate. Indeed, the representation of states on the Poincaré sphere can be applied for any two states of orbital angular momentum [32]. Similarly, the well-known Jones matrices which describe the propagation of polarized light through an optical system have equivalents for the propagation of orbital angular momentum through a system with astigmatic optical elements [33, 34]. There are also joint matrices for light that is both polarized and possesses orbital angular momentum. An alternative to the use of these joint matrices is to apply the spin (Jones) matrix and then the orbital angular momentum matrix separately. This is equivalent to the separation of the spin and orbital components in the hydrogen wavefunction.

That this orbital angular momentum is a true momentum was first demonstrated in optical tweezers [35]. Optical tweezers use the gradient force associated with a tightly focused beam of the light to trap a microscopic dielectric particle [36]. A few milliwatts is all that is required to trap a 5 μm diameter sphere suspended in a liquid medium. Using an LG mode as the trapping beam results in a transfer of angular momentum to the particle causing it to spin about the beam axis.

The similarities in behavior of the two types of angular momenta in a light beam are also shown in optical tweezers when a small, mildly absorptive particle is trapped *on-axis*. When the light is purely circularly polarized, the particle may be made to rotate clockwise or anticlockwise depending on the handedness of the polarization where $\sigma = \pm 1$. When the same trapped particle sees light with $l = \pm 1$, it can also be made to rotate in either direction. Application of light where σ and l have the same sign leads to a faster rotation proportional to $(\sigma + l)$, while if σ and l have opposite signs the particle slows to a halt which arises clearly from $(\sigma - l)$. This demonstrates the mechanical equivalence of spin and orbital angular momentum [37]. In other words, the spin angular momentum can be added to or subtracted from the orbital component, consistent with the statement that the optical angular momentum of a light beam is $(l + \sigma)\,\hbar$. This statement is in agreement with the theory of angular momentum flux. It is observed that the center of mass in the on-axis case does not move and both the spin and orbital angular momentum contribute to making the sphere rotate about its own axis. *Off-axis*, such a particle behaves rather differently. It responds to orbital angular momentum by orbiting the axis of the beam with an angular velocity proportional to the local intensity of the beam. It also spins, because of σ, about its own axis. Again the velocity depends on the local intensity but otherwise spin and orbital manifest themselves in that case in different ways – highlighting the intrinsic and extrinsic nature of optical angular momentum [38]. These various studies in optical tweezers have spawned significant work, worldwide, where the induced rotation of the particles acts as a microfluidic pump [39, 40] or other optically driven micromachine [41–44].

It should be observed that spin, σ, is said to be *intrinsic* because it is independent of the choice of axis about which it is calculated. However, orbital, l, depends upon the choice of axis. Nevertheless, when there is a direction, z, for which the transverse linear momentum of the beam is zero, both l and σ are invariant under a shift of axis and the orbital component might be said to be *quasi-intrinsic*. For off-axis apertures in cylindrically symmetric beams the transverse linear momentum is nonzero and l is *extrinsic*.

Closely related to the use of LG beam in optical tweezers is their interaction with cold atoms [45]. In many cases, rather than the helical wave fronts, it is the on-axis intensity zero that enables the confinement of blue-detuned atoms

[46]. Residual scattering can additionally lead to guiding along the length of the singularity. Cooling the atoms further to create a Bose–Einstein condensate (BEC) results in yet more interesting interactions between matter and the orbital angular momentum of light. This includes an optically induced rotation of the BEC [47, 48].

The interaction of light carrying orbital angular momentum with an in resonance atom has also been investigated [49]. It is found that the frequency shift of a resonant transition in an atom moving with an angular velocity Ω through a polarized beam with orbital angular momentum is Ωl while the torque on the center of mass of the atom is $\hbar l\Gamma$ and independent of σ [50]. There appear to be no torques on the atom's center of mass that depend on $(l+\sigma)$.

This frequency shift is an example of an angular Doppler effect readily observed when a light beam is rotated at angular frequency Ω about its own axis. This is not to be confused with the transverse Doppler shift observed when an emitter moves toward or away from the source. For the spinning beam, the frequency of the light is shifted for spin by $\delta\omega' = \Omega\sigma$, for orbital angular momentum by $\delta\omega'' = \Omega l$ [51] and for total angular momentum by $\delta\omega''' = \Omega\left(\sigma + l\right)$ [52]. For combined beams with the same polarization but different total orbital angular momenta, a spectrum of shifted components $\delta\omega_1 = \Omega\left(\sigma + l_1\right)$, $\delta\omega_2 = \left(\sigma + l_2\right)$, and so on, is produced. This is one of the effects found to depend upon the sum of the spin and orbital components. The phenomenon can be understood by the realization that time evolution of a helical phase front is indistinguishable from rotation about the beam axis. A full rotation of the beam changes the phase of the light by $l+\sigma$ cycles. Such phase and associated frequency shifts also extend to polychromatic light, where all spectral components are frequency shifted by the same amount [53].

Attempts have been made to see if analogs to electron spin-orbit interactions common in atoms exists in light. The only evidence so far is that in the dissipative force on a moving atom there is a term proportional to σl. It is, however, small and only comparable in size to terms which are usually ignored, of order $\left(1/k^2\right)$ [54].

Second harmonic generation or up-conversion in a nonlinear crystal can produce second harmonic generation for helically phased modes, where

$$\omega_{\text{Shg}} = \omega_{\text{In}} + \omega_{\text{In}} = 2\omega_{\text{In}} \text{ and } l_{\text{Shg}} = l_{\text{In}} + l_{\text{In}} = 2l_{\text{In}} \tag{1.15}$$

This is in contrast to the spin angular momentum which can only be unity, at most. Here, there is another difference between orbital and spin angular momentum. There is no potential upper bound to l_{Shg} and we see that up-conversion may be used to change the order of the mode [55]. There is no equivalent change of polarization mode. This conversion of l_{Shg} arises through strict phase matching and because the wavevectors and Poynting vector of the fundamental and second-harmonic helical beams are collinear. This implies that when the wave-number doubles then l must also double [56]. Such a process is consistent with the conservation of orbital angular momentum within the light fields. This work is a precursor of work on down-conversion where one input photon creates two photons of lower energy. This has important implications for the higher order entanglement possible with orbital angular momentum [57]. In down-conversion, correlation of orbital angular momentum can be achieved with a pair of holograms that determine l_{Idler} and

l_{Signal} for a given l_{Pump} [58, 59]. Although spin is limited to ± 1, there is a wide range of l_{Idler} and l_{Signal} for a given l_{Pump}. The high-dimensionality of the Hilbert space and information content [60] combined with techniques for sorting single photons [61–63] creates opportunities in, for example, quantum information processing [64]. It is the study of the down converted beams and a violation of a Bell inequality [65] that illustrates that orbital angular momentum is a meaningful concept at the quantum level and hence a true photon property.

For spin angular momentum and circularly polarized light, the light source need not be either temporally or spatially coherent. For orbital angular momentum the situation is more complicated. Orbital angular momentum is a meaningful concept across the full electromagnetic spectrum [66] and has been considered ranging from radio frequency [67] to X ray regimes [68]. As orbital angular momentum is associated with the phase cross section of the beam, there is no restriction on its temporal coherence; each spectral component can have a perfect $\exp(-il\phi)$ phase structure. Beams with such multispectral components can be generated using the normal forked diffraction grating, but with its spectral dispersion compensated by a prism [69] or second grating [70]. These beams have the exact anticipated orbital angular momentum to cause microscopic objects to rotate about the axis of the beam [71]. Perfect helical wave fronts imply a complete spatial coherence. Degrading the spatial coherence destroys the fidelity of the on-axis phase singularity and the on-axis intensity zero. If the beam has some degree of spatial coherence then when transmitted through a spiral phase plate or diffracted from a forked diffraction grating, the resulting beam can be decomposed into an incoherent sum of different modes having a finite average value of orbital angular momentum. These beams have been termed *Rankine vortices* [72]. Spiral phaseplates built into telescopes have been shown to be useful astronomical filters, which could suppress the light from a point-star so that an off-axis source of light from a planet might be detected [73, 74].

Another aspect of clear distinction between spin and orbital angular momentum is the existence of a Fourier relationship for orbital angular momentum and angular position [75], and a related uncertainty relationship. The uncertainty relationship was originally discussed for measurements of linear position and linear momentum. In the case of orbital angular momentum, a similar expression can be written for small uncertainties in angular position, $\Delta\phi\Delta l = \hbar/2$ [76]. No equivalent expression exists for spin. The uncertainty associated with the measurement of orbital angular momentum may prove to be a limitation to the evident virtues of orbital angular momentum as a means of exploiting entanglement, and so on.

One marked difference in the literature, since light beams possessing orbital angular momentum have been realized and understood, arises because, in order to exploit their dependence on space, the formal way in which light interacts with atoms has had to be developed. It is no longer sufficient to investigate the interaction of atoms with plane waves. The traditional semiclassical approach is still in the main appropriate, but it must now be applied to specifically structured Gaussian beams.

Various reviews have been written, which summarize the development of the field of the last 15 years [3, 77–80] and many aspects of the current state of this work is discussed in later chapters of this book.

References

1. Poynting, J.H. (1909) The wave motion of a revolving shaft, and a suggestion as to the angular momentum in a beam of circularly polarised light. *Proc. R. Soc. London Ser. A*, **82**, 560–567.
2. Allen, L., Beijersbergen, M.W., Spreeuw, R.J.C., and Woerdman, J.P. (1992) Orbital angular momentum of light and the transformation of Laguerre-Gaussian laser modes. *Phys. Rev. A*, **45**, 8185–8189.
3. Allen, L., Padgett, M.J., and Babiker, M. (1999b) *The Orbital Angular Momentum of Light*, Progress in Optics, vol. XXXIX, Elsevier Science Publishers B V, Amsterdam.
4. Barnett, S.M. and Allen, L. (1994) Orbital angular momentum and nonparaxial light beams. *Opt. Commun.*, **110**, 670–678.
5. Padgett, M.J. and Allen, L. (1995) The Poynting vector in Laguerre-Gaussian laser modes. *Opt. Commun.*, **121**, 36–40.
6. Courtial, J. and Padgett, M.J. (2000) Limit to the orbital angular momentum per unit energy in a light beam that can be focussed onto a small particle. *Opt. Commun.*, **173**, 269–274.
7. Berry, M.V. and McDonald, K.T. (2008) Exact and geometrical optics energy trajectories in twisted beams. *J. Opt. A: Pure Appl. Opt.*, **10**, 7.
8. Ganic, D., Gan, X.S., and Gu, M. (2003) Focusing of doughnut laser beams by a high numerical-aperture objective in free space. *Opt. Express*, **11**, 2747–2752.
9. Monteiro, P.B., Neto, P.A.M., and Nussenzveig, H.M. (2009) Angular momentum of focused beams: beyond the paraxial approximation. *Phys. Rev. A*, **79**, 033830.
10. Nieminen, T.A., Stilgoe, A.B., Heckenberg, N.R., and Rubinsztein-Dunlop, H. (2008) Angular momentum of a strongly focused Gaussian beam. *J. Opt. A: Pure Appl. Opt.*, **10**, 6.
11. Barnett, S.M. (2002) Optical angular momentum flux. *J. Opt. B: Quantum Semiclassical Opt.*, **4**, S7–S16.
12. McGloin, D. and Dholakia, K. (2005) Bessel beam: diffraction in a new light, contemporary. *Physics*, **46**, 15–28.
13. Gutierrez-Vega, J.C., Iturbe-Castillo, M.D., and Chavez-Cerda, S. (2000) Alternative formulation for invariant optical fields: Mathieu beams. *Opt. Lett.*, **25**, 1493–1495.
14. Bandres, M.A. and Gutierrez-Vega, J.C. (2004) Ince-Gaussian beams. *Opt. Lett.*, **29**, 144–146.
15. Harris, M., Hill, C.A., and Vaughan, J.M. (1994) Optical helices and spiral interference fringes. *Opt. Commun.*, **106**, 161–166.
16. Padgett, M., Arlt, J., Simpson, N., and Allen, L. (1996) An experiment to observe the intensity and phase structure of Laguerre-Gaussian laser modes. *Am. J. Phys.*, **64**, 77–82.
17. Soskin, M.S., Gorshkov, V.N., Vasnetsov, M.V., Malos, J.T., and Heckenberg, N.R. (1997) Topological charge and angular momentum of light beams carrying optical vortices. *Phys. Rev. A*, **56**, 4064–4075.
18. Beijersbergen, M.W., Allen, L., van der Veen, H., and Woerdman, J.P. (1993) Astigmatic laser mode converters and transfer of orbital angular momentum. *Opt. Commun.*, **96**, 123–132.
19. Bazhenov, V.Y., Vasnetsov, M.V., and Soskin, M.S. (1990) Laser beams with screw dislocations in their wave-fronts. *JETP Lett.*, **52**, 429–431.
20. Nye, J.F. and Berry, M.V. (1974) Dislocation in wave trains. *Proc. R. Soc. London Ser. A: Math. Phys. Eng. Sci.*, **336**, 165–190.

21. Coullet, P., Gil, L., and Rocca, F. (1989) Optical vortices. *Opt. Commun.*, **73**, 403–408.
22. Desyatnikov, A.S., Kivshar, Y.S., and Torner, L. (2005) *Optical Vortices and Vortex Solitons*, Progress in Optics, Vol. 47, Elsevier Science Publishers B V, Amsterdam.
23. Furhapter, S., Jesacher, A., Bernet, S., and Ritsch-Marte, M. (2005) Spiral interferometry. *Opt. Lett.*, **30**, 1953–1955.
24. Bernet, S., Jesacher, A., Furhapter, S., Maurer, C., and Ritsch-Marte, M. (2006) Quantitative imaging of complex samples by spiral phase contrast microscopy. *Opt. Express*, **14**, 3792–3805.
25. Beijersbergen, M.W., Coerwinkel, R.P.C., Kristensen, M., and Woerdman, J.P. (1994) Helical-wave-front laser-beams produced with a spiral phaseplate. *Opt. Commun.*, **112**, 321–327.
26. Turnbull, G.A., Robertson, D.A., Smith, G.M., Allen, L., and Padgett, M.J. (1996) Generation of free-space Laguerre-Gaussian modes at millimetre-wave frequencies by use of a spiral phaseplate. *Opt. Commun.*, **127**, 183–188.
27. Leach, J., Keen, S., Padgett, M.J., Saunter, C., and Love, G.D. (2006b) Direct measurement of the skew angle of the Poynting vector in a helically phased beam. *Opt. Express*, **14**, 11919–11924.
28. Masajada, J. and Dubik, B. (2001) Optical vortex generation by three plane wave interference. *Opt. Commun.*, **198**, 21–27.
29. O'Holleran, K., Padgett, M.J., and Dennis, M.R. (2006) Topology of optical vortex lines formed by the interference of three, four, and five plane waves. *Opt. Express*, **14**, 3039–3044.
30. O'Holleran, K., Dennis, M., Flossmann, F., and Padgett, M.J. (2008) Fractality of light's darkness. *Phys. Rev. Lett.*, **100**, 053902.
31. O'Holleran, K., Dennis, M., and Padgett, M. (2009) Topology of Light's Darkness. *Phys. Rev. Lett.*, **102**, 143902.
32. Padgett, M.J. and Courtial, J. (1999) Poincare-sphere equivalent for light beams containing orbital angular momentum. *Opt. Lett.*, **24**, 430–432.
33. Allen, L., Courtial, J., and Padgett, M.J. (1999a) Matrix formulation for the propagation of light beams with orbital and spin angular momenta. *Phys. Rev. E*, **60**, 7497–7503.
34. Allen, L. and Padgett, M. (2007) Equivalent geometric transformations for spin and orbital angular momentum of light. *J. Mod. Opt.*, **54**, 487–491.
35. He, H., Friese, M.E.J., Heckenberg, N.R., and Rubinszteindunlop, H. (1995) Direct observation of transfer of angular-momentum to absorptive particles from a laser-beam with a phase singularity. *Phys. Rev. Lett.*, **75**, 826–829.
36. Ashkin, A., Dziedzic, J.M., Bjorkholm, J.E., and Chu, S. (1986) Observation of a single-beam gradient force optical trap for dielectric particles. *Opt. Lett.*, **11**, 288–290.
37. Simpson, N.B., Dholakia, K., Allen, L., and Padgett, M.J. (1997) Mechanical equivalence of spin and orbital angular momentum of light: an optical spanner. *Opt. Lett.*, **22**, 52–54.
38. O'neil, A.T., Macvicar, I., Allen, L., and Padgett, M.J. (2002) Intrinsic and extrinsic nature of the orbital angular momentum of a light beam. *Phys. Rev. Lett.*, **88**, 053601.
39. Ladavac, K. and Grier, D.G. (2004) Microoptomechanical pumps assembled and driven by holographic optical vortex arrays. *Opt. Express*, **12**, 1144–1149.
40. Leach, J., Mushfique, H., Di Leonardo, R., Padgett, M., and Cooper, J. (2006) An optically driven pump for microfluidics. *Lab Chip*, **6**, 735–739.
41. Galajda, P. and Ormos, P. (2001) Complex micromachines produced and driven by light. *Appl. Phys. Lett.*, **78**, 249–251.
42. Paterson, L., Macdonald, M.P., Arlt, J., Sibbett, W., Bryant, P.E., and Dholakia, K. (2001) Controlled rotation of optically trapped microscopic particles. *Science*, **292**, 912–914.
43. Grier, D.G. (2003) A revolution in optical manipulation. *Nature*, **424**, 810–816.
44. Knoner, G., Parkin, S., Nieminen, T.A., Loke, V.L.Y., Heckenberg, N.R., and

Rubinsztein-Dunlop, H. (2007) Integrated optomechanical microelements. *Opt. Express*, **15**, 5521–5530.

45. Barreiro, S. and Tabosa, J.W.R. (2003) Generation of light carrying orbital angular momentum via induced coherence grating in cold atoms. *Phys. Rev. Lett.*, **90**, 133001.
46. Ozeri, R., Khaykovich, L., and Davidson, N. (1999) Long spin relaxation times in a single-beam blue-detuned optical trap. *Phys. Rev. A*, **59**, R1750–R1753.
47. Andersen, M.F., Ryu, C., Clade, P., Natarajan, V., Vaziri, A., Helmerson, K., and Phillips, W.D. (2006) Quantized rotation of atoms from photons with orbital angular momentum. *Phys. Rev. Lett.*, **97**, 170406.
48. Wright, K.C., Leslie, L.S., and Bigelow, N.P. (2008) Optical control of the internal and external angular momentum of a Bose-Einstein condensate. *Phys. Rev. A*, **77**, 041601.
49. Allen, L., Babiker, M., Lai, W.K., and Lembessis, V.E. (1996) Atom dynamics in multiple Laguerre-Gaussian beams. *Phys. Rev. A*, **54**, 4259–4270.
50. Allen, L., Babiker, M., and Power, W.L. (1994) Azimuthal Doppler-shift in light-beams with orbital angular-momentum. *Opt. Commun.*, **112**, 141–144.
51. Courtial, J., Dholakia, K., Robertson, D., Allen, L., and Padgett, M.J. (1998) Measurement of the rotational frequency shift imparted to a rotating light beam possessing orbital angular momentum. *Phys. Rev. Lett.*, **80**, 3217–3219.
52. Courtial, J., Robertson, D., Dholakia, K., Allen, L., and Padgett, M.J. (1998) Rotational frequency shift of a light beam. *Phys. Rev. Lett.*, **81**, 4828–4830.
53. Nienhuis, G. (2006) Polychromatic and rotating beams of light. *J. Phys. B*, **39**, S529–S544.
54. Allen, L., Lembessis, V.E., and Babiker, M. (1996b) Spin-orbit coupling in free-space Laguerre-Gaussian light beams. *Phys. Rev. A*, **53**, R2937–R2939.
55. Dholakia, K., Simpson, N.B., Padgett, M.J., and Allen, L. (1996) Second-harmonic generation and the orbital angular momentum of light. *Phys. Rev. A*, **54**, R3742–R3745.
56. Courtial, J., Dholakia, K., Allen, L., and Padgett, M.J. (1997b) Second-harmonic generation and the conservation of orbital angular momentum with high-order Laguerre-Gaussian modes. *Phys. Rev. A*, **56**, 4193–4196.
57. Arnaut, H.H. and Barbosa, G.A. (2000) Orbital and intrinsic angular momentum of single photons and entangled pairs of photons generated by parametric down-conversion. *Phys. Rev. Lett.*, **85**, 286–289.
58. Mair, A., Vaziri, A., Weihs, G., and Zeilinger, A. (2001) Entanglement of the orbital angular momentum states of photons. *Nature*, **412**, 313–316.
59. Oemrawsingh, S.S.R., Ma, X., Voigt, D., Aiello, A., Eliel, E.R.T., Hooft, G.W., and Woerdman, J.P. (2005) Experimental demonstration of fractional orbital angular momentum entanglement of two photons. *Phys. Rev. Lett.*, **95**, 240501.
60. Gibson, G., Courtial, J., Padgett, M.J., Vasnetsov, M., Pas'ko, V., Barnett, S.M., and Franke-Arnold, S. (2004) Free-space information transfer using light beams carrying orbital angular momentum. *Opt. Express*, **12**, 5448–5456.
61. Leach, J., Courtial, J., Skeldon, K., Barnett, S.M., Franke-Arnold, S., and Padgett, M.J. (2004) Interferometric methods to measure orbital and spin, or the total angular momentum of a single photon. *Phys. Rev. Lett.*, **92**, 013601.
62. Marrucci, L., Manzo, C., and Paparo, D. (2006) Optical spin-to-orbital angular momentum conversion in inhomogeneous anisotropic media. *Phys. Rev. Lett.*, **96**, 163905.
63. Deng, L.P., Wang, H.B., and Wang, K.G. (2007) Quantum CNOT gates with orbital angular momentum and polarization of single-photon quantum logic. *J. Opt. Soc. Am. B: Opt. Phys.*, **24**, 2517–2520.
64. Molina-Terriza, G., Torres, J.P., and Torner, L. (2002) Management of the angular momentum of light: preparation of photons in multidimensional vector states of angular momentum. *Phys. Rev. Lett.*, **88**, 013601.
65. Leach, J., Jack, B., Romero, J., Ritsch-Marte, M., Boyd, R.W., Jha, A.K., Barnett, S.M., Franke-Arnold, S.,

and Padgett, M.J. (2009) Violation of a Bell inequality in two-dimensional orbital angular momentum state-spaces. *Opt. Express*, **17**, 8287–8293.

66. Harwit, M. (2003) Photon orbital angular momentum in astrophysics. *Astrophys. J.*, **597**, 1266–1270.
67. Thide, B., Then, H., Sjoholm, J., Palmer, K., Bergman, J., Carozzi, T.D., Istomin, Y.N., Ibragimov, N.H., and Khamitova, R. (2007) Utilization of photon orbital angular momentum in the low-frequency radio domain. *Phys. Rev. Lett.*, **99**, 087701.
68. Sasaki, S. and Mcnulty, I. (2008) Proposal for generating brilliant x-ray beams carrying orbital angular momentum. *Phys. Rev. Lett.*, **100**, 124801.
69. Leach, J. and Padgett, M.J. (2003) Observation of chromatic effects near a white-light vortex. *New J. Phys.*, **5**, 7.
70. Mariyenko, I.G., Strohaber, J., and Uiterwaal, C. (2005) Creation of optical vortices in femtosecond pulses. *Opt. Express*, **13**, 7599–7608.
71. Wright, A., Girkin, J., Gibson, G.M., Leach, J., and Padgett, M.J. (2008) Transfer of orbital angular momentum from a super-continuum, white-light beam. *Opt. Express*, **16**, 9495–9500.
72. Swartzlander, G.A. and Hernandez-Aranda, R.I. (2007) Optical Rankine vortex and anomalous circulation of light. *Phys. Rev. Lett.*, **99**, 160901.
73. Foo, G., Palacios, D.M., and Swartzlander, G.A. (2005) Optical vortex coronagraph. *Opt. Lett.*, **30**, 3308–3310.
74. Lee, J.H., Foo, G., Johnson, E.G., and Swartzlander, G.A. (2006) Experimental verification of an optical vortex coronagraph. *Phys. Rev. Lett.*, **97**, 053901.
75. Yao, E., Franke-Arnold, S., Courtial, J., Barnett, S., and Padgett, M.J. (2006) Fourier relationship between angular position and optical orbital angular momentum. *Opt. Express*, **14**, 9071–9076.
76. Franke-Arnold, S., Barnett, S., Yao, E., Leach, J., Courtial, J., and Padgett, M.J. (2004) Uncertainty principle for angular position and angular momentum. *New J. Phys.*, **6**, 103.
77. Allen, L., Barnett, S.M., and Padgett, M.J. (2003) *Optical Angular Momentum*, IoP Publishing. ISBN-0 7503 0901 6.
78. Padgett, M., Courtial, J., and Allen, L. (2004) Light's orbital angular momentum. *Phys. Today*, **57**, 35–40.
79. Molina-Terriza, G., Torres, J.P., and Torner, L. (2007) Twisted photons. *Nat. Phys.*, **3**, 305–310.
80. Franke-Arnold, S., Allen, L., and Padgett, M. (2008) Advances in optical angular momentum. *Laser Photonics Rev.*, **2**, 299–313.

2
Vortex Flow of Light: "Spin" and "Orbital" Flows in a Circularly Polarized Paraxial Beam

Aleksandr Bekshaev and Mikhail Vasnetsov

2.1
Introduction

Rotational properties of light attract a steady and growing interest in current literature in optics (see reviews in [1–6]). In general, these properties are associated with the circulatory flows of energy in the plane orthogonal to the beam propagation axis. One of the most impressive examples of such energy flows is represented by the well-known "optical vortices" that became a key concept of a new chapter of modern physical optics called *singular optics* [1]. Scalar or linearly polarized fields are transpierced with "threads of darkness," that is, curves of zero amplitude. The wave front structure around these lines generates phase vortices with associated energy vortices, which propagate with the beam and may emerge, annihilate, and interact in accordance with trajectories of the zero-amplitude lines [1]. In such situations, the local energy circulation associated with separate vortices are of importance. In this chapter, we restrict ourselves to beams with relatively simple configuration with (possibly) only one zero line coinciding with the propagation axis.

The rotational characteristics of such beams are generally expressed by the mechanical angular momentum (AM) of the optical field. The AM can be transmitted to other objects, for example, microparticles [6–10]. Two sorts of AM are commonly accepted based on the nature and origination of the considered rotational properties [2, 11]. The spin AM is inherent in light beams with circular or elliptic polarization and depends on the field vector rotations that take place in every point of the beam cross-section; the orbital AM is attributed to the "macroscopic" energy circulation caused by the beam's spatial configuration (e.g., the screw wave front dislocations as a phase frame of optical vortices [1–5, 12]). The orbital and spin AMs can be treated as classical as well as quantum light property [2]. Although there exist some theoretical subtleties concerning the validity of separating the total AM of the electromagnetic field into the spin and orbital parts in the general case [2, 13], the notions of spin and orbital AM are suitable and physically consistent in many practical situations.

Twisted Photons: Applications of Light with Orbital Angular Momentum.
Edited by Juan P. Torres and Lluis Torner

ISBN: 978-3-527-40907-5

2.2 Spin and Orbital Flows: General Concepts

In the last few years, considerable attention has been paid to mutual exchange of the spin and orbital AM in the AM-carrying light beams; in particular, the spin-to-orbital AM conversion induced by the beam transformations causing its strong transverse inhomogeneity [14–21]. Such transformations – for example, sharp focusing [14–18] or transmitting through small apertures [19–21] – are always accompanied by essential deviations from the paraxial character of the beam propagation. Under nonparaxial conditions, the unambiguous separation of the beam AM into the spin and orbital parts is impossible [2, 13]; however, one still can separate the contribution associated with the beam polarization state and the contribution owing to the beam spatial inhomogeneity [11, 16, 22]. Namely, the energy flow density (the Poynting vector time-averaged over the oscillation period) of a monochromatic optical beam can be presented in the form

$$\mathbf{S} = \mathbf{S}_C + \mathbf{S}_O \tag{2.1}$$

where $\mathbf{S}_C$ and $\mathbf{S}_O$ are the so-called *spin and orbital flow densities* (spin and orbital currents) that have recently been studied in detail [22–24]. By using the Gaussian system of units and denoting the light velocity as c and the wave number as k, the summands of Eq. (2.1) are represented by the expressions

$$\mathbf{S}_C = \frac{c}{16\pi k}\,\mathrm{Im}\left[\nabla \times \left(\mathbf{E}^* \times \mathbf{E}\right)\right], \quad S_O = \frac{c}{8\pi k}\,\mathrm{Im}\left[\mathbf{E}^* \cdot (\nabla)\,\mathbf{E}\right] \tag{2.2}$$

Here, $\mathbf{E}$ is the complex electric field (the true electric field strength equals $\mathrm{Re}\left[\mathbf{E}\exp\left(-i\omega t\right)\right]$, where the oscillation frequency $\omega = ck$), $\left[\mathbf{E}^* \cdot (\nabla)\,\mathbf{E}\right]$ is the invariant Berry notation [22] of the vector differential operation that in Cartesian coordinates reads as

$$\left[\mathbf{E}^* \cdot (\nabla)\,\mathbf{E}\right]_j = E_x^* \frac{\partial E_x}{\partial j} + E_y^* \frac{\partial E_y}{\partial j} + E_z^* \frac{\partial E_z}{\partial j}$$

with j standing for x, y, and z. In agreement with Eqs. (2.1) and (2.2), the electromagnetic AM of the beam with respect to a certain reference point with radius-vector $\mathbf{R}_0$ can also be represented as a sum of two terms corresponding to summands of Eq. (2.1),

$$\mathcal{L} = \frac{1}{c^2}\,\mathrm{Im}\int \left[(\mathbf{R} - \mathbf{R}_0) \times S\right] d^3R = \mathcal{L}_C + \mathcal{L}_O \tag{2.3}$$

which can be reduced to the forms

$$\begin{aligned} \mathcal{L}_C &= \frac{1}{8\pi\omega}\,\mathrm{Im}\int \left(\mathbf{E}^* \times \mathbf{E}\right) d^3R, \\ \mathcal{L}_O &= \frac{1}{8\pi\omega}\,\mathrm{Im}\int (\mathbf{R} - \mathbf{R}_0) \times \left[\mathbf{E}^* \cdot (\nabla)\,\mathbf{E}\right] d^3R \end{aligned} \tag{2.4}$$

Here, $\mathbf{R}$ is the radius-vector of the current point in 3D space, the integration is performed over the whole space, and it is supposed that $\mathbf{E} \to 0$ rapidly enough at $|\mathbf{R}| \to \infty$.

As is seen from Eq. (2.4), the term $\mathcal{L}_C$, in contrast to $\mathcal{L}_O$, essentially involves the vector nature of the light wave and does not depend on the position of the reference point, whose properties it shares with the spin AM of a paraxial beam [1]. Moreover, in case of a paraxial beam propagating, say, along axis z, the expression of $\mathcal{L}_C$ following from Eq. (2.4) coincides with the usual spin AM definition [1, 4, 25]. Hence, it can be referred to as the *nonparaxial spin AM*. The similar but reverse arguments allow the term $\mathcal{L}_O$ to be considered as the orbital AM of a nonparaxial beam. When a paraxial beam is tightly focused, its total AM (Eq. (2.3)) is conserved but the initial well-defined paraxial spin and orbital AM are generally redistributed between the nonparaxial spin and orbital AMs (Eq. (2.4)) of the focused beam. This effect is commonly treated as the spin-to-orbital AM conversion.

In experiments, the spin and orbital AM, both in the paraxial and nonparaxial versions, can be distinguished from each other according to the motion of particles suspended within the field of the tested light beam. Under the spin AM action, a particle can only rotate near its own axis, regardless of its position within the beam cross-section, while in the optical field with orbital AM, particles shifted from the beam axis can exhibit orbital motion around it [7–9]. Observation of such orbital motion is the main experimental evidence that the spin-to-orbital AM conversion takes place in strongly focused beams [15, 16].

However, this deduction loses sight of the fact that the spin AM, *per se*, can also induce the orbital motion of a particle, even in the paraxial case. This conclusion readily follows from the recent analyses of energy flows in light beams [22, 23]. In this work, we intend to accentuate this fact and demonstrate its possible manifestations in the usual experimental approaches designed to perform optically induced rotations (optical spanners) [7–10].

2.3
Transverse Energy Flows in Circularly Polarized Paraxial Beams

Let us consider a paraxial light beam propagating along axis z. The electric vector distribution of this beam can be represented as [23, 24]

$$\mathbf{E} = \mathbf{E}_\perp + \mathbf{e}_z E_z = \exp(ikz)\left(\mathbf{u} + \frac{i}{k}\mathbf{e}_z \operatorname{div}\mathbf{u}\right) \tag{2.5}$$

where the slowly varying vector complex amplitude $\mathbf{u} = \mathbf{u}(x, y, z)$ is related to complex amplitudes of orthogonal polarization components of the field (Eq. (2.5)) and $\mathbf{e}_z$ is the unit vector of longitudinal direction. On the basis of circular polarization

$$\mathbf{e}_\sigma = \frac{1}{\sqrt{2}}\left(\mathbf{e}_x + i\sigma\mathbf{e}_y\right)$$

($\mathbf{e}_x, \mathbf{e}_y$ are unit vectors of the transverse coordinates, $\sigma = \pm 1$ is the photon spin number, or helicity),

$$\mathbf{u} = \mathbf{e}_{+1}u_{+1} + \mathbf{e}_{-1}u_{-1} \tag{2.6}$$

$u_\sigma \equiv u_\sigma(x, y, z)$ is the scalar complex amplitude of the corresponding circularly polarized component. Note that in the component with $\sigma = 1$, the electric vector rotates counterclockwise when seen against the beam propagation (left polarization in the terminology of optics [26]). "Partial" intensity and phase distributions of each polarization component equal

$$I_\sigma(x, y, z) = \frac{c}{8\pi} \left| u_\sigma(x, y, z) \right|^2 \tag{2.7}$$

and

$$\varphi_\sigma = \frac{1}{2i} \ln \frac{u_\sigma}{u_\sigma^*} \tag{2.8}$$

The spin flow density (Eq. (2.2)) of the paraxial field (Eq. (2.5)) reduces to

$$\mathbf{S}_C = \frac{1}{2k} \left[\mathbf{e}_z \times \nabla (I_{-1} - I_{+1}) \right] = \frac{1}{2k} \operatorname{rot} \left[\mathbf{e}_z (I_{+1} - I_{-1}) \right] = \frac{1}{2k} \operatorname{rot} (\mathbf{e}_z s_3) \tag{2.9}$$

[23] where s_3 is the fourth Stokes parameter characterizing the degree of circular polarization [27]. Equation (2.9) means that, although in transversely uniform beams the circular polarization produces no macroscopic energy current [4, 28, 29], the specific energy flow occurs in beams with inhomogeneous s_3. In particular, this flow is of circulatory character near extrema of the function $s_3(x, y)$ [23, 24].

The situation becomes especially suitable for analysis in the widespread case of a beam with uniform circular polarization and a circular intensity profile. Then $s_3 = \sigma I_\sigma$ and in the polar frame

$$r = \sqrt{x^2 + y^2}, \quad \phi = \arctan(y/x)$$

the corresponding spin flow (Eq. (2.9)) is expressed by the formula

$$\mathbf{S}_C = -\frac{\sigma}{2k} \left(-\mathbf{e}_r \frac{1}{r} \frac{\partial}{\partial \phi} + \mathbf{e}_\phi \frac{\partial}{\partial r} \right) I_\sigma \tag{2.10}$$

where the unit vectors of polar coordinates are introduced in agreement with equations

$$\mathbf{e}_x = \mathbf{e}_r \cos\phi - \mathbf{e}_\phi \sin\phi, \quad \mathbf{e}_y = \mathbf{e}_r \sin\phi + \mathbf{e}_\phi \cos\phi$$

For comparison, the orbital flow density (transverse part of the second expression (2.2)) of the same beam, in accordance with Eqs. (2.5)–(2.8), is given by the equation [23]

$$\mathbf{S}_O = \frac{1}{k} I_\sigma \nabla \varphi_\sigma = \frac{1}{k} I_\sigma \left(\mathbf{e}_\phi \frac{1}{r} \frac{\partial}{\partial \phi} + \mathbf{e}_r \frac{\partial}{\partial r} \right) \varphi_\sigma \tag{2.11}$$

For paraxial beams, it is natural to consider the AM with respect to the propagation axis z and to characterize it by the linear density (AM per unit length of the beam) [1, 4, 12] which is expressed by the proper modification of Eq. (2.3) [4, 23]

$$\mathcal{L}' = \frac{1}{c^2} \operatorname{Im} \int [\mathbf{r} \times \mathbf{S}] \, d^2 r = \frac{1}{c^2} \operatorname{Im} \int S_\phi r^2 \, dr \, d\phi$$

where $\mathbf{r}$ is the transverse radius-vector, S_ϕ is the Poynting vector azimuthal component and the integration is performed over the whole cross-section of the beam. With allowance for Eqs. (2.10) and (2.11), the spin and orbital AM linear densities for a paraxial beam can be written in the well-known forms [23]

$$\mathcal{L}'_C = -\frac{\sigma}{2\omega c}\int_0^\infty r^2 dr \int_0^{2\pi} \frac{\partial I_\sigma}{\partial r} d\phi = \frac{\sigma}{\omega c}\int_0^\infty rdr \int_0^{2\pi} I_\sigma d\phi \tag{2.12}$$

$$\mathcal{L}'_O = \frac{1}{\omega c}\int_0^\infty rdr \int_0^{2\pi} I_\sigma \frac{\partial \varphi_\sigma}{\partial \phi} d\phi \tag{2.13}$$

(in the second Eq. (2.12), the fact that $I_\sigma(r,\phi) \to 0$ when $r \to \infty$ has been employed).

One can notice a great degree of similarity between Eqs. (2.10) and (2.11): both $\mathbf{S}_C$ and $\mathbf{S}_O$ originate from the beam transverse inhomogeneity and their components are directly related to the azimuthal and radial derivatives of the beam profile parameters. However, while the orbital flow is mainly "produced" by the phase gradient (and the variable intensity can only modify it due to factor I_σ), the spin flow is completely dependent on the amplitude inhomogeneity of a circularly polarized beam. There also exists a difference in the interrelations between the streamline patterns of $\mathbf{S}_C$ ($\mathbf{S}_O$) and the spatial derivatives of the corresponding "master" parameter I_σ (φ_σ): while $\mathbf{S}_O$ is always directed *along* the *phase* gradient, $\mathbf{S}_C$ is *orthogonal* to the *intensity* gradient. Nevertheless, in what concerns the action on suspended microparticles, both flows are expected to be almost equivalent, provided the quantitative characteristics of the flow patterns are commensurate. Now consider the detailed characterization of the spin and orbital flows in some simple examples.

For a Gaussian beam in the waist cross-section (beam waist radius b) the intensity (Eq. (2.7)) and phase (Eq. (2.8)) distributions appear in the forms

$$\varphi_\sigma = 0,\ I_\sigma = I_{\sigma 0}\exp\left(-\frac{r^2}{b^2}\right) \tag{2.14}$$

The wave front of this beam is flat and the orbital flow (Eq. (2.11)) vanishes; the spin flow is determined by the last term of Eq. (2.10)

$$\mathbf{S}_C = -\sigma \mathbf{e}_\phi \frac{1}{2k}\frac{\partial I_\sigma}{\partial r} \tag{2.15}$$

which due to Eq. (2.14) gives

$$\mathbf{S}_C = \sigma \mathbf{e}_\phi \frac{r}{kb^2} I_{\sigma 0}\exp\left(-\frac{r^2}{b^2}\right) \tag{2.16}$$

(in Figure 2.1 the spin flow pattern in the left-polarized beam, $\sigma = 1$, is presented).

A slightly more complicated situation occurs in Laguerre–Gaussian (LG) beams which, along with the spin helicity (circular polarization), possess the "orbital helicity" – the screw wave front dislocations giving rise to optical vortices of the lth order ($|l| > 1$ is the integer azimuthal index) [2, 4, 12]. Restricting ourselves, for simplicity, to beams with zero radial index, let us again consider the waist cross-section where

$$\varphi = l\phi,\ I_\sigma = \frac{1}{|l|!} I_{\sigma 0}\left(\frac{r}{b}\right)^{2|l|}\exp\left(-\frac{r^2}{b^2}\right) \tag{2.17}$$

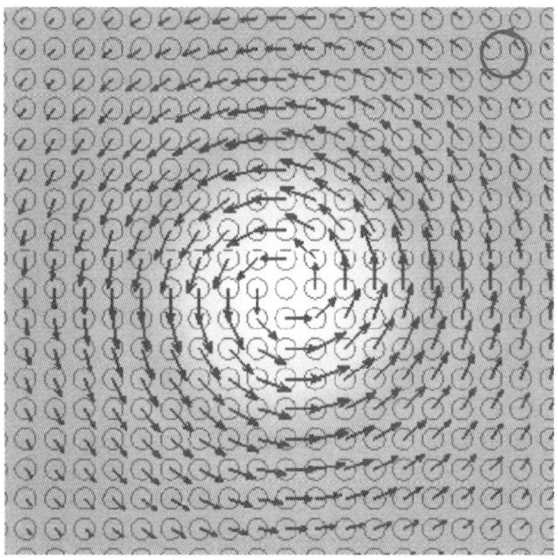

Figure 2.1 Map of the spin flow density of Eq. (2.16) for a left-polarized Gaussian beam ($\sigma = 1$, polarization handedness is shown in the upper right corner); lengths of arrows correspond to relative flow density, the intensity distribution and polarization ellipses (circles) are shown in the background, the beam is viewed against the propagation axis. (Please find a color version of this figure on the color plates.)

The normalization constant $(|l|!)^{-1}$ warrants that the beam's total power for every l is the same. With allowance for Eq. (2.17) the last term of Eq. (2.10) gives

$$\mathbf{S}_C = -\mathbf{e}_\phi \sigma I_{\sigma 0} \frac{1}{|l|!} \frac{1}{kb} \left(\frac{r}{b}\right)^{2|l|-1} \left(|l| - \frac{r^2}{b^2}\right) \exp\left(-\frac{r^2}{b^2}\right) \tag{2.18}$$

and, following Eq. (2.11), the orbital flow is found to be

$$\mathbf{S}_O = \mathbf{e}_\phi I_{\sigma 0} \frac{1}{|l|!} \frac{1}{kb} \left(\frac{r}{b}\right)^{2|l|-1} l \exp\left(-\frac{r^2}{b^2}\right) \tag{2.19}$$

These equations stipulate a simple relation between the spin and orbital flows of the circularly polarized beams that are being considered:

$$\mathbf{S}_C = -\frac{\sigma}{l} \left(|l| - \frac{r^2}{b^2}\right) \mathbf{S}_O \ (l \neq 0) \tag{2.20}$$

The derived dependencies are illustrated by Figure 2.2a–d. In contrast to the spin and orbital AM densities of Eqs. (2.12) and (2.13), which usually coincide [2, 4] with the transverse intensity distribution of circularly polarized LG beams (curves I), the corresponding transverse energy flows (curves S_C and S_O) behave differently. At any l, the circulatory energy flows vanish on the axis ($r = 0$); of course, far from the axis ($r \to \infty$), they vanish also. In the intermediate region, absolute values of the spin and orbital flows possess extrema. The orbital flow magnitude $|\mathbf{S}_O|$ (Eq. (2.19)) has the maximum at

$$\frac{r}{b} = \sqrt{\frac{2|l| - 1}{2}} \quad (|l| > 0) \tag{2.21}$$

and extremum points of the spin flow density (Eq. (2.18)) satisfy the condition

$$\left(\frac{r}{b}\right)^2 = |l| + \frac{1}{4} \pm \frac{\sqrt{16|l| + 1}}{4} \tag{2.22}$$

which corresponds to maxima of $|\partial I_\sigma / \partial r|$ on the inner and outer sides of the bright ring of the "donut" mode pattern (Eq. (2.17)); at $l = 0$, the inner

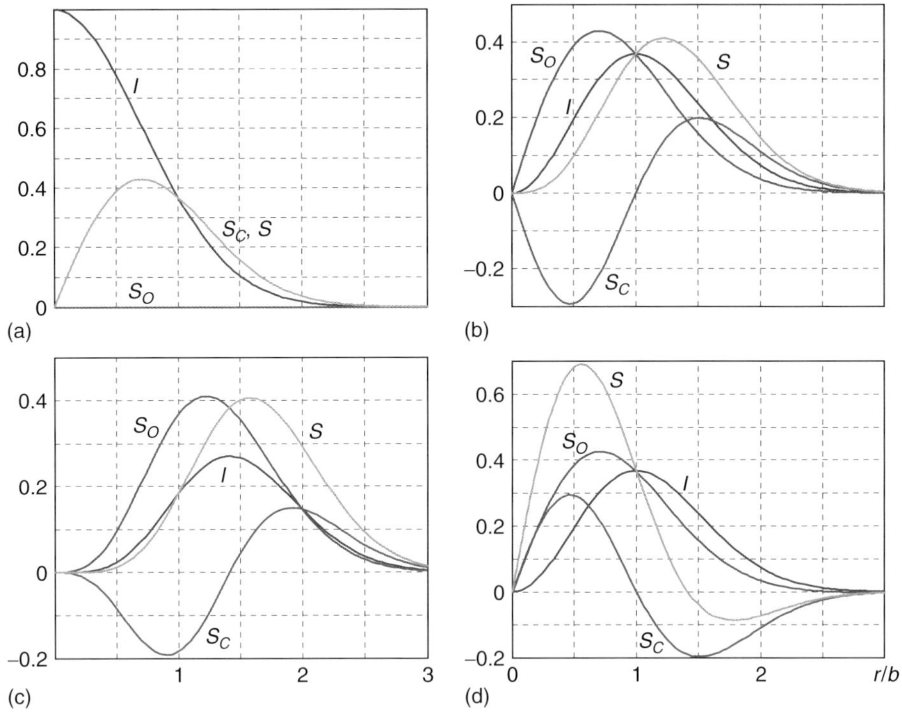

Figure 2.2 Radial profiles of (I) intensity (Eq. (2.17)) in units of $I_{\sigma 0}$, (S_C) spin flow density (Eq. (2.18)), (S_O) orbital flow density (Eq. (2.19)), and (S) total transverse flow density (Eq. (2.24)) (all in units of $I_{\sigma 0}/kb$), for the circularly polarized LG beams with zero radial index and the following sets of parameters: (a) $\sigma = 1, l = 0$ (Gaussian beam of Figure 2.1), (b) $\sigma = 1, l = 1$, (c) $\sigma = 1, l = 2$, (d) $\sigma = -1, l = 1$. (Please find a color version of this figure on the color plates.)

extremum disappears and the only maximum of the absolute spin flow density occurs at

$$r = b\big/\sqrt{2} \tag{2.23}$$

The expected zero spin flow takes place at the "brightest" line of the ring where I_σ is maximal.

Equations (2.18) and (2.19) and Figure 2.2 show that in many cases the magnitudes of spin and orbital flow densities are of the same order. Consequently, they are expected to have similar experimental manifestations. In particular, since the orbital flow, due to associated mechanical momentum, can force the orbital rotation of particles [7, 8], the same effect can be caused by the spin flow. This must be taken into account in experiments on the spin-to-orbital AM conversion [15, 16]. In real situations, it is the total transverse energy flow

$$\mathbf{S} = \mathbf{S}_C + \mathbf{S}_O = \left(1 - \sigma\frac{|l|}{l} + \frac{\sigma}{l}\frac{r^2}{b^2}\right)\mathbf{S}_O \tag{2.24}$$

(Figure 2.2, curves marked S), with associated mechanical momentum $\mathbf{P} = \mathbf{S}/c^2$, that is likely to be the motivating factor for orbital rotation of the probing particles. The spin and orbital contributions may support as well as suppress each other (Figure 2.2). In the region $r/b < l$, which is the most important physically because it contains prevailing part of the beam power, the orbital flow dominates; otherwise (at the beam periphery) the spin contribution is more intensive.

An interesting situation occurs in the near-axis region $r/b \ll 1$ where, due to Eq. (2.20), absolute magnitudes of the spin and orbital flows are almost identical. Then, if signs of l and σ coincide (i.e., handedness of the macroscopic optical vortex of the LG beam and handedness of the circular polarization are the same), the total transverse energy circulation is zero at small $r \ll b$ (Figure 2.2b). That the spin flow can be directed oppositely to the polarization handedness, seems, at first sight, counterintuitive but can be simply explained by the "cell model" of the spin flow formation [4, 29]. Formally, this follows immediately from the fact that the spin flow handedness is determined not only by σ, but also by the sign of $\partial I_\sigma/\partial r$ (Eq. (2.15)).

On the contrary, if the polarization handedness is opposite to the orbital circulation, the spin and orbital flows add constructively and enable the maximum local values of the total rotational energy flow available for circularly polarized LG beams with given l, as is seen from Figure 2.2d, curve S.

The flow maps presented in Figure 2.3 are in full agreement with the data of Figure 2.2d. For considered beams, the orbital flow density possesses the same handedness in the whole cross-section (compare Figure 2.3a and curve S_O); however, the spin and the total flows may reverse. Regions of opposite circulations in Figure 2.3b,c are separated by contours, where the relevant energy flow constituent vanishes, corresponding to sign alterations in curves S_C and S.

Note that in calculation of the full spin AM over the whole cross-section (e.g., by first formula (Eq. (2.12))), the "opposite" spin flow of the near-axis region is compensated for by the periphery contribution where the spin flow reverses. As a

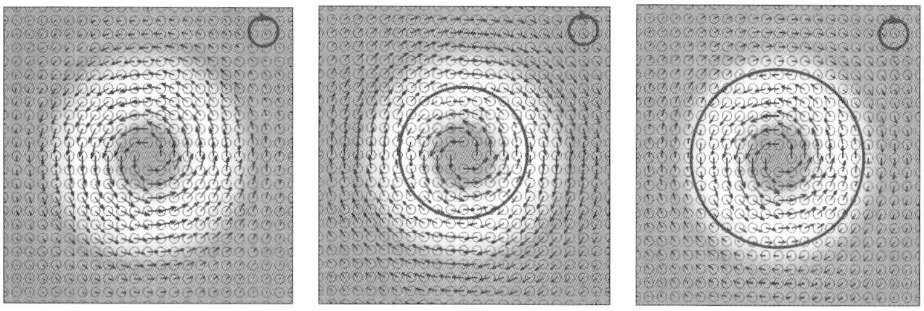

(a) (b) (c)

Figure 2.3 Maps of the (a) orbital $\mathbf{S}_O$, (b) spin $\mathbf{S}_C$, and (c) total $\mathbf{S}$ transverse energy flows in the cross-section of a right-polarized LG beam (Eq. (2.17)) with $l = 1, \sigma = -1$ (case of Figure 2.2d). At every point, polarization is the same as shown in the upper right corners; circular contours in panels (b) and (c) are contours where the corresponding flow component vanishes. (Please find a color version of this figure on the color plates.)

result, the handedness of the total spin AM of the considered uniformly polarized beam always coincides with σ, which is seen from the second Eq. (2.12) where this compensation is ensured automatically.

2.4 Orbital Rotation without Orbital Angular Momentum

Now let us dwell upon the peculiarities of the spin flow as a factor inducing the orbital motion of suspended particles and experimental conditions enabling unambiguous manifestation of the spin flow. Its action would be especially expressive in case of the Gaussian beam (Eq. (2.14)) where the orbital flow is absent. In the most general features, the idea of the experiment does not differ much from that devised for the orbital AM demonstration [9, 10, 15]. The tested beam falls normally onto the cell with suspended particles which are situated off-axially with respect to the beam axis. They experience the motive force proportional to the local energy flow density expressed, for example, by Eqs. (2.15), (2.16), (2.18), and (2.19). In fact, this force is directed tangentially and, if it is the only force acting on the particles, they move centrifugally. In order to get the closed orbital motion, some additional steps should be taken to keep the particles in a fixed circular trajectory. In experiments with beams carrying the orbital AM, this can be realized by special configuration of the beam itself. If it possesses a (multi-)ring-like transverse profile (e.g., LG beam with at least one nonzero index, Bessel beam), the particles experience the gradient force due to the optical field inhomogeneity, and tend to be confined within rings of high or low intensity depending on their optical properties. In studies of the spin flow, this technique is inappropriate because in regions of the intensity extrema the spin flow vanishes (Eq. (2.15)). This forces us to look for other solutions (Figure 2.4). For example, the cell with suspended particles may contain a ring-like channel or cuvette of the proper mean radius corresponding to the maximum spin flow (Figure 2.4a,b), for example, what is dictated by Eqs. (2.22) or (2.23). In the channel, the particles are kept mechanically, for example, due to the special shape of the cell bottom. Such a mechanical trapping may be inconvenient because the particles' orbital motion is hampered by friction at the channel boundaries. Otherwise, the channel can be formed by a sort of ring-like optical trap, for example, by an auxiliary light beam with ring-like intensity profile. The intensity of the auxiliary field must be sufficient to form the perceptible peak or gap in the resulting intensity distribution (on account of the driving beam whose spin flow is analyzed); besides, the auxiliary beam should be free from additional rotatory action (i.e., possess no orbital AM).

Interesting possibilities open up due to variable handedness of the transverse energy circulation, as Figure 2.3c displays. This pattern means that direction of the tangential force applied to a particle depends on its radial position so the speed and direction of the orbital rotation can be switched by changing the driving beam radius or the ring-like trap radius. Another expected peculiarity of the motion caused by the spin flow is that the particles absorbing a part of the incident circularly polarized

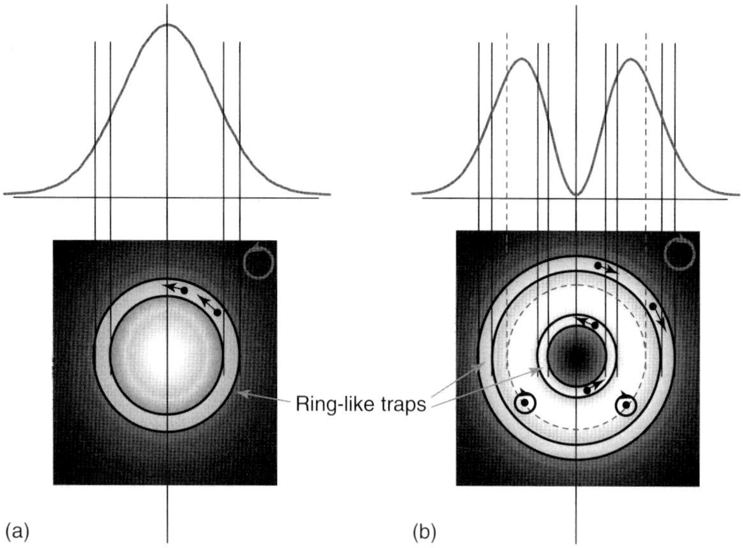

Figure 2.4 Possible schemes of orbital motion of the absorbing suspended particles confined in the ring-like traps within the circularly polarized field of (a) Gaussian beam of Figure 2.1 and (b) LG beam with $l = 1$, $\sigma = -1$ of Figure 2.3. Top row: diametric sections of the intensity profiles with boundaries of the ring-like traps, bottom row: views of the beam cross sections with the trap traces (polarization handedness is indicated in the upper right corners). Circles with arrows indicate the expected orbital motion of the trapped particles and dashed lines in panel (b) specify locations where orbital motion is not excited (see the circular contour in Figure 2.3c). Particles situated at this contour perform only the spinning motion (shown by the arrow loops); in all other positions the spinning motion is not shown but is also expected in addition to the orbital one. (Please find a color version of this figure on the color plates.)

light will thus be set in rotation about their own axes, in addition to the orbital motion around the driving beam axis (Figure 2.4b). Handedness of this spinning motion is the same over the whole cross-section of the homogeneously polarized beam, although its rate will generally vary in accordance with the inhomogeneous intensity.

2.5 Conclusion

The relatively simple examples of light beams with internal rotation considered in this work illustrate the main properties and potentiality of the spin and orbital flows in problems of optical manipulation. Even if not employed, the spin constituent of the transverse energy circulation must be taken into account in experiments involving the optically induced orbital rotation of microparticles. Interesting applications of circularly polarized beams may arise from the possibility of combining the orbital and spinning motion of the same particle. A more rich variety of particle motion and new possibilities of their control can be expected in case of more complicated driving beams; for example, those with inhomogeneous polarization. The presented analysis constitutes the starting point and outlines the way in which these complicated cases can be studied further.

References

1. Soskin, M.S. and Vasnetsov, M.V. (2001) Singular optics. *Prog. Opt.*, **42**, 219–276.
2. Allen, L., Padgett, M.J., and Babiker, M. (1999) Orbital angular momentum of light. *Prog. Opt.*, **39**, 291–372.
3. Franke-Arnold, S., Allen, L., and Padgett, M. (2008) Advances in optical angular momentum. *Laser Photon. Rev.*, **2** (4), 299–313.
4. Bekshaev, A., Soskin, M., and Vasnetsov, M. (2008) *Paraxial Light Beams with Angular Momentum*, Nova Science Publishers, New York.
5. Dennis, M.R., O'Holleran, K., and Padgett, M.J. (2009) Singular optics: optical vortices and polarization singularities. *Prog. Opt.*, **53**, 293–364.
6. Nieminen, T.A., Higuet, J., Knoner, G., Loke, V.L.Y., Parkin, S., Singer, W., Heckenberg, N.R., and Rubinsztein-Dunlop, H. (2006) Optically driven micromachines: progress and prospects. *Proc. SPIE*, **6038**, 237–245.
7. He, H., Friese, M.E.J., Heckenberg, N.R., and Rubinsztein-Dunlop, H. (1995) Direct observation of transfer of angular momentum to absorptive particles from a laser beam with a phase singularity. *Phys. Rev. Lett.*, **75**, 826–829.
8. Simpson, N.B., Dholakia, K., Allen, L., and Padgett, M.J. (1997) Mechanical equivalence of spin and orbital angular momentum of light: an optical spanner. *Opt. Lett.*, **22**, 52–54.
9. O'Neil, A.T., MacVicar, I., Allen, L., and Padgett, M.J. (2002) Intrinsic and extrinsic nature of the orbital angular momentum of a light beam. *Phys. Rev. Lett.*, **88**, 053601.
10. Garces-Chavez, V., McGloin, D., Summers, M.D., Fernandez-Nieves, A., Spalding, G.C., Cristobal, G., and Dholakia, K. (2004) The reconstruction of optical angular momentum after distortion in amplitude, phase and polarization. *J. Opt. A: Pure Appl. Opt.*, **6**, S235–S238.
11. Jackson, J.D. (1999) *Classical Electrodynamics*, 3rd edn, John Wiley & Sons, Inc., New York.
12. Allen, L., Beijersbergen, M.V., Spreeuw, R.J.C., and Woerdman, J.P. (1992) Orbital angular momentum of light and the transformation of Laguerre-Gaussian modes. *Phys. Rev. A*, **45**, 8185–8189.
13. Barnett, S.M. (2002) Optical angular momentum flux. *J. Opt. B: Quantum Semiclass. Opt.*, **4**, S1–S10.
14. Bomzon, Z., Gu, M., and Shamir, J. (2006) Angular momentum and geometrical phases in tight-focused circularly polarized plane waves. *Appl. Phys. Lett.*, **89**, 241104.
15. Zhao, Y., Edgar, J.S., Jeffries, G.D.M., McGloin, D., and Chiu, D.T. (2007) Spin-to-orbital angular momentum conversion in a strongly focused optical beam. *Phys. Rev. Lett.*, **99**, 073901.
16. Nieminen, T.A., Stilgoe, A.B., Heckenberg, N.R., and Rubinsztein-Dunlop, H. (2008) Angular momentum of a strongly focused Gaussian beam. *J. Opt. A: Pure Appl. Opt.*, **10**, 115005.
17. Monteiro, P.B., Maia Neto, P.A., and Nussenzveig, H.M. (2009) Angular momentum of focused beams: beyond the paraxial approximation. *Phys. Rev. A*, **79**, 033830.
18. Chen, B. and Pu, J. (2009) Tight focusing of elliptically polarized vortex beams. *Appl. Opt.*, **48**, 1288–1294.
19. Vuong, L.T., Adam, A.J.L., Brok, J.M., Planken, P.C.M., and Urbach, H.P. (2010) Electromagnetic spin-orbit interactions via scattering of sub-wavelength apertures. *Phys. Rev. Lett.*, **104**, 083903.
20. Schouten, H.F., Visser, T.D., and Lenstra, D. (2004) Optical vortices

near sub-wavelength structures. *J. Opt. B*, **6**, S404–S409.

21. Schouten, H.F., Visser, T.D., Gbur, G., Lenstra, D., and Blok, H. (2004) Connection between phase singularities and the radiation pattern of a slit in a metal plate. *Phys. Rev. Lett.*, **93**, 173901.
22. Berry, M. (2009) Optical currents. *J. Opt. A: Pure Appl. Opt.*, **11**, 094001 (12 pp).
23. Bekshaev, A.Ya. and Soskin, M.S. (2007) Transverse energy flows in vectorial fields of paraxial beams with singularities. *Opt. Commun.*, **271**, 332–348.
24. Bekshaev, A. and Soskin, M. (2007) Transverse energy flows in vectorial fields of paraxial light beams. *Proc. SPIE*, **6729**, 67290G.
25. Berry, M.V. (1998) Paraxial beams of spinning light. *Proc. SPIE*, **3487**, 6–11.
26. Born, M. and Wolf, E. (1970) *Principles of Optics*, 4th edn, Pergamon Press, Oxford.
27. Shurkliff, W.A. (1962) *Polarized Light*, Harvard University Press, Cambridge, MA.
28. Allen, L. and Padgett, M.J. (2000) The Poynting vector in Laguerre-Gaussian beams and the interpretation of their angular momentum density. *Opt. Commun.*, **184**, 67–71.
29. Bekshaev, A.Ya. (2006) Spin angular momentum of inhomogeneous and transversely limited light beams. *Proc. SPIE*, **6254**, 56–63.

3
Helically Phased Beams, and Analogies with Polarization

Miles Padgett

3.1
Introduction

Helically phased beams are characterized in terms of an azimuthal phase dependence of $\exp(i\ell\theta)$, which implies that the beam axis is a phase singularity around which the optical phase changes by $2\pi\ell$. Optical realizations of such beams have been considered at least since 1979, when Vaughan and Willets noted the interference patterns in the output from a Krypton ion laser and by analogy with superfluids [1], Coullet coined the phrase "optical vortex" [2]. Helically phased beams were created with high purity, directly from a laser designed by Tamm and Weiss [3] and using diffraction gratings designed by Soskin *et al.* [4]. However, it was not until 1992 when Allen *et al.*, working within Woerdman's group, recognized that helical phasefronts, and their associated azimuthal component of the Poynting vector gave an angular momentum corresponding to $\ell\hbar$ per photon [5]. This orbital angular momentum (OAM) is independent of the polarization state of the beam and hence unrelated to the spin of individual photons. We see that the phrase "optical vortex" in fact applies to the vortex of the transverse energy and momentum flow around the singularity.

Although terms such as phase singularity, optical vortex, and OAM are often used interchangeably, they are, of course, distinct phenomena. Within a cross section of the field, the point of phase singularity has no intensity and hence, carries no energy or momentum. This point is, however, surrounded by a region of higher intensity, and the helical phasefronts give rise to an azimuthal component of the energy and momentum flow – an optical vortex [6]. This momentum flow can be expressed in terms of an OAM directed parallel to the beam axis.

The 1992 study used cylindrical lenses to convert the Hermite–Gaussian (HG) output of a laser into a Laguerre–Gaussian (LG) mode, which is the most obvious example of a mode having helical phasefronts. In addition to the azimuthal index, that is ℓ, the radial profile is characterized by a radial index p, giving the number of radial nodes. The operating principles of these cylindrical lens mode converters were also detailed by the same group [7]. In essence, the astigmatic focus between

Twisted Photons: Applications of Light with Orbital Angular Momentum.
Edited by Juan P. Torres and Lluis Torner

ISBN: 978-3-527-40907-5

the lenses introduces a Gouy phase that depends upon the indices, n and m, of the incident HG mode. For an HG mode, rotated by 45° with respect to the lens axis, the phasing between modes is such that the result is a perfect LG mode, where $\ell = n - m$ and $p = \min(m, n)$. Of particular interest to us here, is the recognition that the action of the lenses was mathematically analogous to the change in phase introduced by a waveplate for the transformation of the polarization state. The mode converter can be designed to either transform the HG mode into an LG mode (refer to the transformation of a linear polarization at 45° into circularly polarized light by a quarter-waveplate) or reverse the handedness of an LG mode (in analogy with a half-waveplate).

3.2
Representation of Helically Phased Beams

Polarization states are describable within a two-dimensional state space and hence, all possible polarization states can be represented on the surface of a Bloch-sphere, in this case, termed the *Poincaré sphere*. For describing polarization, the poles of the sphere correspond to circularly polarized states and the equator to linear states at various angles. Any polarization state can be expressed as an appropriately weighted and phased superposition of right- and left-handed states, where right- and left-handed circularly polarized light corresponds to a spin angular momentum (SAM) of the photon of $\pm\hbar$. In contrast to the two orthogonal states of SAM, ℓ can take any integer value for OAM and hence the state space is unbounded. However, for an OAM subspace described by the superposition of any two modes, a similar sphere can still be constructed [8]. Most obviously, this applies to the modes $LG_{(l=+1,p=0)}$ and $LG_{(l=-1,p=0)}$. Placing these modes at the north and south poles of the sphere gives perfect $HG_{(n=1,m=0)}$ modes around the equator, with an orientation depending upon longitude; that is, the relative phase at which the two LG beams are added together. The representation can be extended for the superposition of any pair of LG modes $\pm l$, in which case the beams at the equator comprise $2l$ petals, see Figure 3.1. Similar representations can also by applied to other beam types characterized by helical phase fronts [9, 10]. On this Poincaré-type sphere, we see that the LG modes are analogous to circular polarization and the HG modes analogous to linear polarization. The cylindrical lens modes converters described in the previous section give identical state transformation between these modes, as waveplates do for polarization.

Of course this Poincaré sphere representation is limited to the description of superpositions of two orthogonal states. For polarization states, an equivalent representation to the Poincaré sphere are the Jones matrices, where any polarization state is described by a two-element column-vector and any transformation of state by a two-by-two matrix. Such a matrix approach can be extended to represent any number of orthogonal states. Both $LG_{(l,p)}$- and $HG_{(m,n)}$ modes are characterized in terms of their mode order, $N = \ell + 2p = m + n$, where modes of the same order

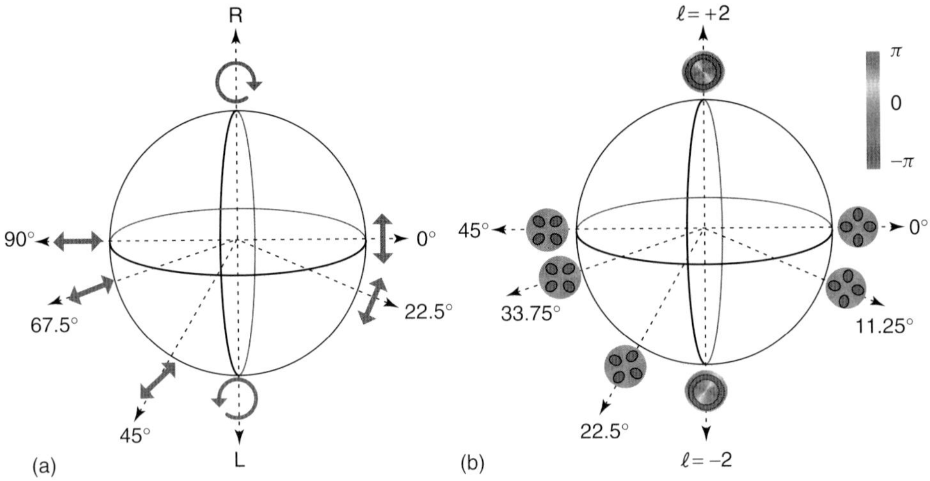

Figure 3.1 (a) The Poincaré sphere representation of polarization state. (b) An equivalent representation for beams formed from the superposition of Laguerre–Gaussian modes $\ell = 2, p = 0$ and $\ell = -2, p = 0$. (Please find a color version of this figure on the color plates.)

experience the same Gouy phase and expand upon propagation at the same rate. For a mode order N, there are $N + 1$ orthogonal modes that can be represented by an $N + 1$ element column-vector and transformation between modes of the same order by an $N + 1$ element matrix [11]. In all cases, the transformation matrices can describe the action of mode converters as well as rotations or mirror inversion of any mode. This analogy between modal and polarization states is pertinent to a number of research areas.

3.3 Exploiting the Analogous Representations of Spin and Orbital Angular Momentum

3.3.1 Rotational Doppler Shifts and Geometrical Phase

In 1979, Arnold and Garetz demonstrated that a half-waveplate, spinning with respect to a second waveplate, introduces a frequency shift to circularly polarized light, which can be completely characterized in terms of Jones matrices [12]. In terms of the instantaneous electric field, it is rotated by twice the angle between it and the optic axis of the half-waveplate. For incident circular polarization, this results in a reversal of the handedness of the transmitted light. A second half-waveplate returns the handedness of polarization to the original sense. A single rotation of either waveplate causes the field to undergo two additional rotations in the opposite direction, explaining why the observed frequency shift is twice the rotation rate of the spinning waveplate. For a beam with circular intensity, this rotation of the polarization is identical to a rotation of the beam itself, hence the frequency shift $\Delta\omega$, can be expressed in terms of the beam rotation frequency Ω, as $\Delta\omega = \sigma\Omega$ where $\sigma = \pm 1$ for right- and left-handed circularly polarized light respectively.

When the polarizations transformation is traced on the surface of the Poincaré sphere, the shift can also be interpreted in terms of an evolving geometric or "Berry" phase [13]. The direct analogy between polarization and modal states was appreciated in the original work on the cylindrical lens mode converter [7] and it was soon recognized that mode converters too, could be considered to introduce a geometric phase change between the transmitted mode components [14], and a corresponding frequency shift for a spinning mode converter [15]. Indeed, the geometric phases are prevalent throughout such mode transformations [16]. Originally, the equivalent mode converter to a half-waveplate was a pair of cylindrical lenses separated by twice their focal length; this acts to invert the incident image about the major axis of the lenses. A more common form of the image inverter is, however, a Dove prism that, in our case, transforms an LG mode from $+l$ to $-l$.

Creating a noncircular beam spinning about its own axis is harder than one might imagine. Rather than rotating the polarization state with a spinning waveplate, one now needs to rotate a complex image, without introducing any associated translation or tilt. At optical frequencies this has proved extremely challenging, since any mode converter based on cylindrical lenses or Dove prisms suffers from slight misalignment. However, it is also possible to produce helically phased beams at much longer wavelengths, where the relative mechanical precision of the various components and their alignment is much better.

In the microwave region of the electromagnetic spectrum, helical phasefronts are readily produced using a spiral phase plate inserted into the free-space propagating beam [17]. It is similarly possible to fabricate lenses, prisms, waveplates, and so on, also operating at these quasi-optical frequencies. The longer wavelength means that standard machining tolerances are sufficient to ensure that any misalignment is well below the wavelength of the beam and it is possible to make a rotating mode converter that spins the transmitted beam without any significant translation or tilt. The other advantage of these comparatively low frequencies is that rather than being deduced from interferometric techniques, the frequency can be counted directly.

Using a pair of Dove prisms to spin a helically phased beam gives a measured frequency shift of $\Delta\omega = l\Omega$ [18].

Of course, an interesting question is what happens when the beam carries both OAM and SAM. Is the result two separate frequency shifts or do the two components of angular momentum combine to give a shift proportional to the total angular momentum? Plotting the instantaneous field cross section of a circularly polarized, helically phased beam shows a $(l + \sigma)$-fold rotational symmetry, meaning that a single rotation of the beam about its propagation axis advances or retards the phase by $(l + \sigma)$ cycles. Spinning the beam gives a frequency shift of $\Delta\omega = (l + \sigma)\,\Omega$,

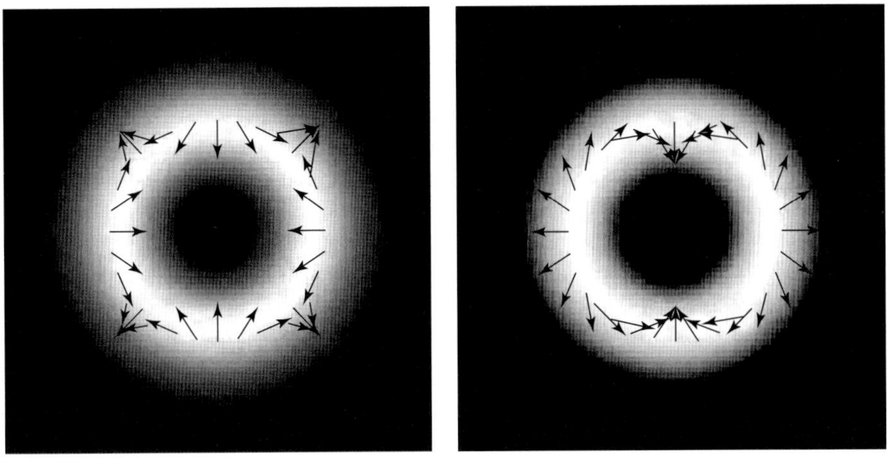

(a) (b)

Figure 3.2 Cross sections showing the orientation of the electric field of circularly polarized helically phased beams. (a) $l = 3, \sigma = 1$, revealing a $l + \sigma = 4$-fold rotational symmetry and (b) $l = -3, \sigma = 1$, revealing a $l + \sigma = 2$-fold rotational symmetry.

where the total angular momentum of that photon is $J = \hbar\left(l + \sigma\right)$ (Figure 3.2). Note that in this case, the SAM and the OAM act indistinguishably and the induced frequency shift could be used as a direct measure of the beam's total angular momentum [19].

An interesting comparison between the translational and rotational Doppler shifts is made by expressing both in terms of the linear and angular momentum per photon as $p_0 = \hbar k_0$ and $J = \hbar\left(l + \sigma\right)$ respectively. The translational shift can now be written as $\Delta\omega = v\left(p_0/\hbar\right)$ and the rotational shift as $\Delta\omega = \Omega\left(J/\hbar\right)$. For both SAM and OAM, the energy exchange implied by the frequency shift can be identified with the reversal of handedness and the associated azimuthal component of the optical momentum. Rotation of the optical component against this reaction force transfers energy to or from the transmitted beam [20].

Finally, this rotational Doppler shift should not be confused with the normal Doppler shift seen when a body rotates about an axis, orientated to give a component of velocity toward or away from the observer. The rotational Doppler shift arising from the light's angular momentum is maximal in a direction parallel to the angular rotation vector, along which the translational shift is zero.

3.3.2
Mode Sorting using Geometric Phase

The first experiments to investigate the quantum implications of OAM were those by Mair *et al.* [21]. They used the classic forked hologram to selectively couple a target mode into a single-mode fiber and by interchanging the hologram, they were able to test for specific modes, one at a time. However, this is inherently lossy, yielding efficiency that is no greater than the reciprocal of the number of states to be tested [22]. Certainly, if the full potential of the large Hilbert space offered by OAM is to be realized, then it is desirable to be able to measure the projection of a single photon onto many dimensions in OAM Hilbert space simultaneously [23].

Subsequent to the work on rotational frequency shifts, Courtial recognized that the geometric phase associated with the rotation of a helically phased beam could be used as the basis of a mode sorter, allowing the angular momentum of individual photons to be identified [24]. Rather than spinning the Dove prism to produce an evolving phase and hence a frequency shift, it suffices to place Dove prisms in the arms of a Mach–Zehnder interferometer. The relative rotation angle between them introduces a fixed geometric phase that depends upon the index l of the incident mode. For an angular displacement of the Dove prisms of α, the phase shift between the arms is given by $\Delta\phi = 2\alpha l$. For $\alpha = \pi/2$, this gives a relative phase shift of 0 for even–l and π for odd–l, resulting in even–l and odd–l beams being coupled to different output ports. Thus, a single interferometer can be used to separate odd and even modes with 100% efficiency. The approach can be extended by cascading interferometers with different values of α such that, for example, the even modes can be further separated (using $\alpha = \pi/4$) into those modes $l = 2 + 4n$ and $l = 4n(n = 0, \pm 1, \pm 2, \ldots)$. A separation into N different modes requires $N - 1$ interferometers [25].

The same interferometric approach could be adapted to separate polarization states, where the Dove prism would be replaced by half-waveplates – the interferometric equivalent to a polarizing beam splitter. However, as with the rotational frequency shift, the concept can be further extended by combining Dove prisms and half-waveplates so that the induced phase shift is given by the total angular momentum, that is $\Delta\phi = 2\alpha\left(l + \sigma\right)$.

Unfortunately, although the interferometric approach does show that it is, in principle, possible to sort orbital angular momentum states with 100% efficiency, the technical challenges of keeping multiple interferometers aligned are great. An alternative approach that eliminates the need for precise alignment remains an exciting opportunity for ongoing research.

Distinct from this sorting of OAM states is the switch between OAM states by changing the spin state of the beam. These mode switchers are based on birefringent material, whose orientation varies within the azimuthal position in the beam such that the spatially varying geometric phase imparts a helical phase structure to the transmitted light [26–28]. Such optical elements can be used for both the efficient generation and measurement of angular momentum states.

3.3.3 Entanglement of Spatial Modes

Another application of the analogous representation of SAM and OAM is in the area of the quantum entanglement of OAM states. As already mentioned,

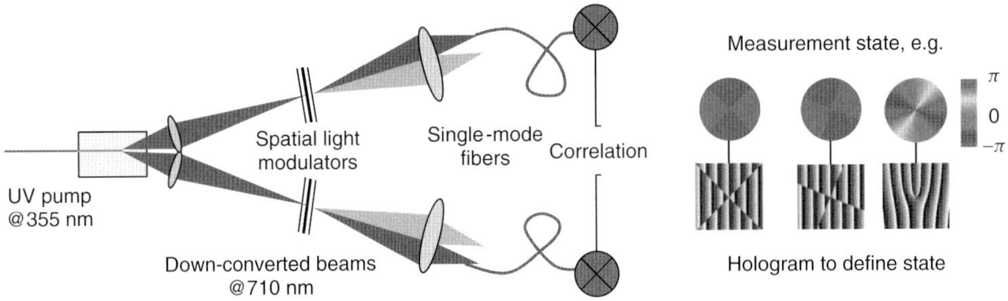

Figure 3.3 Observing correlations in the orbital angular momentum of down-converted beams. The use of spatial light modulators to define various holograms allows the measurement of an arbitrary spatial mode, including orbital angular momentum states analogous to those of polarization on the Poincaré sphere. (Please find a color version of this figure on the color plates.)

the first examination of the truly quantum aspects of OAM was performed by Zeilinger *et al.* [21, 29]. They used forked holograms to record the correlations in the measured OAM states of the down-converted photons (Figure 3.3). However, correlations between parallel and orthogonal states can have a classical explanation. The key to observing the uniquely quantum properties is that such correlations persist in a complementary measurement basis, or more generally that partial correlations can be observed between various superpositions of the states. In the original work, these superpositions were measured by deliberate displacement of the measurement axis.

The key demonstration of quantum entanglement was, of course, by Aspect *et al.*, who observed the measured correlations in the polarization states of two separated photons [30]. That quantum mechanics was both nonlocal and indescribable by a hidden variable theory is revealed through a violation of a Bell inequality. For a Hilbert space of two dimensions, the Bell inequality relates to the variation in the correlation as the angle between the two measurement states is varied. For polarization, this angle simply maps to the relative angle between the two measurement polarizers. As one polarizer is rotated with respect to the other, the measured correlations should vary sinusoidally. In terms of the Poincaré sphere, this corresponds to holding one measurement state fixed on the equator and scanning the other around the equator. In terms of OAM states, the equator corresponds to an equally weighted superposition of right- and left-handed helically phased beams, giving a "petal"-type intensity distribution with $2l$-fold rotational symmetry. Changing the phase of the superposition gives a rotation of the petals. A phase change of 2π corresponds to one rotation of the state around the equator and the rotation of the beam by the angle between neighboring petals [31]. This equivalent representation of polarization and helically phased beams means that the original demonstration of a violation for polarization states can be repeated exactly for helically phased beams both for discrete [32] and continuous [33] variables. The directness of the analogy between SAM and OAM means that the robust arguments applied to polarization can be applied to helically phased beams and spatial modes.

3.3.4
Photon Drag and the Mechanical Faraday Effect

That the speed of light is independent of the frame of reference is, of course, a cornerstone of relativity. However, this is not the case when the refractive index of the moving medium is no longer unity. A moving medium changes the velocity, an effect known as *photon drag* or *ether drag*. When the medium has a velocity parallel to the propagation direction, the "Fresnel" drag results in a retardation or advance of the optical phase, which can be observed interferometrically. When the medium has a velocity that is transverse to the propagation direction, the transmitted light is laterally displaced. For a medium of thickness L, refractive index n, and moving a velocity v, the lateral displacement is given by $\Delta x = \left(n - l/n\right) vL/c$. The analysis and observation of this photon drag effect was extensively investigated by Jones *et al.* [34–36].

A closely related effect is for rotational motion where the plane of polarization is rotated by a spinning object through an angle $\Delta\theta = \left(n - l/n\right) \Omega L/c$. This rotational effect was also investigated by Jones *et al.* [37, 38] and has been subsequently referred to as a *mechanical Faraday effect* [39]. The rotation of a plane of linear polarization is equivalent to a phase change between the right- and left-handed circular polarization states, which prompts the question "what influence does a rotating medium have on the relative phase of helically phased beams, and how might this be observed?" The annular intensity cross section of OAM modes means that the rotation drag can be analyzed as a linear drag acting at a radius vector on the skew rays describing the mode, which gives identical expressions for the phase change associated with both SAM and OAM (Figure 3.4). However, whereas

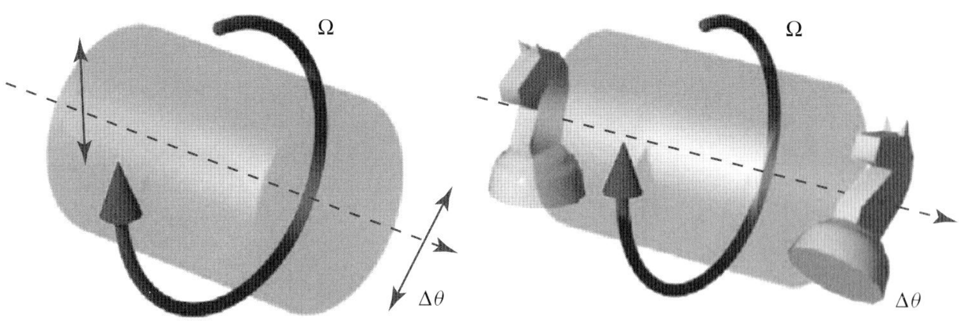

Figure 3.4 A spinning medium is predicted to slightly rotate both the polarization state and the image of the transmitted light. (Please find a color version of this figure on the color plates.)

the phase change between the SAM states gives a rotation of the polarization state, the phase change between OAM states gives a rotation of the resulting beam. Since OAM states form a complete basis set, capable of describing any cross section, this rotation of the beam is manifested as the rotation of an image [40]. So it would appear that a spinning window should slightly rotate the image of any scene, through an angle $\Delta\theta = (n - 1/n)\,\Omega L/c$ [41]. This effect has been confirmed by observing the delay of a spinning image at it is passed through a stationary window [42], but the true observation of the rotational transformation of an image that is observed through a spinning window remains an experimental challenge.

3.4 Conclusions

Beyond being a convenient representation, the use of a Poincaré sphere or Jones matrix description of helically phased beams, and hence their analogy to polarization, gives insight into the behaviors of orbital as compared to spin angular momentum. Mode transformations and geometrical phase both have analogous behaviors for spin, orbital, and total angular momenta, leading to frequency shifts, mode sorting, and quantum phenomena. Subtler than these transformations are analogies between birefringence and optical activity, and the corresponding effects of helically phased beams. As was recognized in the initial work on mode converters, birefringence, which imparts a phase change between orthogonal states, is akin to astigmatic focusing of helically phased beams. The equivalent to optical activity is actually image rotation, a fact revealed by experiments on photon drag as light passes through rotating or translating media.

Despite these numerous areas where SAM and OAM exhibit equivalent behavior, there is certainly not a general equivalence. Nowhere is this distinction more apparent than in the interaction of angular momentum carrying light and microscopic objects – optical spanners. In the special case of the object being bigger than the optical beam, partial absorption leads to an equivalent torque from both spin and orbital components. More generally, for asymmetric scatter, or birefringently induced polarization changes, or when the object is smaller than the optical beam, SAM and OAM have distinct effects on particle motion. For a circularly polarized, helically phased annular beam, a small birefringent particle falling into the ring will experience a torque from the circular polarization, causing it to spin round its own axis and a scattering recoil force from the helical phasefronts causing it to orbit around the beam axis. On an atomic scale, the circular polarization can excite a $\sigma = \pm 1$ Zeeman transition, whereas the helical phase fronts can induce a complicated recoil shift of the atomic center of mass. More complicated might be the interaction between a helically phased beam and an extended atom-like object such a Rydberg atom or loosely bound exciton. Under these circumstances, whether the spin and orbital components behave in an equivalent fashion remains a point of conjecture.

References

1. Vaughan, J.M. and Willetts, D.V. (1979) Interference properties of a light-beam having a helical wave surface. *Opt. Commun.*, **30**, 263–267.
2. Coullet, P., Gil, L., and Rocca, F. (1989) Optical vortices. *Opt. Commun.*, **73**, 403–408.
3. Tamm, C. and Weiss, C.O. (1990) Bistability and optical switching of spatial patterns in a laser. *J. Opt. Soc. Am. B: Opt. Phys.*, **7**, 1034–1038.
4. Bazhenov, V.Y., Vasnetsov, M.V., and Soskin, M.S. (1990) Laser-beams with screw dislocations in their wave-fronts. *JETP Lett.*, **52**, 429–431.
5. Allen, L., Beijersbergen, M.W., Spreeuw, R.J.C., and Woerdman, J.P. (1992) Orbital angular-momentum of light and the transformation of Laguerre-Gaussian laser modes. *Phys. Rev. A*, **45**, 8185–8189.
6. Padgett, M.J. and Allen, L. (1995) The Poynting vector in Laguerre-Gaussian laser modes. *Opt. Commun.*, **121**, 36–40.
7. Beijersbergen, M.W., Allen, L., Vanderveen, H., and Woerdman, J.P. (1993) Astigmatic laser mode converters and transfer of orbital angular-momentum. *Opt. Commun.*, **96**, 123–132.
8. Padgett, M.J. and Courtial, J. (1999) Poincare-sphere equivalent for light beams containing orbital angular momentum. *Opt. Lett.*, **24**, 430–432.
9. Soares, W.C., Caetano, D.P., and Hickmann, J.M. (2006) Hermite-Bessel beams and the geometrical representation of nondiffracting beams with orbital angular momentum. *Opt. Express*, **14**, 4577–4582.
10. Wang, W., Dennis, M.R., Ishijima, R., Yokozeki, T., Matsuda, A., Hanson, S.G., and Takeda, M. (2007) Poincare sphere representation for the anisotropy of phase singularities and its applications to optical vortex metrology for fluid mechanical analysis. *Opt. Express*, **15**, 11008–11019.
11. Allen, L., Courtial, J., and Padgett, M.J. (1999) Matrix formulation for the propagation of light beams with orbital and spin angular momenta. *Phys. Rev. E*, **60**, 7497–7503.
12. Garetz, B.A. and Arnold, S. (1979) Variable frequency-shifting of circularly polarized laser-radiation via a rotating half-wave retardation plate. *Opt. Commun.*, **31**, 1–3.
13. Bretenaker, F. and Le Floch, A. (1990) Energy exchanges between a rotating retardation plate and a laser-beam. *Phys. Rev. Lett.*, **65**, 2316–2316.
14. van Enk, S.J. (1993) Geometric phase, transformations of Gaussian light-beams and angular-momentum transfer. *Opt. Commun.*, **102**, 59–64.
15. Nienhuis, G. (1996) Doppler effect induced by rotating lenses. *Opt. Commun.*, **132**, 8–14.
16. Galvez, E.J., Crawford, P.R., Sztul, H.I., Pysher, M.J., Haglin, P.J., and Williams, R.E. (2003) Geometric phase associated with mode transformations of optical beams bearing orbital angular momentum. *Phys. Rev. Lett.*, **90**, 203901.
17. Turnbull, G.A., Robertson, D.A., Smith, G.M., Allen, L., and Padgett, M.J. (1996) Generation of free-space Laguerre-Gaussian modes at millimetre-wave frequencies by use of a spiral phaseplate. *Opt. Commun.*, **127**, 183–188.
18. Courtial, J., Dholakia, K., Robertson, D., Allen, L., and Padgett, M.J. (1998a) Measurement of the rotational frequency shift imparted to a rotating light beam possessing orbital angular momentum. *Phys. Rev. Lett.*, **80**, 3217–3219.
19. Courtial, J., Robertson, D., Dholakia, K., Allen, L., and Padgett, M.J. (1998b) Rotational frequency shift of a light beam. *Phys. Rev. Lett.*, **81**, 4828–4830.
20. Padgett, M.J. (2004) The mechanism for energy transfer in the rotational frequency shift of a light beam. *J. Opt. A: Pure Appl. Opt.*, **6**, S263–S265.
21. Mair, A., Vaziri, A., Weihs, G., and Zeilinger, A. (2001) Entanglement of the orbital angular momentum states of photons. *Nature*, **412**, 313–316.
22. Gibson, G., Courtial, J., Padgett, M.J., Vasnetsov, M., Pasko, V., Barnett, S.M.,

and Franke-Arnold, S. (2004) Free-space information transfer using light beams carrying orbital angular momentum. *Opt. Express*, **12**, 5448–5456.

23. Molina-Terriza, G., Torres, J.P., and Torner, L. (2002) Management of the angular momentum of light: preparation of photons in multidimensional vector states of angular momentum. *Phys. Rev. Lett.*, **88**, 013601.
24. Leach, J., Padgett, M.J., Barnett, S.M., Franke-Arnold, S., and Courtial, J. (2002) Measuring the orbital angular momentum of a single photon. *Phys. Rev. Lett.*, **88**, 257901.
25. Leach, J., Courtial, J., Skeldon, K., Barnett, S.M., Franke-Arnold, S., and Padgett, M.J. (2004) Interferometric methods to measure orbital and spin, or the total angular momentum of a single photon. *Phys. Rev. Lett.*, **92**, 013601.
26. Biener, G., Niv, A., Kleiner, V., and Hasman, E. (2002) Formation of helical beams by use of Pancharatnam-Berry phase optical elements. *Opt. Lett.*, **27**, 1875–1877.
27. Marrucci, L., Manzo, C., and Paparo, D. (2006) Optical spin-to-orbital angular momentum conversion in inhomogeneous anisotropic media. *Phys. Rev. Lett.*, **96**, 163905.
28. Calvo, G.F. and Picon, A. (2007) Spin-induced angular momentum switching. *Opt. Lett.*, **32**, 838–840.
29. Vaziri, A., Weihs, G., and Zeilinger, A. (2002) Experimental two-photon, three-dimensional entanglement for quantum communication. *Phys. Rev. Lett.*, **89**, 240401.
30. Aspect, A., Dalibard, J., and Roger, G. (1982) Experimental test of Bell inequalities using time-varying analyzers. *Phys. Rev. Lett.*, **49**, 1804–1807.
31. Franke-Arnold, S., Leach, J., Padgett, M.J., Lembessis, V.E., Ellinas, D., Wright, A.J., Girkin, J.M., Ohberg, P., and Arnold, A.S. (2007) Optical ferris wheel for ultracold atoms. *Opt. Express*, **15**, 8619–8625.
32. Leach, J., Jack, B., Romero, J. *et al.* (2009) Violation of a Bell inequality in two-dimensional orbital angular momentum state-spaces. *Opt. Express*, **17**, 8287–8293.
33. Lassen, M., Leuchs, G., and Andersen, U.L. (2009) Continuous variable entanglement and squeezing of orbital angular momentum states. *Phys. Rev. Lett.*, **102**, 163602.
34. Jones, R.V. (1972) Fresnel aether drag in a transversely moving medium. *Proc. R. Soc. London, Ser. A: Math. Phys. Sci.*, **328**, 337–352.
35. Jones, R.V. (1975) Ether-drag in a transversely moving medium. *Proc. R. Soc. London, Ser. A: Math. Phys. Eng. Sci.*, **345**, 351–364.
36. Player, M.A. (1975) Dispersion and transverse Aether drag. *Proc. R. Soc. London, Ser. A: Math. Phys. Eng. Sci.*, **345**, 343–344.
37. Jones, R.V. (1976) Rotary Aether drag. *Proc. R. Soc. London, Ser. A: Math. Phys. Eng. Sci.*, **349**, 423–439.
38. Player, M.A. (1976) Dragging of plane of polarisation of light propagating in a rotating medium. *Proc. R. Soc. London, Ser. A: Math. Phys. Eng. Sci.*, **349**, 441–445.
39. Nienhuis, G., Woerdman, J.P., and Kuscer, I. (1992) Magnetic and mechanical faraday effects. *Phys. Rev. A*, **46**, 7079–7092.
40. Allen, L. and Padgett, M. (2007) Equivalent geometric transformations for spin and orbital angular momentum of light. *J. Mod. Opt.*, **54**, 487–491.
41. Padgett, M., Whyte, G., Girkin, J., Wright, A., Allen, L., Ohberg, P., and Barnett, S.M. (2006) Polarization and image rotation induced by a rotating dielectric rod: an optical angular momentum interpretation. *Opt. Lett.*, **31**, 2205–2207.
42. Leach, J., Wright, A., Gotte, J. *et al.* (2008) "Aether drag" and moving images. *Phys. Rev. Lett.*, **100**, 153902.

4
Trapping and Rotation of Particles in Light Fields with Embedded Optical Vortices

Michael Mazilu and Kishan Dholakia

4.1
Introduction

Modern photonics has been at the heart of a large range of advances in the natural sciences. One of the areas of success of photonics has been optical micromanipulation, which has sustained its importance and impact during that time. The advent of the laser heralded a myriad of breakthroughs that have had repercussion across all of the sciences. The light–matter interaction is at the core of micromanipulation and has emerged as one of the key advances seen and indeed has enabled truly groundbreaking studies: this includes the topics of laser cooling, Bose–Einstein condensation, and – central to this chapter – the topic of manipulating and trapping particles from the nanometric size scale right up to the size of a single cell. We restrict ourselves here to a discussion of the forces and torques upon mesoscopic particles and cells. Pertinent to the topic of this book, we shall emphasize and explain the role of optical vortex fields, namely Laguerre–Gaussian (LG) light modes as well as other light fields, in optical micromanipulation. One of the key drivers of the area has been the ability to exploit the orbital angular momentum (OAM), inherent to certain forms of these modes, as well as the shaped wave front and dark vortex core.

The output of a laser cavity is electromagnetic fields that form solutions of the wave equations and naturally satisfy appropriate boundary conditions. These solutions may form a complete and orthogonal set of functions [1] and are termed the *higher modes of propagation* in textbooks. Essentially, any transverse light field distribution can be decomposed in terms of these orthogonal set of modes. In the literature, these higher-order transverse modes are often referred to as *nonzero-order-* or *sculpted* light fields. Such nonzero-order light fields are of importance in optical micromanipulation as they enable forms of motion and tailored trajectories of trapped objects that would be difficult if not impossible to achieve with combinations of standard Gaussian fields. Light fields may impart linear momentum to trapped objects, which is the core to understanding the basic remit of the field of micromanipulation. It is instructive to consider how objects are

Twisted Photons: Applications of Light with Orbital Angular Momentum.
Edited by Juan P. Torres and Lluis Torner

ISBN: 978-3-527-40907-5

confined and held in traditional Gaussian beams before progressing to the impact and role of optical vortex fields.

The key point about implementing optical vortices in trapping, in particular the LG modes, is that angular as well as linear momentum may be transferred to trapped objects. While the origin of the linear momentum of light is established, the angular momentum content is due to the helical or inclined wave fronts of the field. The consequence is an unusual trajectory to the Poynting vector for the light field: this denotes the energy flow that has an azimuthal component and this leads to the notion of OAM [2].

The OAM takes on discrete values $l\hbar$, where l is the azimuthal index (integer) of the beam; in such light beams, the OAM offers itself as a useful tool in the exertion of torques on particles on a toroidal trap by simply increasing the azimuthal index in contrast to the optical torque arising from spin angular momentum which varies with optical power [3] and is limited to $\hbar$ per photon. A number of light modes may be generated with embedded optical vortices. The most prominent for the purposes of this chapter is the LG mode or beam although we mention others such as Bessel and Mathieu light beams. We commence our discussion of the LG light mode with an emphasis on their relevant parameters for optical micromanipulation. We then progress to a theoretical discussion of optical forces and then finally describe some of the main experimental uses of optical vortex fields for rotation and trapping of particles.

4.2 Laguerre–Gaussian Light Beams

If we explore the topic of transverse laser modes, the circularly symmetric Laguerre–Gaussian (LG) laser modes form a complete orthonormal basis set for paraxial light beams. Any light field may thus be expanded and represented as a sum of these modes with appropriate weight. An LG mode is typically described as LG_p^l, where l and p are the two integer indices that describe the mode. The azimuthal index l is the number of complete 2π phase cycles around the mode circumference. A given mode will have $p+1$ radial modes and we typically deal with $p = 0$ modes of the form of an annulus. The azimuthal index l is often referred to as the *topological charge* of the optical vortex.

The azimuthal phase term $\exp(-il\phi)$ of LG modes gives rise to a well-defined OAM of $l\hbar$ per photon [2]. As this integer may exceed unity, the angular momentum is greater than that associated with the spin state of the field. The azimuthal phase term implies an inclined phasefront and an azimuthal component to the energy flow (Poynting vector): this, thus, gives us a physical interpretation of the origin of OAM [2, 4]. In the paraxial approximation, this form of angular momentum in the LG beam can be decoupled from spin angular momentum arising from its polarization state [5]. The scalar field of an LG mode $u(\mathrm{LG}_p^p)$ of indices l and p may be written as

$$u(\mathrm{LG}_l^p) \propto \exp\left(\frac{-ikr^2 z}{2(z^2+z_r^2)}\right)\exp\left(\frac{-r^2}{\omega^2}\right)\exp\left(-i(2p+l+1)\arctan\left(\frac{z}{z_r}\right)\right)$$
$$\exp(-il\phi)(-1)^p\left(\frac{r\sqrt{2}}{\omega}\right)^l L_p^l\left(\frac{2r^2}{\omega^2}\right) \tag{4.1}$$

where z denotes the distance from the beam waist, z_r is the Rayleigh range, k is the wave number, ω is the radius at which the Gaussian term $\exp(-r^2/\omega^2)$ falls to $1/e$ of its on-axis value, r is the radial distance, ϕ is the azimuthal angle, and L_p^l is the generalized Laguerre polynomial. The term $(2p+l+1)\arctan(z/z_r)$ is the Guoy phase of the LG mode, which will vary with the mode indices.

In terms of optical trapping experiments, efficient generation of the LG laser modes and other beams such as Bessel or Mathieu beams with embedded optical vortices is crucial. Intracavity generation of LG modes from within the laser resonator cavity is possible if the cavity has a certain asymmetry [6]; however, this is problematic in most instances and certainly in the field of optical micromanipulation, modes with optical vortices, such as the LG or Bessel modes, are usually generated externally to the trapping laser. One of the first techniques of LG mode generation made use of the output of a higher order Hermite–Gaussian (HG) transverse mode (for example, by inserting an intracavity cross-wire) from a laser cavity. The resultant HG mode may then be converted to an LG laser mode using a mode converter [7]. A mode converter comprises two cylindrical lenses of focal length f, canonically disposed with respect to one another. When placed at a distance of $\sqrt{2}f$ between the two cylindrical lenses, this system of lenses introduces a Guoy phase shift of $\pi/2$ on an incident HG mode laser beam of indices m and n and transforms it to an LG mode of indices $l=(m-n)$ and $p=\min(m,n)$. Notably, we may obtain a high-purity LG mode by this method, but this requires a careful selection of the HG laser mode desired and avoidance of any undesired astigmatism in the optical system. Such a mode of generation was used for one of the early experiments in optical rotation with LG beams where Simpson and colleagues rotated trapped absorptive Teflon particles in three-dimensional optical tweezers [8].

Ideally, we would wish to take a fundamental TEM_{00} Gaussian beam and sculpt the output mode and embed vortices directly within the beam profile. Diffractive optical elements have enabled this to become a mainstream technology in recent years. They may be dynamic or static in nature. For the case of LG beams, the two most powerful methods are the use of a spiral phase element or the use of a computer-generated hologram.

A spiral phase element is typically a high refractive index substrate that is shaped into the spiral phase ramp [9, 10]. The incident Gaussian beam is not deviated in direction but is directly converted to an LG beam. With the advances seen in recent microfabrication techniques, the spiral phase element has been miniaturized [11, 12] and even generated in microfluidic geometries.

A recent study fabricated the hologram in polydimethylsiloxane (PDMS) using soft-lithography methods [13]. The exact output of the spiral plate is not a pure LG mode but is rather a superposition of LG modes [9]. Such plates are now commercially available and can result in a high conversion efficiency typically in excess of 80%.

The computer-generated holographic generation method requires little in the way of "fabrication process" as one can mathematically encode the spiral phase with a given input field at an angle onto a computer-generated pattern. Importantly, this may be achieved with an on-axis or off-axis hologram; the off-axis method is useful to avoid any "zeroth-order" overlap and is also key in the case of broadband vortex generation. The transmission function of the off-axis hologram T required to generate a single-ringed ($p = 0$) LG beam of azimuthal order $\pm l$ may be represented as

$$T = \frac{1}{2}(1 - \cos(k_x x \pm l\phi)) \tag{4.2}$$

where k_x defines the periodicity of the grating along the x axis, l is the azimuthal order, and ϕ is $\tan^{-1}(y/x)$. Propagation of a Gaussian beam, $G(r, \phi, z)$, through the hologram T will yield a mathematical equation that will have three functions that contribute to the zeroth- and the two conjugate orders. We find that the topological orders from the two diffraction orders are equal but opposite in charges (positive and negative).

$$\begin{aligned} G(r,\phi,z)\,T &= G(r,\phi,z)\left(\frac{1}{2}(1 - \cos(k_x x \pm l\phi))\right) \\ &= G(r,\phi,z)\left(\frac{1}{2} - \frac{e^{i(k_x x \pm l\phi)} + e^{-i(k_x x \pm l\phi)}}{4}\right) \\ &= \left(\frac{1}{2}G(r,\phi,z) - \frac{1}{4}G(r,\phi,z)e^{i(k_x x \pm l\phi)} - \frac{1}{4}G(r,\phi,z)e^{-i(k_x x \pm l\phi)}\right) \end{aligned} \tag{4.3}$$

This holographic technique has gained more popularity because of the ease and versatility of LG beam generation, and, in particular, has been used with dynamic elements such as the spatial light modulator. This element is an array of liquid crystal droplets that can be electrically or optically addressed and be used to generate holograms in real time, obviating the need to swap between static diffractive optical elements in a setup. The OAM of these light fields can be seen by careful consideration of the helical wave fronts of an LG beam, and is related to the azimuthal phase l seen in Eq. (4.1). The inclined helical wave front leads to a Poynting vector that moves in a corkscrew-like manner [2]. This angular momentum is therefore linked with the azimuthal component of the Poynting vector as already stated.

4.3 Origin of Optical Torques and Forces

4.3.1 Intuitive Picture of Optical Forces

The propagation of light corresponds to the flow of electromagnetic energy through space. This flow of energy is interrupted when light is incident on a surface of a transparent body. Indeed, some of the light gets reflected while some gets refracted and all together the flow of energy is modified. In mechanical terms, the variation of the optical energy flow can be seen as a force acting on the flow. This force originates from the surface of the transparent body where light changes the propagation direction. Because of the action–reaction principle, the force of the surface acting on the light is equal and opposite in sign to the force of the light field acting on the surface. This force is the optical force.

In intuitive terms, the flow of electromagnetic energy is equivalent to the momentum of the light field. When a light beam scatters from a body, its momentum is changed. As the total momentum of the system is conserved, the difference between the initial and final momentum is transferred to the scattering body (Figure 4.1), which is then subject to an optical force.

It is relatively easy to calculate the optical forces acting on the objects, which are small in comparison to the wavelength of light. In general, the forces acting on these objects can be decomposed into two parts that only depend on the light field intensity distribution and its energy flow. The gradient part of the field makes the dielectric particle seek regions of the optical beam that have the highest intensity. Its origin is the induced electric dipole in the small particles. The second part of optical force is the scattering force. This scattering force simply pushes small particles in the direction of the energy flow and it is due to the overall linear momentum transfer in an isotropic scattering case. In optical tweezers, these two force components are arranged to counteract each other so as to form a stable trapping position. This is the case for the trapping of microparticles in a tightly focused laser beam. Indeed, a microparticle is pushed along the beam, in the direction of propagation of the beam. In certain conditions the gradient force can cancel this scattering force to create stable trapping positions.

The optical forces mentioned above correspond to the linear momentum of the light field and its translation to forces. Another conserving quantity that is transferred from the optical field to the scattering bodies is the angular momentum. In classical mechanics, the angular momentum is associated with the rotational state of a rigid body. An isolated physical system conserves its angular momentum, meaning that its rotation continues "in the same way" for as long as no external torque acts upon it. The torque $\boldsymbol{\tau}$ defines the amount of rotation, with respect to a point, induced on a body by a force $\mathbf{F}$

$$\boldsymbol{\tau} = \mathbf{r} \times \mathbf{F} \tag{4.4}$$

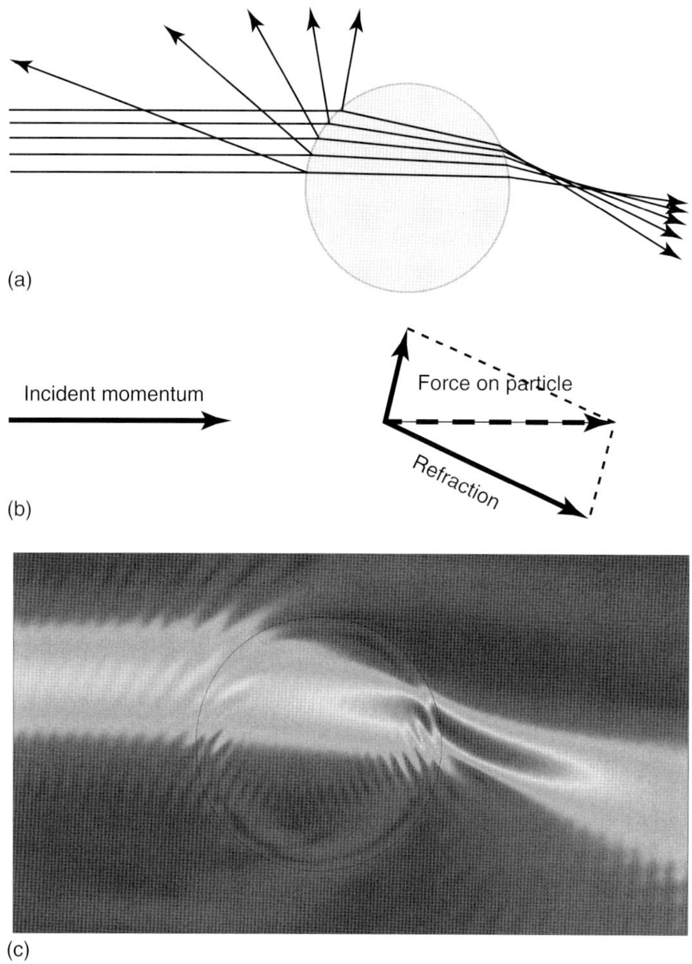

Figure 4.1 Optical forces arising from the conservation of momentum. (a) Bundle of rays being reflected and refracted by a transparent sphere. (b) Total momentum before and after the optical interaction. (c) Refraction and reflection of a light beam by a dielectric sphere. (Please find a color version of this figure on the color plates.)

where **r** is the position vector linking the center of the rotation and the point of application of the force.

In a similar way, the electromagnetic angular momentum of a light beam can be defined as its ability to induce torque on an optically scattering object. There are two distinct mechanisms through which optical torque is transported by the electromagnetic wave. The first one is spin angular momentum that is transmitted by the polarization of the light and interacts with birefringent materials, for example. Indeed, a quarter-wave optical plate that converts linearly polarized light into circularly polarized light experiences a mechanical torque in the process. The second transfer mechanism is OAM, which corresponds to the locally skewed motion of the energy flow of the light beam with respect to the average energy flow of the light beam. The OAM of light can be observed in the case of optical micromanipulation of microparticles in LG beams, for example.

4.3.2 Angular Momentum within Geometric Optics

The most intuitive way to calculate the angular momentum of light is by considering the momentum transfer occurring during the refraction and reflection of rays by a dielectric particle whose size is large compared to the wavelength. Within this approach, the change of direction of a ray corresponds to a momentum transfer to the reflecting or refracting surface. The direction and amplitude of the momentum transfer is given by the vectorial momentum conservation relation between each of the rays involved (Figure 4.1). The incident, refracted, and reflected rays carry a momentum proportional to their intensities and having a direction given by their propagation. The difference between the incident total momentum and the scattered one determines the force acting on the surface.

The total force acting on a large spherical particle can be calculated using simple geometrical optical ray tracing, which involves integration of propagation of all the rays using the below shown formula [14, 15]:

$$\mathbf{F} = \frac{1}{c}\int_S dS\, I(\mathbf{r})\cos(\theta_i)\left(\hat{\mathbf{u}}_i - R\hat{\mathbf{u}}_r - T^2\sum_{k=1}^{N} R^{k-1}\hat{\mathbf{u}}_{tk}\right) \tag{4.5}$$

where θ_i is the angle of the incident ray (Figure 4.2) with respect to the normal at the point of incidence and the unit vectors $\hat{\mathbf{u}}_i$, $\hat{\mathbf{u}}_r$, and $\hat{\mathbf{u}}_{tk}$ correspond, respectively, to the propagation direction of the incident, reflected, and transmitted beams where the latter takes into account k internal reflections. The coefficients T and

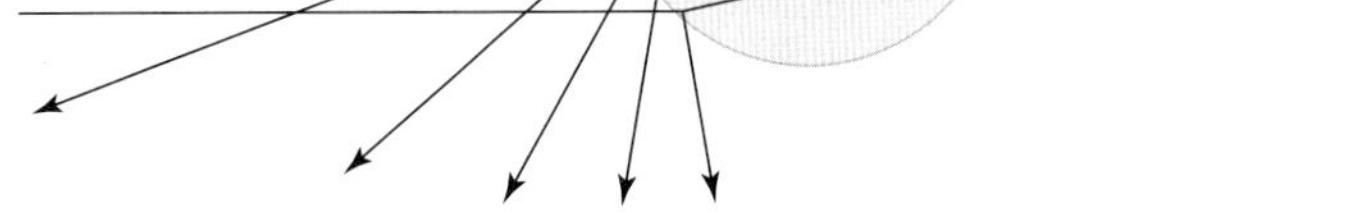

Figure 4.2 Rays reflecting and refracting from a transparent dielectric sphere. The various terms are defined in the text. (Please find a color version of this figure on the color plates.)

R denote Fresnel's transmission and reflection coefficients averaged over the two different polarization directions. Within this ray optics picture, the total optical force depends on the way an incident beam decomposes into its constituent rays and amplitudes. This optical forces also give rise to a torque that can be calculated using Eqs (4.4) and (4.5). In the case of a higher order Bessel beam, this torque transfers OAM from the beam to the particle.

4.3.3 Paraxial Beams

The optical ray optics formula shown above delivers a good force and torque approximation in the case of large optical particles and noninterfering beams. Both polarization effects and angular spin transfer to particles are more difficult to describe within this formula. To deal with these limitations one needs to take into account the wave nature and propagation properties of the light field. Fortunately, in most experiments, it is sufficient to consider the propagation of the beams within the paraxial approximation. This approximation considers optical beams that propagate along an optical axis and whose amplitude varies slowly compared to the wavelength of the carrier wave. Here, we consider a carrier wave propagating in the z direction defined by $\exp(ik_z z - i\omega t)$ where $k_z = \omega/c$ is the wave-vector with ω and c representing the optical frequency and speed of light in vacuum. The paraxial scalar field $u(x, y, z)$ is then a solution of the paraxial equation

$$\frac{\partial^2 u}{\partial x^2} + \frac{\partial^2 u}{\partial y^2} + 2ik_z \frac{\partial u}{\partial z} = 0 \tag{4.6}$$

The electromagnetic vector fields associated with these scalar paraxial fields depend additionally on the polarization state of the vector field and are defined by

$$\begin{aligned} \mathbf{E} &= -\nabla V - \partial_t \mathbf{A} \\ &= -\frac{c^2}{i\omega}\nabla(\nabla \cdot \mathbf{A}) + i\omega \mathbf{A} \\ \mathbf{B} &= \nabla \times \mathbf{A} \end{aligned} \tag{4.7}$$

where we use the vector potential $\mathbf{A} = (a_x, a_y, 0)u(x, y, z)\exp(ik_z z - i\omega t)$ and the Lorenz gauge condition $c^2\nabla \cdot \mathbf{A} + \partial_t V = 0$ to define the electric field $\mathbf{E}$ and the magnetic flux $\mathbf{B}$. These two fields are complex valued and their real part corresponds to the real fields. The electric scalar potential denoted by V is also assumed to oscillate at the same frequency as the carrier wave. The complex amplitudes a_x and a_y correspond to the field amplitude along the x and y directions. The relative magnitude and phase of these two amplitudes give rise to different polarization states of light. For example, $a_x = ia_y$ corresponds to circularly polarized light while $a_x = a_y$ corresponds to linearly polarized light. The general case of two different amplitudes having a phase delay between them gives rise to elliptical polarization.

In the case of small particles, we can decompose the optical force into a gradient and a scattering force. The gradient force simply pushes the small particles toward the regions of high intensity while the scattering force is related to the energy flow and polarization of the beam. This scattering force is proportional to the time-averaged Poynting vector given by [2, 16]

$$< \mathbf{E} \times \mathbf{H} > = \frac{(|a_x|^2 + |a_y|^2)}{2\mu_0}(i\omega(u^*\nabla u - u\nabla u^*) - 2\omega k|u|^2\hat{\mathbf{z}}) + i\omega\frac{(a_x a_y^* - a_x^* a_y)}{2\mu_0}\nabla|u^2| \times \hat{\mathbf{z}} \quad (4.8)$$

where $\mathbf{H} = \mathbf{B}/\mu_0$ defines the magnetic field. The brackets $< \cdot >$ denote the time average over an optical cycle $T = 2\pi/\omega$ and $\hat{\mathbf{z}}$ is the unit vector in the z direction. The first part of this equation is polarization independent while the second part depends on the polarization state of the light. This term is zero for linearly polarized light and nonzero for circularly polarized light.

Using the scattering force, we can now calculate the optical scattering torque on a small particle in the case of an LG beam as defined by Eq. (4.1). If we consider the torque in the direction of propagation z we have

$$\boldsymbol{\tau} \cdot \hat{\mathbf{z}} \propto (\mathbf{r} \times < \mathbf{E} \times \mathbf{H} >) \cdot \hat{\mathbf{z}} \propto \frac{\omega l}{\mu_0}|u(\mathrm{LG}_l^p)|^2 - \frac{\omega r i(a_x a_y^* - a_x^* a_y)}{2\mu_0}\partial_r|u(\mathrm{LG}_l^p)|^2 \quad (4.9)$$

where $r = \sqrt{x^2 + y^2}$ is the radial distance from the optical axis. We can now identify the two terms in this torque component as the OAM, which is proportional to the vortex charge, l, of the LG beam. The second term, which is polarization or spin dependent, is zero for linearly polarized light and changes sign depending on the handedness of the circular or elliptical polarization of the incident beam. This term is best described through the definition of the spin in z direction, $\sigma_z = i(a_x a_y^* - a_x^* a_y)$. The total torque acting on a scatterer depends on the relative sign between the vortex charge l and spin density of the light field σ_z. Same sign enhances the torque while opposite sign decreases the torque on the scatterer (Figure 4.3).

4.3.4 Maxwell's Stress Tensor

From a fundamental perspective, the origin of the optical forces is the momentum transfer from the optical fields to a scattering object. The ray optics approach introduced in Section 4.3.2 takes this transfer into account by considering the momentum transported by each ray and its change in the refraction and reflection process. This approach breaks down when considering small objects with respect to the wavelength. In addition, the ray optics framework cannot account for coherence, interference, and polarization effects. The methods that we discussed in Section 3.3 are based on the paraxial approximation. Within this approximation, it is possible to describe coherent laser beams propagating along an optical axis and displaying only slow intensity variations over the wavelength. Beams that are tightly focused or the interaction with objects smaller than the wavelength cannot be treated within

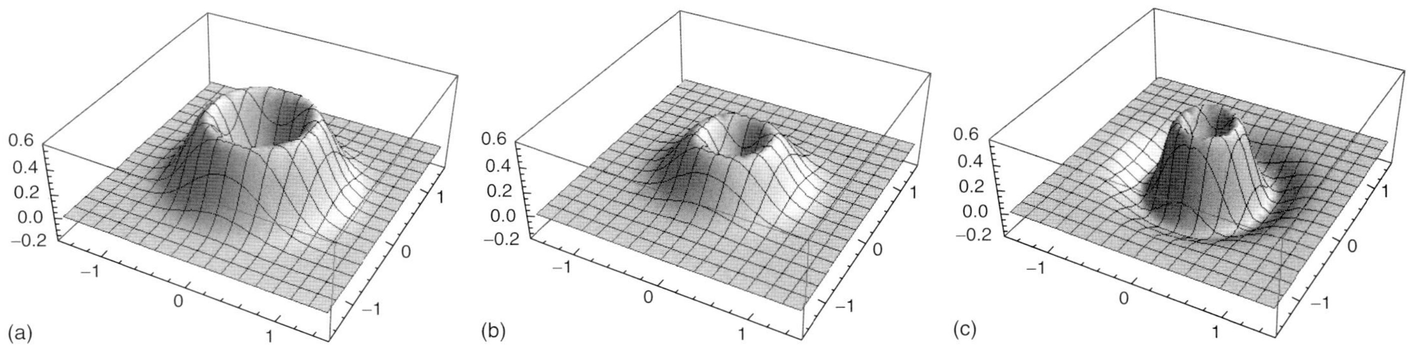

Figure 4.3 Optical torque arising in the transverse cross section of an LG beam ($l = 1$, $p = 0$) as defined by Eq. (4.9). (a) Spin and vortex charge are of the same sign. (b) Linearly polarized beam. (c) Spin and vortex charge are of opposite signs. The total integrated torque is double in the same sign case compared to the linearly polarized case and averages to zero in the opposite sign case. (Please find a color version of this figure on the color plates.)

this approximation without further corrections. It is specifically these cases that are interesting in trapping, tweezing, and rotating of small particles by focused vortex beams. To treat these cases, we need to work with the full vectorial solution of Maxwell's equations.

In this section, we introduce the complete definition of the electromagnetic momentum flux. This momentum flux is described by a 3×3 tensor or matrix, which when applied to the unit normal to a surface gives the vector associated with the momentum flux per unit of area across this surface. This matrix is called *Maxwell's stress tensor* and is physically part of the energy–momentum conservation relation. To define Maxwell's stress tensor and the forces and torques associated with it, we start from the time-dependent version of Maxwell's equations and show that its divergence gives indeed the variation of the energy flow density just as the divergence of this flow density gives the variation of the energy density. These two relations are the conservation relations in the absence of any scattering. Finally, we consider the case of corresponding monochromatic waves giving rise to the optical forces and torques for most experimental cases discussed later.

The propagation of light in vacuum is defined by Maxwell's equations.

$$\begin{aligned} \nabla \cdot \epsilon_0 \mathbf{E}_t &= 0 \\ \nabla \cdot \mu_0 \mathbf{H}_t &= 0 \\ \nabla \times \mathbf{E}_t &= -\mu_0 \frac{\partial \mathbf{H}_t}{\partial t} \\ \nabla \times \mathbf{H}_t &= \epsilon_0 \frac{\partial \mathbf{E}_t}{\partial t} \end{aligned} \tag{4.10}$$

where $\mathbf{E}_t$ and $\mathbf{H}_t$ are the electric and magnetic fields. The subscript t indicates the time dependence of the fields. The vacuum permittivity and permeability, ϵ_0 and μ_0, define the speed of propagation of the electromagnetic waves through the relation $c = 1/\sqrt{\epsilon_0 \mu_0}$.

For electromagnetic waves in the absence of charges and currents, that is, in vacuum, the flux of momentum density defined by Maxwell's stress tensor $\widetilde{\boldsymbol{\sigma}}$ is as follows [17]:

$$\widetilde{\boldsymbol{\sigma}} = \frac{c^2}{2}\left(2\epsilon_0 \mathbf{E}_t \otimes \mathbf{E}_t + 2\mu_0 \mathbf{H}_t \otimes \mathbf{H}_t - (\epsilon_0 \mathbf{E}_t \cdot \mathbf{E}_t + \mu_0 \mathbf{H}_t \cdot \mathbf{H}_t)\widetilde{\mathbf{I}}\right)$$

where $\otimes$ corresponds to the tensor product defined by $(\mathbf{A} \otimes \mathbf{B})_{ij} = A_i B_j$ and where $\widetilde{\mathbf{I}}_{ij} = \delta_{ij}$ is the identity tensor. Here, the indices indicate the three different coordinate components.

To verify the conservation relation associated with Maxwell's stress tensor, we determine its divergence. This can be simplified into two steps. The first is

$$\begin{aligned} \nabla \cdot \widetilde{\boldsymbol{\sigma}} &= \frac{c^2}{2}\nabla \cdot \left(2\epsilon_0 \mathbf{E}_t \otimes \mathbf{E}_t + 2\mu_0 \mathbf{H}_t \otimes \mathbf{H}_t - (\epsilon_0 \mathbf{E}_t \cdot \mathbf{E}_t + \mu_0 \mathbf{H}_t \cdot \mathbf{H}_t)\widetilde{\mathbf{I}}\right) \\ &= \frac{c^2}{2}(2\epsilon_0((\nabla \cdot \mathbf{E}_t)\mathbf{E}_t + (\mathbf{E}_t \cdot \nabla)\mathbf{E}_t) + 2\mu_0((\nabla \cdot \mathbf{H}_t)\mathbf{H}_t + (\mathbf{H}_t \cdot \nabla)\mathbf{H}_t) \\ &\quad -2\epsilon_0((\mathbf{E}_t \cdot \nabla)\mathbf{E}_t + \mathbf{E}_t \times (\nabla \times \mathbf{E}_t)) - 2\mu_0((\mathbf{H}_t \cdot \nabla)\mathbf{H}_t + \mathbf{H}_t \times (\nabla \times \mathbf{H}_t)) \end{aligned}$$

where we use the vector calculus identities $\nabla \cdot (\mathbf{A} \otimes \mathbf{B}) = (\nabla \cdot \mathbf{A})\mathbf{B} + (\mathbf{A} \cdot \nabla)\mathbf{B}$ and $\nabla((\mathbf{A} \cdot \mathbf{B})\widetilde{\mathbf{I}}) = (\mathbf{A} \cdot \nabla)\mathbf{B} + (\mathbf{B} \cdot \nabla)\mathbf{A} + \mathbf{A} \times (\nabla \times \mathbf{B}) + \mathbf{B} \times (\nabla \times \mathbf{A})$. In a second step, we use Maxwell's equations in vacuum to further simplify

$$\begin{aligned} \nabla \cdot \widetilde{\sigma} &= -c^2(\epsilon_0 \mathbf{E}_t \times (\nabla \times \mathbf{E}_t) + \mu_0 \mathbf{H}_t \times (\nabla \times \mathbf{H}_t)) \\ &= c^2 \epsilon_0 \mu_0 (\mathbf{E}_t \times \partial_t \mathbf{H}_t - \mathbf{H}_t \times \partial_t \mathbf{E}_t) \\ &= \partial_t (\mathbf{E}_t \times \mathbf{H}_t) \end{aligned} \tag{4.11}$$

making it possible to introduce the Poynting vector $\mathbf{S} = \mathbf{E}_t \times \mathbf{H}_t$ where $\times$ corresponds to the vector product. Physically, the Poynting vector determines the flow of energy and its divergence corresponds to the variation of the energy density. Using the same procedure outlined above, we can determine the conservation relation of the energy

$$\nabla \cdot \mathbf{S} + \partial_t \mathcal{E} = 0 \tag{4.12}$$

where $\mathcal{E} = 1/2(\epsilon_0 \mathbf{E}_t \cdot \mathbf{E}_t + \mu_0 \mathbf{H}_t \cdot \mathbf{H}_t)$.

In the following, without any loss of generality, we consider monochromatic fields of the form $\mathbf{E}_t = \mathbf{E}\exp(-i\omega t)$ and $\mathbf{H}_t = \mathbf{H}\exp(-i\omega t)$, where ω is the optical frequency and $\mathbf{E}$ and $\mathbf{H}$ are complex, frequency-dependent vector fields. The introduction of these complex vector fields greatly facilitates the discussion of circularly polarized fields and their associated intrinsic angular momentum. Maxwell's equations for these monochromatic fields read then as

$$\begin{aligned} \nabla \cdot \epsilon_0 \mathbf{E} &= 0 \\ \nabla \cdot \mu_0 \mathbf{H} &= 0 \\ \nabla \times \mathbf{E} &= i\omega\mu_0 \mathbf{H} \\ \nabla \times \mathbf{H} &= -i\epsilon_0 \omega \mathbf{E} \end{aligned} \tag{4.13}$$

The general, time-dependent fields can be retrieved through a superposition of the monochromatic solutions corresponding to an inverse Fourier transform of the monochromatic field amplitudes. For monochromatic fields, Maxwell's stress tensor, the Poynting vector, and the energy density can be written as

$$\begin{aligned} \widetilde{\sigma} &= \frac{c^2}{2}\left(\epsilon_0(\mathbf{E} \otimes \mathbf{E}^* + \mathbf{E}^* \otimes \mathbf{E}) + \mu_0(\mathbf{H}^* \otimes \mathbf{H} + \mathbf{H} \otimes \mathbf{H}^*) - 2\mathcal{E}\widetilde{\mathbf{I}}\right) \\ \mathbf{S} &= \frac{1}{2}(\mathbf{E} \times \mathbf{H}^* + \mathbf{E}^* \times \mathbf{H}) \\ \mathcal{E} &= \frac{1}{2}(\epsilon_0 \mathbf{E} \cdot \mathbf{E}^* + \mu_0 \mathbf{H} \cdot \mathbf{H}^*) \end{aligned}$$

where the asterisk corresponds to the complex conjugate. The conservation relations of these quantities simplifies to

$$\begin{aligned} \nabla \cdot \widetilde{\sigma} &= 0 \\ \nabla \cdot \mathbf{S} &= 0 \end{aligned} \tag{4.14}$$

stating simply that for monochromatic waves, solutions of Maxwell's equations (Eq. (4.13)) have a time-independent energy density and energy flux density.

4.3.5 Linear Momentum Transfer

The conservation relations (Eq. (4.14)) defined in the previous section hold true for any linear superposition of solutions of Maxwell's equation (Eq. (4.13)). In the presence of any sources or scattering objects, such as charges or electric dipoles, the conservation relations are no longer fulfilled. From the electromagnetic wave point of view, the interference between the incident field and the scattered field is at the origin of this loss of conservation. On the scatterer's side, this interference manifests itself through the transfer of linear momentum in the form of a force acting on the scattering object.

In the following, without any loss of generality, we consider monochromatic fields as defined by Eq. (4.13) and electric polarization of the form $\mathbf{P}_t = \mathbf{P}\exp(-i\omega t)$. In this case, Maxwell's equation reads as

$$\begin{aligned}\nabla\cdot\epsilon_0\mathbf{E} &= \nabla\cdot\mathbf{P}\\ \nabla\cdot\mu_0\mathbf{H} &= 0\\ \nabla\times\mathbf{E} &= i\omega\mu_0\mathbf{H}\\ \nabla\times\mathbf{H} &= -i\epsilon_0\omega\mathbf{E} - i\omega\mathbf{P}\end{aligned} \tag{4.15}$$

where we can define the local electric charge density $\rho = \nabla\cdot\mathbf{P}$ and the associated current $\mathbf{j} = -i\omega\mathbf{P}$. This gives rise to the forces acting on the scattering object as a function of the induced charges and currents:

$$\nabla\cdot\boldsymbol{\sigma} + \partial_t\mathbf{S} = \frac{1}{2}(\rho\mathbf{E}^* + \rho^*\mathbf{E} + \mu_0\mathbf{j}^*\times\mathbf{H} + \mu_0\mathbf{j}\times\mathbf{H}^*)$$

where we recognize the Coulomb and Lorentz force.

It is also possible to express the total momentum transfer to the scattering object by integrating the momentum flux density on a surface surrounding the object. This total optical force is given by [18, 19]

$$< \mathbf{F} > = \frac{1}{2}\int_S \boldsymbol{\sigma}^{(p)}\cdot\mathbf{n}dS \tag{4.16}$$

$$= \Re\left(\int_S (\epsilon_0\mathbf{E}\cdot\mathbf{n})\mathbf{E}^* + \mu_0(\mathbf{H}\cdot\mathbf{n})\mathbf{H}^* - \frac{1}{2}(\epsilon_0\mathbf{E}\cdot\mathbf{E}^* + \mu_0\mathbf{H}\cdot\mathbf{H}^*)\mathbf{n}dS\right) \tag{4.17}$$

where S and dS are the surface and element of the surface surrounding the particle. The unit vector $\mathbf{n}$ corresponds to the normal pointing outward from this surface. The brackets $< \cdot >$ denote the time average over an optical cycle $T = 2\pi/\omega$ and $\Re()$ to the real part of a complex number.

4.3.6
Angular Momentum Transfer

Similar to the definition of a torque by Eq. (4.4), we can introduce a flux density of the angular momentum by considering the following tensor:

$$\sigma_{ij}^{(a)} = \varepsilon_{ikl}\sigma_{jk}x_l \tag{4.18}$$

where ε_{ikl} is the Levi-Civita totally antisymmetric pseudotensor of third rank defined by $\varepsilon_{ijk} = \text{sgn}(i-j)\text{sgn}(j-k)\text{sgn}(k-i)$ with the function sgn representing the sign function. The vector $(x_1, x_2, x_3) = \mathbf{r} = (x, y, z)$ describes the position vector with respect to the point around which the angular momentum is determined. This pseudotensor corresponds to the vector product at tensor level and is equivalent to the definition given by Jackson [17]. As a consequence of this definition, the divergence of the angular momentum tensor is intrinsically linked to the divergence of the linear momentum tensor. Indeed, we have

$$\nabla \cdot \boldsymbol{\sigma}^{(a)} = \mathbf{r} \times \nabla \cdot \boldsymbol{\sigma} \tag{4.19}$$

where we used the symmetry property of the linear momentum tensor $\sigma_{jk} = \sigma_{kj}$. This relation shows that when the linear momentum is divergence free the angular momentum tensor is also divergence free. This means that if there is no linear momentum transfer to a small particle, then there is no angular momentum transfer.

As with the linear angular momentum, we can define a time-averaged angular momentum transfer to a scattering object. This gives rise to a total torque with respect to the origin of the coordinates system, given by

$$\begin{aligned} <\boldsymbol{\Gamma}> &= \frac{1}{2}\int_S \boldsymbol{\sigma}^{(a)} \cdot \mathbf{n}dS \\ &= \Re\left(\int_S (\epsilon_0 \mathbf{E}\cdot\mathbf{n})(\mathbf{r}\times\mathbf{E}^*) + \mu_0(\mathbf{H}\cdot\mathbf{n})(\mathbf{r}\times\mathbf{H}^*)\right. \\ &\quad \left. -\frac{1}{2}(\epsilon_0\mathbf{E}\cdot\mathbf{E}^* + \mu_0\mathbf{H}\cdot\mathbf{H}^*)(\mathbf{r}\times\mathbf{n})dS\right) \end{aligned} \tag{4.20}$$

which can be decomposed into an orbital and a spin angular momentum in the following way:

$$<\boldsymbol{\Gamma}> = \mathbf{r}\times <\mathbf{F}> + <\boldsymbol{\Gamma}_s> \tag{4.21}$$

The OAM, the first term in Eq. (4.21), corresponds to the torque due to the total linear momentum. The extrinsic spin angular momentum, the second term, gives rise to the rotation of the scattering object around its axis.

4.3.7
Polarization Spin Momentum

Contrary to the extrinsic spin angular momentum, which arises from the inhomogeneity of the optical linear momentum, the intrinsic spin momentum arises as a result of a local torque originating in the polarization state of the light [5, 20]. Its density and flux are associated with the difference between the intensity of the left and right polarized light fields. These are defined as

$$\begin{aligned} \mathbf{S}^{(s)} &= ic(\epsilon_0 \mathbf{E} \times \mathbf{E}^* + \mu_0 \mathbf{H}^* \times \mathbf{H}) \\ \mathcal{E}^{(s)} &= \frac{i}{c}(\mathbf{H} \cdot \mathbf{E}^* - \mathbf{E} \cdot \mathbf{H}^*) \end{aligned} \tag{4.22}$$

where $\mathcal{E}^{(s)}$ is the spin energy of the electromagnetic wave while $\mathbf{S}^{(s)}$ its optical flux. In vacuum, these two quantities are related through a conservation relation [21]. In the presence of an optical scatterer that changes the polarization state of the light field this is no longer true. We can deduce the total spin transfer by considering the spin flux across a surface surrounding the particle in the same way as for the momentum transfer in Eq. (4.16).

4.4 Optical Vortex Fields for the Rotation of Trapped Particles

4.4.1 Studies of Rotation of Trapped Objects using Optical Vortex Fields

Rotation of trapped particles has been a topic of immense interest over the last 15 years. This interest has been generated from a number of quarters and has resulted in numerous experiments being performed using rotating light patterns, spin angular momentum and OAM transfer to the trapped particles, and asymmetrical scattering. In this chapter, all these experiments are not reviewed, we rather concentrate on the results of experiments where light fields with optical vortices have played a dominant role. The reason for initiating the rotation particles may differ in several respects: first, quantifying and understanding the manner of rotation is central to a deeper understanding of the angular momentum of light *per se* and its very nature. The trapped particles may map out the spin angular momentum and OAM of a light field. Rotating a particle also leads to potentially interesting applications in the domain of microfluidics. Here, there is interest in pumping very small volumes of fluid as well as looking at local measurements of viscosity: this is a key requirement in biological applications. By applying a torque from an incident light field, spinning an object and recording the maximal rotational velocity, and equating the torque with the maximal torque from the rotational Stokes drag, we may determine local viscosity. Typically, to date, such studies have been achieved using the spin angular momentum of a light field.

Optical forces, as we have seen, will readily localize a particle of higher refractive index than its surroundings close to the beam focus position. A number of mechanisms may transfer angular momentum from a light field to a trapped particle. The first experiments to explore particle rotation through OAM transfer from an optical vortex field explored transfer of OAM from LG beams to absorptive particles. Upon absorption, a transfer of OAM occurred and resulted in a mechanical torque exerted by the LG beam by transfer of its OAM to an absorptive microparticle [3, 8, 22, 23]. The first experiment by He and colleagues [22] trapped absorptive copper oxide particles in two dimensions and set them into rotation using holographically created LG modes: in fact, this experiment constituted one of the first implementations of holographic optical trapping. Trapped particles in two dimensions were set into rotation and the researchers showed that the rotation rate was not a consequence of any asymmetric scattering. The particle was seen to rotate in the opposite sense by reversing the helicity of the 2π azimuthal phase of the trapping beam. In a related work, Simpson and coworkers [8] used a cylindrical lens mode converter to generate a trapping beam of LG modes of single order in the near-infrared. Similar to He *et al.*'s work, absorptive objects were set to rotate in this study, but in this case three-dimensional trapping was seen. The studies experimentally decoupled the spin angular momentum of light from the OAM of light. The spin angular momentum either added to or subtracted from the OAM and by manual adjustment of a waveplate, a "stop–start" rotation of the particle was seen. In analogous but separate studies, Friese *et al.* [3] achieved results using holographically generated LG modes of azimuthal index $l = 3$. In these experiments, in the 1990s, that exploited these LG modes, the particles used were optically absorptive, which was perhaps the simplest way to ensure OAM transfer. Absorption is typically avoided for biological applications, but such works clearly showed the physical properties of light fields with spin angular momentum and OAM.

The very nature of the angular momentum of light can be understood more deeply by examining the motion of particles trapped off axis in optical tweezers created with a vortex light field. Importantly, one may associate certain features of the motion as resulting from distinct contributions of the spin angular momentum and OAM of the light beam. In turn, this allows a distinction to be made between the intrinsic and extrinsic aspects of the angular momentum of light. This study showed that one could actually transfer OAM onto transparent dielectric particles simply by scattering off the inclined wave fronts [24]. The spin angular momentum of a light beam is always noted as intrinsic. Essentially, the z component of the OAM can be described as intrinsic only if the z direction is defined such that the transverse momentum integrated over the whole beam is zero. In contrast, the OAM can be extrinsic or intrinsic. Particles were placed off axis within the circumference of the LG beams and were seen to show differing forms of motion based on whether they were responding to the spin or orbital component of the light field. In turn, this gives insight into the intrinsic and extrinsic nature of spin angular momentum and OAM. Observing both the intrinsic and extrinsic nature of angular momentum simultaneously on a single particle was achieved in 2003 [25]. In this study a birefringent particle was placed in a circularly polarized high azimuthal order Bessel light field. Bessel beams have been of interest for several decades since their identification in the 1980s [26]. These are solutions to the Helmholtz equation that are propagation invariant and thus "nondiffracting" over a given region. While experimentally they do offer such intriguing properties, their transverse profile is a set of concentric rings with a maximum or minimum at beam center so we do compromise by distributing the power equally between the rings. Zeroth-order Bessel modes have been used for creating optical guides and conveyor belts and optical binding. For our discussion, we restrict ourselves to high-order Bessel modes that possess an on-axis vortex. Bessel modes may be created holographically or by the use of diffractive elements known as *axicons* [27]. In the Garces-Chavez *et al.* study, the authors showed that the rotation rate due to spin was inversely proportional to ring radius within which the particle resides, whereas the orbital rotation rate was inversely proportional to the cube of the ring radius. This study showed that a trapped particle could, in principle, be used to map out the angular momentum content of the light field. The high-order Bessel light beam also plays an important role in demonstrating the rotation of low-index particles due to OAM. In this case, particles were held in the dark rings of the beam profile and gained through scattering performed in a circular motion. Azimuthal intensity variations were seen not to be as important as the case for high-index particles in this study [15]. A separate study by Volke-Sepulveda *et al.* considered particle rotation for particles trapped in a high-order Bessel beam [28]. Volke-Sepulveda and colleagues explored transfer of OAM by scattering to dielectric particles in such a field (Figure 4.4). Rotation was observed as expected, varying linearly with power and in accordance with the sense of helicity of the embedded vortex. An important facet of this study was the fact that it was one of the very first studies to bring a theoretical model (using ray optics) together with such experiments to present quantified rotation rates for the trapped objects.

Exploring the total angular momentum transfer to trapped objects was the subject of studies by Parkin *et al.* [29]. They described how to determine the total angular momentum, both spin and orbital, transferred to a particle trapped in optical tweezers. An LG beam with an azimuthal index of 2 with varying degrees of circular

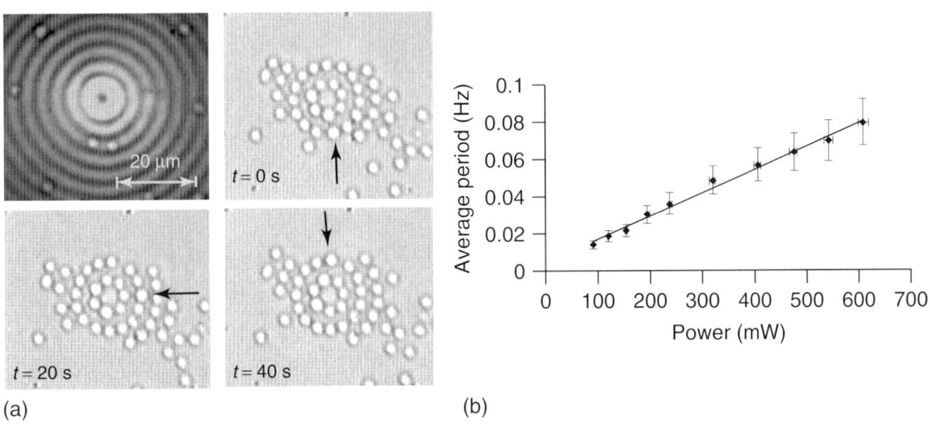

Figure 4.4 (a) The first frame shows the second-order Bessel beam used to trap 3 μm spheres in its rings. The rest of the frames show the trapped spheres rotating in the anticlockwise direction. To see the rotation one sphere in the second ring is highlighted by an arrow. (b) Average period of rotation in the inner ring as a function of the total power of the beam (Reprinted with permission from [28]. © (2002) by IOP Publishing.) (Please find a color version of this figure on the color plates.)

polarization was used to trap and rotate an elongated particle with a well-defined geometry. The method successfully estimates the total optical torque applied to the particle if one assumes negligible absorption. More quantitative work by Zhao *et al.* modeled the forces in an optical vortex trap explicitly accounting for the diffraction of the strongly localized beam and the spherical aberration introduced by the dielectric interface. A multidipole approximation for force calculations agreed well with experimental measurements, for particle displacement of smaller than the radius of the particle [30]. The same team studied the optical spin-to-orbital angular momentum conversion that may occur in a homogeneous and isotropic medium. The studies of this conversion in a tightly focused beam trapping metallic particles showed that the orbital rotation speeds of trapped particles are altered because of this conversion [31]. Curtis and Grier explored [32] the focusing of optical vortex fields that create optical traps. Their studies showed a surprising dependency of the vortex structure and angular momentum flux on the azimuthal index of the trapping light. In fact, they found that the annular radius of the beam scaled linearly with the topological charge (Figure 4.5), which was not expected, and they explored the period of rotation versus the azimuthal index of the LG beam.

If we generalize the "nondiffracting" beam concept, we find that the Mathieu beam represents a more general family of "nondiffracting" modes that may also possess OAM. In fact, in general, they present an elliptical form with a nonuniform intensity variation around the ring. Helical Mathieu beams are fundamental

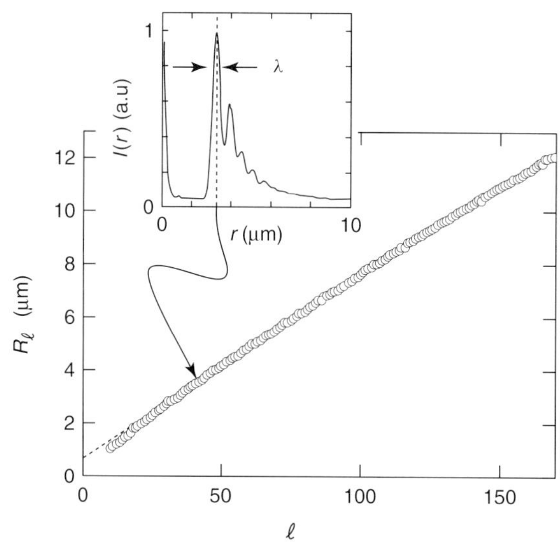

Figure 4.5 Radius of the first vortex ring as a function of the vortex charge. Experimental data points are represented by circles and the dashed lines show theoretical mode. Inset: Azimuthally averaged intensity of the beam (Reprinted with permission from [32]. © (2003) by American Physical Society.)

"nondiffracting" beams, which are solutions of Helmholtz equation in elliptical cylindrical coordinates. The beams are mathematically described by a linear superposition of products of radial and angular Mathieu functions. In contrast to Bessel beams, however, the OAM density of Mathieu beams is not independent of the azimuthal coordinate but it varies with the elliptic angular coordinate of the beam. In 2006, Lopez-Mariscal and colleagues showed how to use such modes to rotate particles [33]. In this case, the mechanism for transfer was again scattering and one could see particle motion around the ring. Interestingly, a competition occurred between the intensity variation and the inclined wave front that drove the particle around the beam circumference, meaning that a nonuniform velocity was recorded for particle motion. The authors saw the transfer of OAM to trapped particles in the azimuthally asymmetric transverse intensity distribution of a helical Mathieu beam. The average rotation rate, instantaneous angular displacement, and terminal velocity of the trapped particles were recorded. They saw that the angular dependence of these parameters agreed well with the variation of the optical gradient force, the transfer of OAM from the Mathieu, beam and the Stokes drag force.

Turning back to LG modes, Jesacher and colleagues explored trapped particles held at an air–water surface [34]. These were seen to orbit in a reverse direction with respect to the OAM of the light field. The observation was explained by two factors: asymmetric particle shape and confinement of the particle at the 2D air–water interface. The experiment showed the effect of the particle shape on the momentum transfer, which is overlooked to some extent in most studies.

Tao and colleagues [35] showed the use of optical vortex beams with fractional topological charges, namely fractional optical vortex beams, to rotate trapped particles. Analogous to the vortex beams with integer topological charges, the fractional optical vortex beams are also capable of rotating particles induced by the transfer of OAM. However, a fractional beam of this type typically shows an intensity discontinuity (low intensity gap) around the beam circumference. In contrast to integer LG beam rotation as described, this may significantly hinder the smooth orbital rotation of the particle. The OAM and radial gap of the fractional vortex beams were exploited in the work to guide and transport microscopic particles.

The scattered light from trapped particles within a vortex field is interesting. Studies by Lee *et al.* explored the far-field diffraction of arrays of trapped particles placed in an optical vortex trapping field [36]. Optically trapped microparticles within such a monochromatic LG beam led to the formation of unique intensity patterns in the far field. This was attributed to multiple interference of the forward scattered light from each particle. Trapped colloids create far-field interference patterns with distinct spiral features that are directly correlated to the helicity of the LG beam. Using two trapped particles, the authors showed the first microscopic version of Young's slit-type experiment to detect the azimuthal phase variation around a singly charged LG trapping beam.

Trapping of metallic nanoparticles may be considered in the intermediate regime between atom trapping and tweezing of micron-sized dielectric beads. Compared to micron-sized dielectric beads, optical tweezing of very small particles poses a significant challenge, as the force necessary to trap, namely the gradient force, decreases with polarizability (directly proportional to particle volume) for fixed material. Intriguingly, metal nanoparticles exhibit substantially increased polarizability when compared to their equivalently sized dielectric counterparts. In turn, this means we can tweeze metallic nanoparticles more readily than nanometric-sized dielectric particles, though, of course, we must not forget the complex refractive index of such particles. This implies absorption by the particles, which can lead to significant heating. In 2008, a study by Dienerowitz *et al.* [37] showed the confinement and transfer of OAM from the LG beam to 100 nm gold nanoparticles. These particles were in fact restricted to the dark vortex core of the LG beam at 514 nm just below the plasmon resonance of the particles. They were seen to rotate around the beam propagation axis and the sense of nanoparticle rotation remains consistent with the handedness of the LG beam used. In this manner, the transfer of OAM of the beam to the particle was observed. These are the smallest particles to date set into rotation by OAM transfer. The particle rotation rate showed a linear increase with respect to laser power, with a maximum rate of 3.6 Hz at 110 mW. We found a linear dependence of the rotation rate on power of about 33 Hz W^{-1}. These results imply that the scattering force is the dominant interaction mechanism between the laser beam and gold nanoparticle, as OAM is typically transferred by scattering. If we now turn to larger metallic particles, scattering may dominate, meaning they are again expelled from the dark region. In general, Rayleigh particles (tens of nanometers in diameter) have been trapped in three dimensions as scattering is comparable to the dielectric case. Studies by O'Neil and Padgett [38] showed that amalgamations of micron-sized metal particles are restricted to an annular region below the beam focus but outside the high-intensity annular region of the LG trapping mode. OAM was seen to induce a bulk motion of the particle around the LG beam propagation axis. The speed and sense of this rotation was confirmed to be due to the OAM with no contribution from the spin angular momentum state of the trapping beam.

Until very recently, studies of OAM transfer to particles have been restricted to monochromatic trapping beams. Two recent studies extended this to look at rotation of particles in broadband white light vortex fields. In a broader context, white light trapping has begun to be of importance for simultaneous studies of trapping and spectroscopy with applications in aqueous and aerosol samples. The interest of using white light vortices was to extend these works to see the potential relevance of coherence and to show that broadband optical vortices still retained the OAM. In fact, as one might expect, the spatial coherence is the crucial aspect to retain the azimuthal phase variation around the beam profile. The temporal coherence here does not play a major role but does confirm that, indeed, all wavelengths may be encoded with equivalent azimuthal phase variations and subsequently transfer this to trapped objects. Studies by Wright *et al.* [39] showed this with a light source with a 125 nm bandwidth. A plot of the rotation rate for three particles orbiting around the beam axis versus l/r^3 showed a linear dependence where r is the measured radius of the annular ring and l is the azimuthal mode index.

The other simultaneous study by Morris *et al.* [40] showed quantitative verification of the rotation rate using a light source with a bandwidth in excess of

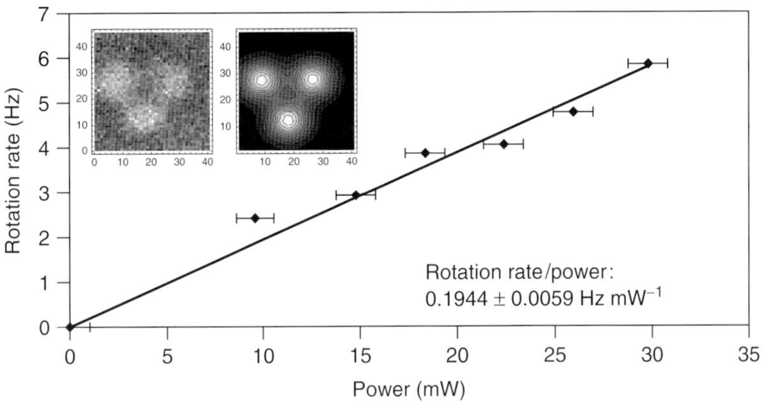

Figure 4.6 Three 1 μm spheres are optically trapped in the first bright annular ring of the focal spot of supercontinuum LG beam ($l = 3$, $p = 0$). The rotation rate is represented as a function of the incident power. The inset shows the trapped spheres and the tracking reconstruction (Reprinted with permission from [49].) (Please find a color version of this figure on the color plates.)

200 nm (Figure 4.6). The numerical modeling explored the electromagnetic linear and angular momentum transfer between the supercontinuum light field and the trapped microparticles in the beam path. This momentum transfer was evaluated using the conservation relation involving Maxwell's stress tensor and its associated angular momentum tensor. A variety of studies and detailed comparison with theory were presented, including the rotation rate directly related to the azimuthal index and power respectively. Broadband light sources for tweezers is a relatively new area and potentially offers trapping combined with spectroscopy and has been used in a variety of studies. In the context of optical vortices, these studies pave the way for more elaborate experiments where the spatial coherence of light and its influence on the OAM content of a vortex field may be explored.

4.5 Optical Vortex Fields for Advanced Optical Manipulation

While the OAM content of LG beams with vortices have been a key to rotation of particles, it is important to note that their very profile itself is of importance for a number of studies in optical trapping. Thus, it is not solely their phase structure that has been of interest but their annular intensity profile as well.

Optical trapping and its physical explanation show that we are reliant on the particle having a higher refractive index than its surroundings. However, this is not always the case. Low-index particles may be found in numerous chemical and biological applications and as one might suspect are in fact repelled from the regions of high light intensity and thus need to be "caged" by light. In 1970

Arthur Ashkin first observed that low refractive index particles are repelled from the high-intensity region of light, while high refractive index particles are drawn into the trap [41]. Building upon this study, he used a high-order mode laser field (TEM_{01}^{*}) to optically levitate low-index glass sphere against gravity [42, 43].

Such tweezers provide the ability to manipulate low refractive index microparticles: all higher-order LG beams ($p = 0$, $l > 1$) possess a smooth annular intensity profile that should allow a low-index particle to be restricted to the dark vortex core. Indeed, this was confirmed by the experiments of Gahagan and Swartzlander [44] who held 20 μm sized hollow spheres and in subsequent studies [45] held both low- and high-index particles in a tweezers geometry. Interestingly, in other studies at approximately the same time, it was realized that the LG beam should offer further improvement in standard optical tweezers, that is, trapping high-index particles. The presence of the vortex should potentially lead to an improvement in the axial confinement: If we consider a Gaussian beam, we see that light coming straight through the center of back aperture of the objective leads to axial scattering forces that act against the gradient forces to destabilize the trap. If the radius of the trapped microparticle is comparable to or greater than the size of the beam waist of the LG beam, high refractive index particles can also be tweezed with an LG beam but will experience a much lower on-axis light scattering force compared to the use of a Gaussian beam because of the LG beam profile. We measured force constants and resonant frequencies for 1–4 μm diameter polystyrene spheres in a single-beam gradient trap using the measurements of backscattered light [46]. This reduces the on-axis scattering force exerted along the axial direction as shown by Simpson *et al.* [47].

Light fields possessing optical vortices allow for the manipulation of droplets where the refractive indices of most liquids are smaller than their surrounding medium. For selective mixing of droplets and motion of droplets, we need careful positioning of the droplets. Illuminating a spiral phase plate creates an on-axis vortex with a uniform annular intensity profile encircling the vortex. Lee *et al.* [48] used this but then intentionally misaligned the plate in a direction orthogonal to the beam propagation direction. This allows for an efficient formation of a stable asymmetrical optical light pattern which may be considered as an off-axis vortex beam. By rotating the spiral plate around the beam axis, a rotating off-axis optical vortex beam was also formed. The beam was used for optical rotation and steering of both high- and low-index particles. In 2007, Lorenz *et al.* [49] adapted these principles and used two such "displaced" LG (optical vortex) beams to controllably fuse two aqueous droplets. This displacement action gradually diminishes the overall annular ring intensity pattern into "crescent-shaped" intensity pattern. In Figure 4.7, we see a vortex-trap-induced fusion of two aqueous droplets in acetophenone.

Further studies by the same group showed the shrinkage and re-expansion of individual femtoliter-volume aqueous droplets that were suspended in an organic medium and held in an optical vortex trap. A change in the volume of aqueous droplets translates into a change in concentration of the dissolved species within the droplets so this may open up further research into chemical processes susceptible

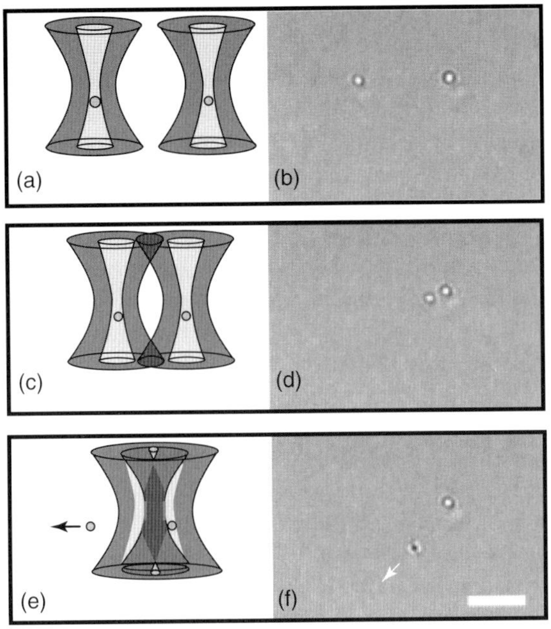

Figure 4.7 Repulsion of two aqueous droplets in a dual vortex trap. (a–d) The two droplets are brought together without overlapping the vortex beams. (e and f) Further overlapping of the vortex beams gives rise to a repulsive force and one of the droplets escapes. The scale bar represents 10 μm [49].

to concentration, such as macromolecular crowding and protein nucleation and crystallization [50].

It is not just droplets that may be manipulated with optical vortex fields. Ultrasound contrast agent is a low-index material and of use for controlled cavitation for possible targeted drug delivery. It may be trapped in the dark core of the vortex and used in combination with exposure to ultrasound for microbubble cavitation and subsequent drug delivery. Such methods for ultrasound agent selection and "sonoporation" were demonstrated by Prentice *et al.* and Garbin *et al.* [51–53].

When we move toward flow in micron-sized channels, fluid flow can be laminar as viscosity dominates over inertia. Methods to induce rapid direction changes within flow channels to control flow or mixing processes of different species of particles or different fluids are of interest in these low Reynolds number environments. Microrheology is thus a key area that may benefit from controlled spinning of optically trapped objects, thereby becoming an important consideration for such studies [54–56]. As mentioned earlier, rotating optically trapped particles also holds immense promise for microrheology where the rotational stokes drag reaches equilibrium with spin angular momentum (circularly polarized fields) rotating a birefringent object permitting a local measurement of viscosity [57] with only picoliter volume of liquid [58]. We do not restrict ourselves to spherical particles as photopolymerized structures [59], which may be set into rotation by asymmetric scattering. Micropumps may be created by simultaneously trapping and rotating the microspheres held by multiple optical tweezers created by acousto-optic modulators [60]. The optical angular momentum of light, be it spin or orbital, can induce controlled torques to trapped particles to generate a rotational motion. Notably, optical torques can be imparted onto particles with the OAM of the LG beams through the mechanisms of scattering or absorption. In terms of applications, optical vortex fields in the form of LG beams too can be extended to the generation of optically driven pumps: Ladavac and Grier [61] exploited holographically created rows of alternating single-ringed LG beams of very high azimuthal index ($l = \pm 21$) to trap and rotate large numbers of microspheres to generate fluid flow. By pushing particles toward the water–air interfaces (away from a hard surface), Jesacher *et al.* [34] observed high rotation rates of particles trapped in holographic optical vortex traps and also demonstrated interactive particle flow steering with arrays of optical vortex pumps. Interestingly, recent work has explored ways to integrate the optical elements required to generate optical OAM and optical vortices into a microdevice. This should permit the rotation of either naturally occurring microparticles or specially fabricated rotors. In the study of Knoner *et al.* [62] two photon photopolymerization was used to fabricate microscopic diffractive optical elements, customized to a wavelength of choice, which may be integrated with micromachines in microfluidic devices. This enables one to have miniaturized diffractive elements to spin particles in sample chambers.

Applying and controlling a small amount of torque on biological particles, that is, cells or chromosomes, often requires that the beams be tailored to the shape of the biological particles. Orientation of particles with optical traps is also a desirable quantity in this respect, and thus in some instances even light fields that have engineered shapes but rotate themselves can be of interest. With higher-order laser modes, Sato *et al.* [63] have demonstrated that the HG modes are capable of rotating elongated biological particles. Optical vortices offer a possibility to apply optical torques onto different types of microparticles. Paterson *et al.* [64] used interfering LG beams with either plane waves or with other LG beams as an alternative beam shaping technique to create rotating light patterns. Chromosomes were controllably oriented and spun by controlled adjustment of the relative optical path length in the LG beam interferometer. Interfering two LG beams of equal but opposite sign azimuthal index can be used to generate an annular array of spots that again may be rotated with careful adjustment of the relative path length between the two arms of the interferometer. Such interferometric patterns can create and rotate three-dimensional cubic structures [65]. Applying the angular Doppler technique to create a frequency shift between the interfering beams, the trapped particle can be spun at a high frequency [66].

The inclined wave front and the annular intensity pattern of an LG beam forms may create an "ideal" toroidal optical trap with a given driving force. In the broader context, a constant driving force along any toroidal optical trap can demonstrate interesting behavior [67–69]. Faucheux *et al.* [68], in their study, used a spinning single-beam trap and saw three different regimes for the particle motion. For small tweezers velocity that was less then 100 μm s^{-1}, the particle was trapped due to the optical forces that moved with the beam. For velocities between 100 μm s^{-1} and 3 mm s^{-1}, the particle was lost from the trap but recaptured as the single-beam tweezers came around again. In the case of large tweezers velocities (larger then 3 mm s^{-1}), the particle diffused around the "time-averaged" annular beam profile but was confined in the radial direction. These observations were interpreted using a model based on the corresponding Fokker–Planck equation. In a more recent experiment, Lutz *et al.* [70] explored up to three particles in a toroidal trap. As a consequence of hydrodynamic interactions, the particles first go through a transient regime and then enter a characteristic limit cycle. Further studies explored how the collective motion of these interacting microparticles changed when a sawtooth potential is applied to the constant driving force. Sawtooth potentials are an important component for the studies of thermal ratchets, which have relevance to the operation of biological motors. Hydrodynamics studies showed that the two-particle clusters exhibit a surprising caterpillar-like motion, which allowed them to surmount the optical potential barriers.

Roichman *et al.* made use of a high-order LG beam ($l = 50$ and 80) and observed the motion of circulating particles in the vortex fields. The particles experienced the same fixed optical intensity pattern, and addition of a small amount of disorder to the system resulted in periodic driving with "strong kicks." The authors showed that this led to a model microscopic system for studying disorder-induced transitions to chaos [71]. Lee and Grier [72] explored a trapped colloidal sphere placed in a corrugated annular field consisting of two opposite (but equal in magnitude charge) optical vortex fields whose interference creates an azimuthally modulated annular trap. The particle dynamics shows it alternating between free motion around the "corrugated" optical vortex (a washboard type potential) and being held in a given local potential energy minima. Fluctuations of velocity were shown to be characterized by a linear Einstein-like diffusion law, but best described by an effective diffusion coefficient for the particle that is enhanced in excess of two orders of magnitude with that expected due to the intermittent trapping, a rather surprising result.

4.6 Conclusions

Light fields with embedded optical vortices have become very important for a host of studies in optical manipulation. In this chapter, we have given an overview of some of the major experiments and theoretical underpinning that has brought this topic to the fore for various researchers over the last two decades in the field of optical micromanipulation of biological and colloidal material. Light fields with embedded vortices have shed light on the OAM of light and enabled the ability to manipulate low-index particles and look at interesting particle dynamics. Undoubtedly, light fields with vortices will play a significant role in micromanipulation for a long time to come.

Acknowledgments

We thank the UK EPSRC for the funding of this work. KD is a Royal Society-Wolfson Merit Award holder.

References

1. Kogelnik, H. and Li, T. (1966) Laser beams and resonators. *Appl. Opt.*, **5** (10), 1550.
2. Allen, L., Beijersbergen, M., and Spreeuw, R. (1992) Orbital angular momentum of light and the transformation of Laguerre-Gaussian laser modes. *Phys. Rev. A*, **45** (11), 8185.
3. Friese, M., Enger, J., and Rubinsztein-Dunlop, H. (1996) Optical angular-momentum transfer to trapped absorbing particles. *Phys. Rev. A*, **54** (2), 1593.
4. Allen, S., Touchette, H., Moukouri, S., and Vilk, Y. (1999) Role of symmetry and dimension in Pseudogap Phenomena. *Phys. Rev. Lett.*, **83** (20), 4128.
5. Barnett, S. (2002) Optical angular-momentum flux. *J. Opt. B*, **4**, S7.
6. Okida, M., Omatsu, T., Itoh, M., and Yatagai, T. (2007) Direct generation of high power Laguerre–Gaussian output from a diode-pumped Nd:YVO_4 1.3-μm bounce laser. *Opt. Express*, **15** (12), 7616.
7. Beijersbergen, M., Allen, L., and der Veen, H.V. (1993) Astigmatic laser mode converters and transfer of orbital angular momentum. *Opt. Commun.*, **96** (1-3), 123.
8. Simpson, N., Dholakia, K., Allen, L., and Padgett, M. (1997) Mechanical equivalence of spin and orbital angular momentum of light: an optical spanner. *Opt. Lett.*, **22** (1), 52.
9. Beijersbergen, M.W., Coerwinkel, R.P.C., Kristensen, M., and Woerdman, J.P. (1994) Helical-wavefront laser beams produced with a spiral phaseplate. *Opt. Commun.*, **112** (5–6), 321–327.
10. Oemrawsingh, S.S.R., van Houwelingen, J.A.W., Eliel, E.R., Woerdman, J.P., Verstegen, E.J.K., Kloosterboer, J.G., and Hooft, G.Wt. (2004) Production and characterization of spiral phase plates for optical wavelengths. *Appl. Opt.*, **43** (3), 688.
11. Cheong, W., Lee, W., Yuan, X., and Zhang, L. (2004) Direct electron-beam writing of continuous spiral phase plates in negative resist with high power efficiency for optical manipulation. *Appl. Phys. Lett.*, **85**, 5784.
12. Lee, W., Yuan, X., and Cheong, W. (2004) Optical vortex beam shaping by use of highly efficient irregular spiral phase plates for optical micromanipulation. *Opt. Lett.*, **29** (15), 1796.
13. Jeffries, G.D.M., Milne, G., Zhao, Y.Q., Lopez-Mariscal, C., and Chiu, D.T. (2009) Optofluidic generation of Laguerre-Gaussian beams. *Opt. Express*, **17** (20), 17555–17562.
14. Gussard, R., Lindmo, T., and Brevik, I. (1992) Calculation of the trapping force in a strongly focused laser-beam. *J. Opt. Soc. Am. B*, **9** (10), 1922–1930.
15. Garces-Chavez, V., Volke-Sepulveda, K., Chavez-Cerda, S., Sibbett, W., and Dholakia, K. (2002) Transfer of orbital angular momentum to an optically trapped low-index particle. *Phys. Rev. A*, **66** (6), 063402.
16. Allen, L., Padgett, M.J., and Babiker, M. (1999) The orbital angular momentum of light. *Prog. Opt.*, **39**, 291–372.
17. Jackson, J.D. (1998) *Classical Electrodynamics*, John Wiley & Sons, Inc.
18. Barton, J., Alexander, D., and Schaub, S. (1989) Theoretical determination of net radiation force and torque for a spherical particle illuminated by a focused laser beam. *J. Appl. Phys.*, **66** (10), 4594–4602.
19. Farsund, O. and Felderhof, B.U. (1996) Force, torque, and absorbed energy for a body of arbitrary shape and constitution in an electromagnetic radiation field. *Phys. A*, **227**, 108–130.

20. Berry, M.V. (2009) Optical currents. *J. Opt. A-Pure Appl. Opt.*, **11** (9), 094001.
21. Mazilu, M. (2009) Spin and angular momentum operators and their conservation. *J. Opt. A-Pure Appl. Opt.*, **11** (9), 094005.
22. He, H., Friese, M., and Heckenberg, N. (1995) Direct observation of transfer of angular momentum to absorptive particles from a laser beam with a phase singularity. *Phys. Rev. Lett.*, **75** (5), 826.
23. Simpson, N., Allen, L., and Padgett, M. (1996) Optical tweezers and optical spanners with Laguerre-Gaussian modes. *J. Mod. Opt.*, **43** (12), 2485.
24. O'Neil, A., MacVicar, I., Allen, L., and Padgett, M. (2002) Intrinsic and extrinsic nature of the orbital angular momentum of a light beam. *Phys. Rev. Lett.*, **88** (5), 053601.
25. Garces-Chavez, V., McGloin, D., Padgett, M.J., Dultz, W., Schmitzer, H., and Dholakia, K. (2003) Observation of the transfer of the local angular momentum density of a multiringed light beam to an optically trapped particle. *Phys. Rev. Lett.*, **91** (9), 093602.
26. Durnin, J., Micelli, J.J., and Eberly, J.H. (1987) Diffraction-free beams. *Phys. Rev. Lett.*, **58** (15), 1499–1501.
27. Mazilu, M., Stevenson, D., Gunn-Moore, F., and Dholakia, K. (2009) Light beats the spread: "non-diffracting" beams. *Laser Photon. Rev.*, doi: 10.1002/lpor.200910019
28. Volke-Sepulveda, K., Garces-Chavez, V., Chavez-Cerda, S., Arlt, J., and Dholakia, K. (2002) Orbital angular momentum of a high-order Bessel light beam. *J. Opt. B*, **4** (2), S82–S89.
29. Parkin, S., Knoner, G., Nieminen, T.A., Heckenberg, N.R., and Rubinsztein-Dunlop, H. (2006) Measurement of the total optical angular momentum transfer in optical tweezers. *Opt. Express*, **14** (15), 6963–6970.
30. Zhao, Y., Milne, G., Edgar, J.S., Jeffries, G.D.M., McGloin, D., and Chiu, D.T. (2008) Quantitative force mapping of an optical vortex trap. *Appl. Phys. Lett.*, **92** (16), 161111.
31. Zhao, Y.Q., Edgar, J.S., Jeffries, G.D.M., McGloin, D., and Chiu, D.T. (2007) Spin-to-orbital angular momentum conversion in a strongly focused optical beam. *Phys. Rev. Lett.*, **99** (7), 073901.
32. Curtis, J.E. and Grier, D.G. (2003) Structure of optical vortices. *Phys. Rev. Lett.*, **90** (13), 133901.
33. Lopez-Mariscal, C., Gutierrez-Vega, J.C., Milne, G., and Dholakia, K. (2006) Orbital angular momentum transfer in helical Mathieu beams. *Opt. Express*, **14** (9), 4182–4187.
34. Jesacher, A., Furhapter, S., Maurer, C., Bernet, S., and Ritsch-Marte, M. (2006) Reverse orbiting of microparticles in optical vortices. *Opt. Lett.*, **31** (19), 2824–2826.
35. Tao, S.H., Yuan, X.C., Lin, J., Peng, X., and Niu, H.B. (2005) Fractional optical vortex beam induced rotation of particles. *Opt. Express*, **13** (20), 7726–7731.
36. Lee, W.M., Garces-Chavez, V., and Dholakia, K. (2006) Interference from multiple trapped colloids in an optical vortex beam. *Opt. Express*, **14** (16), 7436–7446.
37. Dienerowitz, M., Mazilu, M., Reece, P.J., Krauss, T.F., and Dholakia, K. (2008) Optical vortex trap for resonant confinement of metal nanoparticles. *Opt. Express*, **16** (7), 4991–4999.
38. O'Neil, A. and Padgett, M. (2000) Three-dimensional optical confinement of micron-sized metal particles and the decoupling of the spin and orbital angular momentum within an optical spanner. *Opt. Commun.*, **185**, 139.
39. Wright, A.J., Girkin, J.M., Gibson, G.M., Leach, J., and Padgett, M.J. (2008) Transfer of orbital angular momentum from a super-continuum, white-light beam. *Opt. Express*, **16** (13), 9495–9500.
40. Morris, J.E., Carruthers, A.E., Mazilu, M., Reece, P.J., Cizmar, T., Fischer, P., and Dholakia, K. (2008) Optical micromanipulation using supercontinuum Laguerre-Gaussian and Gaussian beams. *Opt. Express*, **16** (14), 10117–10129.
41. Ashkin, A. (1970) Acceleration and trapping of particles by radiation pressure. *Phys. Rev. Lett.*, **24** (4), 156.

42. Ashkin, A. and Dziedzic, J. (1974) Stability of optical levitation by radiation pressure. *Appl. Phys. Lett.*, **24** (12), 586.
43. Ashkin, A. (1997) Optical trapping and manipulation of neutral particles using lasers. *Proc. Natl. Acad. Sci. U.S.A.*, **94**, 4853.
44. Gahagan, K.T. and Swartzlander, G.A. (1996) Optical vortex trapping of particles. *Opt. Lett.*, **21** (11), 827–829.
45. Gahagan, K.T. and Swartzlander, G.A. (1999) Simultaneous trapping of low-index and high-index microparticles observed with an optical-vortex trap. *J. Opt. Soc. Am. B*, **16** (4), 533.
46. Friese, M.E.J., RubinszteinDunlop, H., Heckenberg, N.R., and Dearden, E.W. (1996) Determination of the force constant of a single-beam gradient trap by measurement of backscattered light. *Appl. Opt.*, **35** (36), 7112–7116.
47. Simpson, N.B., McGloin, D., Dholakia, K., Allen, L., and Padgett, M.J. (1998) Optical tweezers with increased axial trapping efficiency. *J. Mod. Opt.*, **45** (9), 1943–1949.
48. Lee, K., Choi, S., and Kim, S. (2005) Radiation of spin waves from magnetic vortex cores by their dynamic motion and annihilation processes. *Appl. Phys. Lett.*, **87**, 192502.
49. Lorenz, R., Edgar, J., Jeffries, G., and Zhao, Y. (2007) Vortex-Trap-Induced fusion of Femtoliter-Volume aqueous droplets. *Anal. Chem.*, **79**, 224.
50. Jeffries, G.D.M., Kuo, J.S., and Chiu, D.T. (2007) Controlled shrinkage and re-expansion of a single aqueous droplet inside an optical vortex trap. *J. Phys. Chem. B*, **111** (11), 2806–2812.
51. Prentice, P., Cuschieri, A., and Dholakia, K. (2005) Membrane disruption by optically controlled microbubble cavitation. *Nat. Phys.*, **1**, 107.
52. Prentice, P.A., MacDonald, M.P., Frank, T.G., Cuschieri, A., Spalding, G.C., Sibbett, W., Campbell, P.A., and Dholakia, K. (2004) Manipulation and filtration of low index particles with holographic Laguerre-Gaussian optical trap arrays. *Opt. Express*, **12**, 593.
53. Garbin, V., Cojoc, D., Ferrari, E., Di Fabrizio, E., Overvelde, M.L.J., van der Meer, S.M., de Jong, N., Lohse, D., and Versluis, M. (2007) Changes in microbubble dynamics near a boundary revealed by combined optical micromanipulation and high-speed imaging. *Appl. Phys. Lett.*, **90** (11), 114103.
54. Bishop, A., Nieminen, T., and Heckenberg, N. (2004) Optical microrheology using rotating laser-trapped particles. *Phys. Rev. Lett.*, **92** (19), 198104.
55. Porta, A.L. and Wang, M. (2004) Optical torque wrench: angular trapping, rotation, and torque detection of quartz microparticles. *Phys. Rev. Lett.*, **92** (19), 190801.
56. Oroszi, L., Galajda, P., Kirei, H., and Bottka, S. (2006) Direct measurement of torque in an optical trap and its application to double-strand DNA. *Phys. Rev. Lett.*, **97**, 058301.
57. Bishop, A., Nieminen, T., and Heckenberg, N. (2003) Optical application and measurement of torque on microparticles of isotropic nonabsorbing material. *Phys. Rev. A*, **68**, 033802.
58. Parkin, S., Knoener, G., and Nieminen, T. (2007) Picoliter viscometry using optically rotated particles. *Phys. Rev. E*, **74**, 041507.
59. Kelemen, L., Valkai, S., and Ormos, P. (2007) Parallel photopolymerisation with complex light patterns generated by diffractive optical elements. *Opt. Express*, **15**, 14488.
60. Terray, A., Oakey, J., and Marr, D. (2002) Microfluidic control using colloidal devices. *Science*, **296**, 1841.
61. Ladavac, K. and Grier, D. (2004) Microoptomechanical pumps assembled and driven by holographic optical vortex arrays. *Opt. Express*, **12**, 1144.
62. Knoner, G., Parkin, S., Nieminen, T.A., Loke, V.L.Y., Heckenberg, N.R., and Rubinsztein-Dunlop, H. (2007) Integrated optomechanical microelements. *Opt. Express*, **15** (9), 5521–5530.
63. Sato, S., Ishigure, M., and Inaba, H. (1991) Optical trapping and rotational manipulation of microscopic particles and biological cells using higher-order mode Nd:YAG laser beamsrecords free of charge. *Electron. Lett.*, **27** (20), 1831.

64. Paterson, L., MacDonald, M., Arlt, J., Sibbett, W., Bryant, P.E., and Dholakia, K. (2001) Controlled rotation of optically trapped microscopic particles. *Science*, **292**, 912.
65. MacDonald, M.P., Paterson, L., Volke-Sepulveda, K., Arlt, J., Sibbett, W., and Dholakia, K. (2002) Creation and manipulation of three-dimensional optically trapped structures. *Science*, **296**, 1101.
66. Arlt, J., MacDonald, M., Paterson, L., Sibbett, W., Dholakia, K., and Volke-Sepulveda, K. (2002) Moving interference patterns created using the angular Doppler-effect. *Opt. Express*, **10**, 844.
67. Faucheux, L., Bourdieu, L., and Kaplan, P. (1995) Optical thermal ratchet. *Phys. Rev. Lett.*, **74** (9), 1504.
68. Faucheux, L., Stolovitzky, G., and Libchaber, A. (1995) Periodic forcing of a Brownian particle. *Phys. Rev. E*, **51** (6), 5239.
69. Harada, T. and Yoshikawa, K. (2004) Fluctuation-response relation in a rocking ratchet. *Phys. Rev. E*, **69**, 031113.
70. Lutz, C., Reichert, M., Stark, H., and Bechinger, C. (2006) Surmounting barriers: the benefit of hydrodynamic interactions. *Europhys. Lett.*, **74** (4), 719.
71. Roichman, Y., Grier, D.G., and Zaslavsky, G. (2007) Anomalous collective dynamics in optically driven colloidal rings. *Phys. Rev. E*, **75** (2), 020401.
72. Lee, S.H. and Grier, D.G. (2006) Giant colloidal diffusivity on corrugated optical vortices. *Phys. Rev. Lett.*, **96** (19), 190601.

5
Optical Torques in Liquid Crystals

Enrico Santamato and Bruno Piccirillo

The interaction between light and liquid crystals (LCs) has been intensely studied for several decades and is still the subject of several applied and fundamental researches. No doubt the great popularity of liquid crystals is largely owed to their very specific electro-optical properties, which have gained growing importance in information technology. Let us consider for instance, liquid crystal displays (LCDs), at present, the leading technology in the information display industry. The electro-optical properties of LCs have also been used to produce photonic devices such as laser beam steering, variable optical attenuators, variable retardation waveplates, tunable-focus lenses, and spatial light modulators (SLMs).

LCs are fluids made up of organic molecules with a specific, anisometric shape, with an arrangement exhibiting a degree of long-range order. The most typical are rod-like molecules or rod-like molecular aggregates (Figure 5.1), which give rise to conventional nematic and smectic phases. Nematic liquid crystals (NLCs) are characterized only by long-range orientational order. The axes of the molecules locally share a common average direction – specified by a unit vector $\mathbf{n}$ called *molecular director* – and the centers of mass of the molecules are randomly spread over space. Smectic liquid crystals (SLCs), on the other hand, are characterized by both the orientational order and a degree of positional order.

The orientational order occurring in LC results in the anisotropy of mechanical, electrical, magnetic, and optical properties [1, 2]. In a sense, LCs, or mesophases, combine the properties of a solid crystal and those of an anisotropic liquid, resulting in very specific optical phenomena, having no counterparts in solids or in isotropic liquids. In what follows we consider only nematic mesophases. NLCs are birefringent and locally uniaxial, the director being coincident with the direction of the optical axis. Most of the specific LC optical effects must be traced back to the reorientation of the director in the macroscopic volume of the material under the influence of an external field or the flow of the liquid. The possibility of reorienting LCs by applying static magnetic or electric fields was known since a long time [3]. The director $\mathbf{n}$ reorients in a static electric (magnetic) field under the action of a dielectric (diamagnetic) torque, which is proportional to the dielectric (diamagnetic)

Twisted Photons: Applications of Light with Orbital Angular Momentum.
Edited by Juan P. Torres and Lluis Torner

ISBN: 978-3-527-40907-5

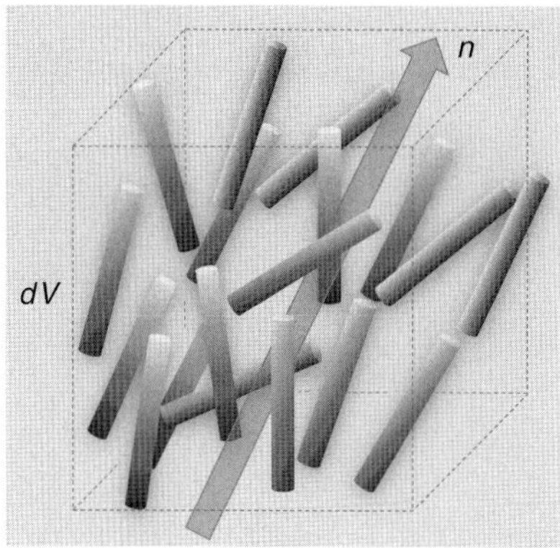

Figure 5.1 Schematic representation of the nematic phase. (Please find a color version of this figure on the color plates.)

anisotropy $\Delta\epsilon = \epsilon_{\parallel} - \epsilon_{\perp}$ ($\Delta\chi = \chi_{\parallel} - \chi_{\perp}$). The corresponding contribution g_E to the density of the nematic free energy is [1]

$$g_E = -\frac{\mathbf{D}\cdot\mathbf{E}}{8\pi} = -\frac{\epsilon_{\perp}E^2}{8\pi} - \frac{\Delta\epsilon\,(E\cdot n)^2}{8\pi} \tag{5.1}$$

that is, the director **n** tends to be aligned along the field, $\mathbf{n} \parallel \mathbf{E}$, if $\Delta\epsilon > 0$ and perpendicular to **n**, $\mathbf{n} \perp \mathbf{E}$, if $\Delta\epsilon < 0$. Analogous expressions and comments hold for magnetic field – induced reorientation as well. Therefore, the molecular reorientation, in a sufficiently slow variation of the external electric (or magnetic) fields, originates as a result of anisotropic electrical properties of the medium, whereas the dynamics of the process depend on both the viscoelastic properties of the mesophase and the initial orientation of the director with respect to the external field. The optical properties of the medium and, in particular, its local optical anisotropy change as a consequence of the molecular reorientation – occurring either locally or throughout the whole sample – and underlie all the known electro-optical effects. In general, the action of an external electric or magnetic field may be described in terms of a torque acting on the molecular director of the medium. The instantaneous orientation of the molecular director is ultimately determined by the balance of both the external torques and internal viscoelastic torques (overdamped motion). The general expressions for the external magnetic and electrical torques, respectively, are

$$\begin{aligned}\tau_m &= \frac{1}{4}\mathbf{B}\times\mathbf{H} = \Delta\chi\,(\mathbf{n}\cdot\mathbf{H})\,(\mathbf{n}\times\mathbf{H})\\ \tau_e &= \frac{1}{4}\mathbf{D}\times\mathbf{E} = \Delta\epsilon\,(\mathbf{n}\cdot\mathbf{E})\,(\mathbf{n}\times\mathbf{E})\end{aligned} \tag{5.2}$$

In order to realize the operation of the reorientation mechanism, let us consider a nematic film, whose director has been prepared beforehand with an alignment parallel to the film surface (uniform or planar alignment), via a suitable substrate coating. Let us suppose now to apply an electric (or magnetic field) along a direction perpendicular to the film surface and consequently it is perpendicular also to the molecular director in the initial configuration. The electrical (or magnetic) torque in Eq. (5.2) tends to enhance any fluctuation of **n** off the initial axis, provided that the dielectric (or diamagnetic) anisotropy is positive. However, only if the applied field is higher than a critical threshold value E_c (or H_c), the surface anchoring forces are overcome by the external field and a director distribution becomes stable with a distortion toward the external field. The extent to which reorientation occurs turns to be a function of the applied field strength. The destabilization of the undistorted state via a static magnetic or electric field, having a strength higher than a critical threshold value, is referred to as *Fréedericksz transition* [3].

NLCs may be reoriented also by the optical field of a laser beam, and an optical counterpart of the Fréedericksz transition also exists. The reorientation with optical fields was discovered only in 1980 by Zel'dovich and Tabiryan [4] and by Zolot'ko *et al.* [5] and later quantitatively discussed by Durbin *et al.* in 1981 [6]. The laser-induced reorientation may be described in terms of an optical torque, in a similar way as its static counterpart. It represents the main contribution to the third-order nonlinearity of LCs, and the effect appears in the form of a refractive index change δn proportional to the laser intensity I. Actually, even with moderate laser intensities, the effect can be so large that δn is no longer linear in I, but appears as a complex nonlinear function of I. It has been largely clarified that molecular correlation is the reason behind the strong reorientation effect [7]. The electronic response certainly also contributes to the third-order nonlinearity of LCs [8–10], but it is not expected to be larger than those of other organic molecules with delocalized electrons.

The existence of a characteristic threshold laser intensity, below which no molecular reorientation can be induced, was demonstrated both theoretically [4] and experimentally [5] in the case of linearly polarized light incident onto an LC initially aligned perpendicular to the film surfaces (homeotropic alignment). The underlying physical mechanism of such an effect, known as the optical Fréedericksz transition (OFT), is essentially the same as in the corresponding dc Fréedericksz transition. The geometry dictates, in fact, that the polarization of the light beam remains linear in traversing the cell, even with molecular reorientation. There are a number of other dc Fréedericksz transitions with different geometries to which an optical analog can be found. In most cases, however, the underlying physical mechanisms of the dc- and optical-field-induced transitions are deeply different, since the beam polarization, as well as the azimuthal structure of the beam transverse wave front, indeed change in the propagation through the medium. This amounts to say that, in some geometries, the optically induced reorientation takes place through an exchange of both spin angular momentum (SAM) and orbital angular momentum (OAM) between the medium and the photons in the optical fields.

In this chapter, we focus our attention exactly on the problem of OAM and SAM transfer from photons to LCs in nonlinear processes. This ability will be recognized to be the distinctive feature of nonlinear optics of LCs. In the following sections, we discuss in some details the optical reorientation in LCs and put special emphasis on the geometries in which the optical reorientation is associated to an SAM and/or OAM transfer from the radiation to the medium. Later, we describe, in general, the processes of angular momentum exchange in terms of the balance among OAM flux, SAM flux, and the momentum flux internal to the fluid. It will be recognized that OAM and SAM, as a matter of facts, act on distinct degrees of freedom in the liquid crystalline fluid. This result reflects onto the possibility of separating OAM and SAM contributions to the total angular momentum of photons also beyond the paraxial approximation. Some of the most relevant dynamical effects observed in the presence of OAM transfer are reviewed. In many cases, the OAM transfer is combined to an SAM transfer from radiation to matter. In such a sense, a suitably reoriented nematic film behaves as an SAM-to-OAM converter. The spin-to-orbital conversion (STOC) observed and described in the literature could be exploited, in future, to realize all-optical tunable converters to be used together with or alternatively to the recently invented q-plates [11–13], to implement OAM technology for quantum optics and optical communications.

5.1
The Optical Reorientation and the Photon Angular Momentum Flux

The laser field can affect the LC molecular orientation in two ways, that is, modifying the orientational distribution (as represented by the orientational order parameter) and changing the direction of average orientation (specified by the director). The former is the only mechanism operating if LCs are in the isotropic liquid phase. The latter dominates in the LC mesophases and is the subject of the present discussion. In the mesophases, molecules are highly correlated and the applied field can no longer affect the correlation significantly. It is, however, easy for the field to alter the average direction $\mathbf{n}$ of orientation of the correlated molecules. This is similar to the case of ferromagnets where the magnetization direction is altered by an applied magnetic field. Director reorientation is the mechanism leading to the so-called giant optical nonlinearity observed in transparent LC mesophases [7]. Changes in the refractive index as large as 0.001–0.1 can be obtained with a laser intensity of several hundreds of watts per square centimeter. Unfortunately, the strong molecular correlation responsible for the giant nonlinearity also leads to a very slow response time, on the order of milliseconds to seconds.

In this section we aim at explaining the actual mechanism through which an optical field may change the original molecular alignment (as represented by the director orientation) imposed on the medium by the boundary conditions through molecular anchoring at the surfaces. In NLC mesophases, each volume element dV, located at a position $\mathbf{r}$ at a time t, is characterized not only by a mass $\rho(\mathbf{r}, t)$, as in conventional isotropic fluid, but also by the orientation of the molecular director $\mathbf{n}(\mathbf{r}, t)$, representing, as above stated, the common average direction of the molecules contained inside dV (Figure 5.1). LCs, therefore, exhibit clearly distinguishable orbital and intrinsic degrees of freedom. Strictly speaking, molecular reorientation refers to the rotation of the director $\mathbf{n}(\mathbf{r}, t)$ in the fixed volume element dV, which is associated with the intrinsic (spin) part of the angular momentum of the material. However, the rotation of the director is dynamically coupled to the rotational motion of the center of mass $\mathbf{r}$ of dV, which is in turn associated with the orbital part of the angular momentum of the medium. As a consequence, the possible effects of an external field on the orbital motion of the molecules reflect on the molecular reorientation. In most cases, because of the high translational viscosity of the fluid, the fluid velocity $\mathbf{v}$ is vanishingly small and the momentum of the external force acting on the centers of mass of the elemental volume dV may be even regarded as an effective source of torque on $\mathbf{n}$.

In what follows, the sources of the optical torques acting in the liquid crystalline material are retraced back in a rigorous way to the angular momentum fluxes carried by a monochromatic radiation in the surrounding medium, which is supposed to be homogenous and isotropic. It will come out quite naturally that the average orientation $\mathbf{n}(\mathbf{r}, t)$ of the molecules contained in dV is directly affected by the spin part of the angular momentum of the optical field, while the center of mass of the volume element dV is directly affected by the orbital part. The problem of the dynamics of NLCs evidently intersects with the problem of the separation of the angular momentum of light in its intrinsic (or spin) and orbital parts, especially beyond the paraxial optics approximation [14–21]. In fact, the separation of the rotational motions inside the liquid crystalline material and the separation of the torques they arise from have repercussions on the corresponding fluxes in the radiation incident from the outside. This makes LCs a very good arena to unambiguously define the OAM and SAM *fluxes* even beyond the paraxial approximation.

5.1.1 Dynamical Equations of Liquid Crystals

We start from the usual equations of motion for the liquid crystalline fluid and for the director $\mathbf{n}$:

$$\rho\dot{\mathbf{v}} = \mathbf{f} = \operatorname{div}\hat{\sigma} - \operatorname{grad}p \tag{5.3}$$

$$\rho\mathbf{r}\times\dot{\mathbf{v}} = \mathbf{r}\times\mathbf{f} = \operatorname{div}\hat{L} - \mathbf{w} \tag{5.4}$$

$$\mathcal{I}\mathbf{n}\times\ddot{\mathbf{n}} = \boldsymbol{\tau} = \operatorname{div}\hat{S} + \mathbf{w} \tag{5.5}$$

where ρ is the fluid density (assumed to be constant); $\mathbf{v}$, the velocity of the fluid (flow of the centers of mass of dV); $\mathbf{f}$, the force per unit volume acting on the fluid; p, the hydrostatic pressure; $\mathcal{I}$, the momentum of inertia per unit volume associated to the rotation of $\mathbf{n}$; $\boldsymbol{\tau}$, the torque density acting on $\mathbf{n}$; $\hat{\sigma}$, the stress tensor; and $\mathbf{w}$, the vector dual to its antisymmetric part, that is, $w_\alpha = \epsilon_{\alpha\beta\gamma}\sigma_{\beta\gamma}$. Elastic ($e$), electromagnetic ($em$), and viscous ($v$) forces contribute, in general, to the torque density $\boldsymbol{\tau} = \boldsymbol{\tau}^e + \boldsymbol{\tau}^{em} + \boldsymbol{\tau}^v$ and to the stress tensor $\hat{\sigma} = \hat{\sigma}^e + \hat{\sigma}^{em} + \hat{\sigma}^v$. The dot in Eq. (5.3) stands for the material derivative. The tensor $\hat{L}$ is defined as $L_{\rho\alpha} = \epsilon_{\alpha\beta\gamma}x_\beta\sigma_{\rho\gamma}$ so that Eq. (5.4) follows from Eq. (5.3). The tensor $\hat{S}$, on its hand, comes from a variational principle based on a suitable free energy functional from which the torque density $\boldsymbol{\tau}$ and the stress tensor $\hat{\sigma}$ can be also deduced. The divergence of a tensor $\hat{T}$ here is defined as $(\text{div}\,\hat{T})_\alpha = \partial_\rho T_{\rho\alpha}$. The terms on the left-hand side in Eqs. (5.3)–(5.5) may be unambiguously interpreted as the densities per unit time of linear momentum, OAM, and intrinsic angular momentum of matter, respectively. The terms on the right-hand side of Eqs. (5.3)–(5.5) can be consistently interpreted as the densities of force, orbital torque, and intrinsic torque acting in the bulk of the medium. Such densities are represented by expressions that contain the divergences of tensors involving the external fields and then make evident their relationship with the fluxes of force and momentum from the outside. It is therefore quite natural referring to the tensors $\hat{L}$and $\hat{S}$ in Eqs. (5.4) and (5.5) as to the orbital and the intrinsic (spin) angular momentum flux densities, respectively. On this definition of fluxes we return later. In the meantime, let us calculate the stress tensor $\hat{\sigma}$, the torque density $\boldsymbol{\tau}$, and the intrinsic angular momentum flux density $\hat{S}$. The elastic and electromagnetic contributions can be deduced applying variational calculus to the free energy functional $\mathcal{F} = \int_V F\,dV = \int_V (F^e + F^{em})\,dV$, where as densities of the elastic and electromagnetic free energy we may take, respectively,

$$F^e = \frac{1}{2}\left[k_1(\text{div}\,\mathbf{n})^2 + k_2(\mathbf{n}\cdot\text{rot}\,\mathbf{n})^2 + k_3(\mathbf{n}\times\text{rot}\,\mathbf{n})^2\right] \tag{5.6}$$

where k_i $(i = 1, 2, 3)$ are the elastic constants for splay, twist, and bend deformations, and

$$F^{em} = \frac{1}{16\pi}(\mathbf{B}^*\cdot\mathbf{H} - \mathbf{D}^*\cdot\mathbf{E}) \tag{5.7}$$

where monochromatic optical fields are assumed and the magnetic and electric inductions $\mathbf{B}$ and $\mathbf{D}$ are related to the corresponding fields by $\mathbf{B} = \hat{\mu}\mathbf{H}$, $\mathbf{D} = \hat{\epsilon}\mathbf{E}$, with magnetic and dielectric tensors related to $\mathbf{n}$ by the uniaxial form: $\hat{\mu} = \mu_0 + \mu_a\mathbf{nn}$, $\hat{\epsilon} = \epsilon_0 + \epsilon_a\mathbf{nn}$. The constants μ_0, ϵ_0, μ_a, ϵ_a characterize the magnetic and electric response of the material. In particular, μ_a and ϵ_a characterize the material anisotropy and they vanish in isotropic media. For monochromatic optical fields, $\mathbf{B}$ and $\mathbf{H}$ can be related to the spatial derivatives of the electric field $\mathbf{E}$, using Maxwell's equation and constitutive relation

$$\mathbf{B} = -(i/k_0)\text{rot}\,\mathbf{E} \qquad \mathbf{H} = \hat{\eta}\mathbf{B} \tag{5.8}$$

with $k_0 = \omega/c$, c being the speed of light in vacuum and ω the optical frequency (cgs units are used), and $\hat{\eta} = \hat{\mu}^{-1} = \eta_0 + \eta_a\mathbf{nn}$. Substituting Eq. (5.8) in Eq. (5.7), the total free energy density $F = F^e + F^{em}$ reduces to a function of the fields $\mathbf{n}(\mathbf{r})$, $\mathbf{E}(\mathbf{r})$, $\mathbf{E}^*(\mathbf{r})$, and of their spatial derivatives. The field equations associated to the total free energy $\mathcal{F}$ are

$$\mathbf{h} = \text{div}\,\hat{\pi} - \partial F/\partial\mathbf{n} = \lambda(\mathbf{r})\mathbf{n} \tag{5.9}$$

$$\boldsymbol{\Lambda} = \text{div}\,\hat{p} - \partial F/\partial\mathbf{E}^* = 0 \tag{5.10}$$

where $\pi_{\rho\gamma} = \partial F/\partial(\partial_\rho n_\gamma)$ and $p_{\rho\gamma} = \partial F/\partial(\partial_\rho E^*_\gamma)$ are the tensors of the generalized momenta associated to the fields $\mathbf{n}$ and $\mathbf{E}^*$, respectively, and $\lambda(\mathbf{r})$ is a Lagrange multiplier accounting for the constraint $\mathbf{n}^2 = 1$. At steady state $(\mathbf{v} = 0, \dot{\mathbf{n}} = 0)$, Eq. (5.9) is equivalent to Eq. (5.5), the sum of the elastic and of the electromagnetic torque densities being given by $\boldsymbol{\tau}^e + \boldsymbol{\tau}^{em} = \mathbf{n} \times \mathbf{h}$. Equation (5.10), on the other hand, is equivalent to Maxwell's equation rot $(\hat{\eta}$ rot $\mathbf{E}) = k_0^2\mathbf{D}$. The elastic and electromagnetic contributions $\hat{\sigma}^e$ and $\hat{\sigma}^{em}$ to the stress tensors $\hat{\sigma}$ in Eq. (5.3) are given by the opposite of the energy-momentum tensors associated to F^e and F^{em}, respectively; that is,

$$\hat{\sigma}^e_{\beta\alpha} = -\pi_{\beta\gamma}\partial_\alpha n_\gamma + \delta_{\beta\alpha}F^e \tag{5.11}$$

$$\hat{\sigma}^{em}_{\beta\alpha} = -p_{\beta\gamma}\partial_\alpha E^*_\gamma + \delta_{\beta\alpha}F^{em} \tag{5.12}$$

It can be easily proved that the electromagnetic force $\mathbf{f}^{em} = \operatorname{div}\hat{\sigma}^{em}$ acting on the unit volume has the right form [22] $f^{em}_\alpha = -(1/16\pi)(E^*_\beta E_\gamma \partial_\alpha \epsilon_{\beta\gamma} + H^*_\beta H_\gamma \partial_\alpha \mu_{\beta\gamma})$. In the LC community $\mathbf{h}$ is known as the molecular field. The explicit expression of the elastic contributions to $\mathbf{h}$ can be found in standard textbooks on the physics of LCs [[1], Eq. (3.22)]. The electromagnetic contribution $\boldsymbol{\tau}^{em}$ to the torque density $\boldsymbol{\tau}$ results in the sum of the optical torque $\boldsymbol{\tau}^o = 1/(8\pi)\Re(\mathbf{D}^* \times \mathbf{E})$ and of the magnetic torque $\boldsymbol{\tau}^{\mathrm{mag}} = 1/(8\pi)\Re(\mathbf{B}^* \times \mathbf{H})$. At optical frequencies, LCs are nonmagnetic, $\mathbf{B}$ and $\mathbf{H}$ are parallel, and $\boldsymbol{\tau}^{\mathrm{mag}}$ vanishes, leaving only the optical torque $\boldsymbol{\tau}^o$. The invariance of $\mathcal{F}$ with respect to a rotation of the coordinate frame may be exploited to split both the elastic and the electromagnetic torque density into the sum of the divergence of a tensor and the antisymmetric part of the corresponding stress tensor

$$\boldsymbol{\tau}^h = \operatorname{div}\hat{S}^h + \mathbf{w}^h, \qquad h = e, em \tag{5.13}$$

The identity in Eq. (5.13) holds true for an arbitrary field $\mathbf{n}$ and a field $\mathbf{E}$ obeying Maxwell's equations (Eq. (5.10)). The tensors $\hat{S}^e$ and $\hat{S}^{em}$ can be regarded as the elastic and the electromagnetic "spin flux densities," respectively.

The contribution σ^v of viscous forces to the overall stress tensor can be deduced on the grounds of phenomenological considerations and can be found in textbooks on the physics of LCs [1], Eqs. (5.31) and (5.32). Adding the electromagnetic field does not change $\hat{\sigma}^v$, since no entropy source is associated to the optical field when light absorption is neglected. In particular, we still have the useful relationship $\boldsymbol{\tau}^v = \mathbf{w}^v$ between the viscous torque density and the antisymmetric part of $\hat{\sigma}^v$. Comparing this relationship with Eq. (5.13), we conclude that no "spin flux density" is associated to viscous torques. From Eq. (5.13) and from the relation $\boldsymbol{\tau}^v = \mathbf{w}^v$, we can see that the last equality on the right-hand side of Eq. (5.5) is a consequence of the rotational invariance of the total free energy of the system.

5.1.2
Angular Momentum Fluxes

Adding Eqs. (5.4) and (5.5) together yields

$$\rho\mathbf{r} \times \dot{\mathbf{v}} + \mathcal{I}\mathbf{n} \times \ddot{\mathbf{n}} = \frac{d}{dt}(\rho\mathbf{r} \times \mathbf{v} + \mathcal{I}\mathbf{n} \times \dot{\mathbf{n}}) = \operatorname{div}(\hat{L} + \hat{S}) = \operatorname{div}\hat{J} \tag{5.14}$$

stating that the conservation of the total (orbital + intrinsic) angular momentum of the system. The fluxes $\hat{L}$ and $\hat{S}$ do not conserve separately, however, because of the presence of the vector $\mathbf{w}$ in Eqs. (5.4) and (5.5). The vector $\mathbf{w}$ is to be interpreted as an internal torque in the volume element dV that couples the L- and S-flux densities. If the total stress tensor was symmetric, then $\mathbf{w}$ would vanish and $\hat{L}$ and $\hat{S}$ would exhibit separate conservation laws. The stress tensor $\sigma_{\rho\alpha}$ and the related L- and S-flux densities, however, are determined up to the following gauge transformations:

$$\begin{aligned}
\sigma_{\rho\alpha} &\rightarrow \sigma'_{\rho\alpha} = \sigma_{\rho\alpha} + \partial_\gamma f_{\gamma\rho\alpha} \\
L_{\rho\alpha} &\rightarrow L'_{\beta\alpha} = \epsilon_{\alpha\beta\gamma} x_\beta \sigma'_{\rho\gamma} \\
S_{\rho\alpha} &\rightarrow S'_{\rho\alpha} = S_{\rho\alpha} + \epsilon_{\alpha\beta\gamma} f_{\rho\beta\gamma}
\end{aligned} \tag{5.15}$$

where $f_{\gamma\rho\alpha} = -f_{\rho\gamma\alpha}$. Equations (5.3)–(5.5) are invariant under the transformations (5.15). The gauge function $f_{\gamma\rho\alpha}$ may be uniquely chosen so to have the components of the spin flux tensor $S'_{\rho\alpha}$ arbitrarily fixed. In particular, the gauge may be fixed so that $S'_{\rho\alpha} = 0$. In this gauge we have $\boldsymbol{\tau} = \mathbf{n} \times \mathbf{h} = \mathbf{w}$, which means that the torque acting on $\mathbf{n}$ is fully determined by the antisymmetric part of the stress tensor. Moreover, in the spinless gauge, the orbital and the total angular momentum flux densities are the same, that is, $\hat{L} = \hat{J}$. This spinless gauge is commonly exploited in the physics of fluids to symmetrize the stress tensor: assuming, in fact, the intrinsic angular momentum to be locally balanced, that is, $\boldsymbol{\tau} = 0$, the antisymmetric part of the stress tensor $\mathbf{w}$ turns to be zero, yielding to a totally symmetric stress tensor $\hat{\sigma}$. The condition of balance of the torques acting on $\mathbf{n}$ entails that the inertial term on the left-hand side of Eq. (5.5) is zero or negligible, as usually assumed in LCs. The last peculiarity was exploited by the Harvard group long ago to describe the hydrodynamics of LCs through a symmetric stress tensor in the small elastic distortion approximation [23]. In block I of Table 5.1, we have reported the stress tensor, its antisymmetric part, and the spin flux density tensor in the spinless gauge. In this gauge $\hat{S} = 0$, by definition, and $\boldsymbol{\tau} = \mathbf{w}$. If we further assume $\boldsymbol{\tau} = 0$, we may retain, in calculating the force density $\mathbf{f}$, only the symmetric part of the total stress tensor $\hat{\sigma}$ reported in block I of Table 5.1. In particular, in the spinless gauge, the electromagnetic part of the force density $\mathbf{f}$ reduces to the divergence of the symmetric part of Maxwell's stress tensor $\hat{\sigma}^M$ as it holds true in ordinary crystals [24]. Though useful to simplify some calculations on slightly distorted LCs, the spinless gauge presents some drawbacks: the dynamical constraint $\boldsymbol{\tau} = 0$ (the local balance of the angular momentum), in fact, is not generally satisfied and, what is worse, the intrinsic and the orbital parts of the angular momentum flux in the material mix so as to become unrecognizable. On the contrary, Ericksen's traditional approach, which is based on the free energy densities in Eqs. (5.6) and (5.7), keeps OAM and SAM separated, and, therefore, seems physically more appropriate, though leading to a nonsymmetric stress tensor [25]. Ericksen's stress tensor, its antisymmetric part, and Ericksen's spin tensor are reported in block II of Table 5.1. It is worth noting that the definitions of OAM and SAM flux along the z axis for a monochromatic field in vacuum proposed in [21] can be brought back just to the L_{33} and S_{33} elements reported in block II of Table 5.1. Here, we derived the same flux densities from a more general Lagrangian approach, exploiting the

Table 5.1 Stress tensor and intrinsic angular momentum flux in different gauges. The flux of the orbital angular momentum is given by $L_{\alpha\beta} = \epsilon_{\beta\mu\nu} x_\mu \sigma_{\alpha\nu}$. The several contributions, for each block, are labeled as S for splay, T for twist, and B for bend, corresponding to the fundamental elastic distortion in nematic liquid crystals. The label E represents the elastic contribution as a whole and em is for the electromagnetic contribution. Finally, we posed $A = \mathbf{n}\cdot\text{rot}\,\mathbf{n}$; $\mathbf{B} = \mathbf{n} \times \text{rot}\,\mathbf{n}$.

		$\sigma_{\alpha\beta}$	$\mathbf{w}$	$S_{\alpha\beta}$
I	S	$n_\alpha h^S_\beta - \delta_{\alpha\beta}(F^S + \mathbf{n}\cdot\mathbf{h}^S)$	$\mathbf{n} \times \mathbf{h}^S$	0
	T	$-n_\beta h^T_\alpha - \delta_{\alpha\beta} F^T$	$\mathbf{n} \times \mathbf{h}^T$	0
	B	$-n_\beta h^B_\alpha - k_3 B_\alpha B_\beta +$ $-(n_\alpha \frac{\partial F^B}{\partial n_\beta} + n_\beta \frac{\partial F^B}{\partial n_\alpha}) + \delta_{\alpha\beta} F^B$	$\mathbf{n} \times \mathbf{h}^B$	0
	em	$\frac{1}{16\pi}\left[(D^*_\alpha E_\beta + B^*_\alpha H_\beta + c.c) +\right.$ $\left.-\delta_{\alpha\beta}(\mathbf{D}^*\cdot\mathbf{E} + \mathbf{B}^*\cdot\mathbf{H})\right]$	$\frac{1}{16\pi}(\mathbf{D}^* \times \mathbf{E} + \mathbf{B}^* \times \mathbf{H} + c.c)$	0
II	S	$-k_1 \partial_\rho n_\rho \partial_\beta n_\alpha + \delta_{\alpha\beta} F^S$	$k_1 \text{rot}\,\mathbf{n}\,\text{div}\,\mathbf{n}$	$k_1 \partial_\rho n_\rho \epsilon_{\alpha\beta\gamma} n_\gamma$
	T	$-k_2 A \epsilon_{\alpha\gamma\rho} n_\rho \partial_\beta n_\gamma + \delta_{\alpha\beta} F^T$	$-k_2 A(\mathbf{B} + \mathbf{n}\,\text{div}\,\mathbf{n})$	$k_2 A(n_\alpha n_\beta - \delta_{\alpha\beta})$
	B	$-k_3(n_\gamma B_\alpha - n_\alpha B_\gamma)\partial_\beta n_\gamma + \delta_{\alpha\beta} F^B$	$-k_3\{[(\mathbf{n}\times\mathbf{B})\cdot\nabla]\mathbf{n} - (\mathbf{n}\times\mathbf{B})\,\text{div}\,\mathbf{n}\}$	$k_3 n_\alpha \epsilon_{\beta\gamma\rho} B_\gamma n_\rho$
	em	$\frac{i}{16\pi k_0}\epsilon_{\alpha\gamma\rho}(H^*_\rho \partial_\beta E_\gamma - c.c) + \delta_{\alpha\beta} F^{em}$	$\frac{i}{16\pi k_0}\left[\mathbf{H}^*\text{div}\,\mathbf{E} - (\mathbf{H}^*\cdot\nabla)\mathbf{E} - c.c\right]$	$\frac{i}{16\pi k_0}(-H^*_\beta E_\alpha + \delta_{\alpha\beta}\mathbf{H}^*\cdot\mathbf{E} - c.c)$
III	E	$-K\left(\partial_\alpha n_\gamma \partial_\beta n_\gamma - \frac{1}{2}\delta_{\alpha\beta}\partial_\gamma n_\rho \partial_\gamma n_\rho\right)$ $+ \delta k_1 \sigma^S_{\alpha\beta} + \delta k_2 \sigma^T_{\alpha\beta} + \delta k_3 \sigma^B_{\alpha\beta}$	$\delta k_1 \mathbf{w}^S + \delta k_2 \mathbf{w}^T + \delta k_3 \mathbf{w}^B$	$K\epsilon_{\beta\gamma\rho} n_\gamma \partial_\alpha n_\rho$ $+ \delta k_1 S^S_{\alpha\beta} + \delta k_2 S^T_{\alpha\beta} + \delta k_3 S^B_{\alpha\beta}$
	em	$-\frac{1}{16\pi k_0^2}\left\{\eta_o\left(\partial_\alpha E_\gamma \partial_\beta E^*_\gamma - \partial_\rho E_\rho \partial_\beta E^*_\alpha + c.c.\right)\right.$ $\left.-\delta_{\alpha\beta}\left[\eta_o\left(\partial_\gamma E_\rho \partial_\gamma E^*_\rho - \partial_\rho E_\rho \partial_\gamma E^*_\gamma\right) - k_0^2 \mathbf{D}^*\cdot\mathbf{E}\right]\right\}$	$-\frac{\eta_o}{16\pi k_0^2}\text{rot}\,\mathbf{E}^*\text{div}\,\mathbf{E} + c.c.$	$\frac{\eta_o}{16\pi k_0^2}\left(\epsilon_{\beta\gamma\rho} E^*_\gamma \partial_\alpha E_\rho\right.$ $\left.- \epsilon_{\alpha\beta\gamma} E^*_\gamma \partial_\rho E_\rho\right) + c.c.$

rotational symmetry of the system. The main drawback of the electromagnetic flux densities $\hat{L}^{em}$ and $\hat{S}^{em}$ derived from the Lagrangian in Eq. (5.7) is that they are not divergence free even in vacuum (only $\hat{J}^{em} = \hat{L}^{em} + \hat{S}^{em}$ is divergence free in vacuum). Having divergenceless fluxes $\hat{L}^{em}$ and $\hat{S}^{em}$ is desirable to have separate conservation laws for the OAM and SAM.

In birefringent media such as LCs, the stress tensor $\hat{\sigma}$ is not symmetric, in general, even in the spinless gauge, and the internal torque $\mathbf{w}$ in Eqs. (5.4) and (5.5) is also present in this case. The presence of the internal torque $\mathbf{w}$ is due to the lack of invariance of the total free energy $\mathcal{F}$ of the system under separate rotation of the center of mass $\mathbf{r}$ and of the components of the fields $\mathbf{n}$ and $\mathbf{E}$. However, the elastic free energy becomes rotationally invariant when all elastic constants k_i $(i = 1, 2, 3)$ become equal, so we may expect that the stress tensor will be symmetric in this limit. Setting $k_i = K$ in Eq. (5.6), F_e reduces to

$$F^0 = \frac{K}{2}[(\text{div}\,\mathbf{n})^2 + (\text{rot}\,\mathbf{n})^2] \tag{5.16}$$

The stress tensor $\hat{\sigma}^0$ derived from F^0 is still nonsymmetric. However, F^0 differs from the free energy density $F^1 = (K/2)\partial_\alpha n_\beta \partial_\alpha n_\beta$ only by divergence terms, so that F^0 and F^1 are equivalent in the bulk [[1], Eq. (3.17)], but the stress tensor $\hat{\sigma}^1$ derived from F^1 is now symmetric. We may write the original elastic free energy density F^e as $F^e = F^0 + \tilde{F}^e$, where $\tilde{F}^e$ is obtained from F^e through the formal substitution $k_i \to (k_i - K)/k_i$ $(i = 1, 2, 3)$. By this choice, when all elastic constants tend to the common value K (this may be the case in LCs near the nematic to isotropic transition), $F^e \to F^0$. Using this decomposition and exploiting the equivalence between F^0 and F^1, we may construct a new stress tensor that, though nonsymmetric in general, becomes symmetric in the one elastic constant approximation $(\delta k_i \to 0)$. This elastic stress tensor and the corresponding spin flux are reported in block III of Table 5.1. A similar argument can be applied to write, within divergence terms, the electromagnetic free energy density F^{em} as $F^1_{em} + \tilde{F}_{em}$, where F^1_{em} generates a symmetric stress tensor in isotropic and homogenous media and $\tilde{F}_{em}$ represents the contribution from the optical anisotropy. Because the electromagnetic stress tensor associated to F^1_{em} is symmetric and $\tilde{F}_{em}$ vanishes in isotropic media, we obtain a stress tensor that reduces to a symmetric one in homogeneous isotropic media. This choice for F^{em} leads to the quantities listed in the last row of block III. We notice that, when this gauge is used, the antisymmetric part $\mathbf{w}$ of the stress tensor is proportional to div $\mathbf{E}$, which is zero in homogeneous and isotropic media. The symmetry of the stress tensor in such media entails that the corresponding flux densities $\hat{L}$ and $\hat{S}$ are both divergence free ($\mathbf{f}$ and $\boldsymbol{\tau}$ are also zero). It is remarkable that there is no gauge function like $f_{\gamma\rho\alpha}$ in Eq. (5.15) settling the crossing between gauges I and II to the last one. This is not surprising, considering that the gauge transformations in Eq. (5.15) are not the most general: we may still add to $\sigma_{\rho\alpha}$ a divergence-free *symmetric* tensor. Assuming now the LC sample to be immersed in a homogenous and isotropic medium, Eqs. (5.4) and (5.5), upon integration over a region V with its border ∂V completely immersed in the surrounding medium respectively, assume the form

$$\int_V \rho(\mathbf{r}\times\dot{\mathbf{v}})\cdot\mathbf{u}d\mathbf{r} = \oint_{\partial V} u\cdot\hat{L}^{em}\cdot\mathbf{u}\,ds - \int_V \mathbf{w}^{em}\cdot\mathbf{u}\,d\mathbf{r} - \int_V \mathbf{w}^{e}\cdot\mathbf{u}\,d\mathbf{r} + \int_V (\mathbf{r}\times\mathbf{f}^{v})\cdot\mathbf{u}\,d\mathbf{r} \tag{5.17.1}$$

$$\int_V I(\mathbf{n}\times\ddot{\mathbf{n}})\cdot\mathbf{u}d\mathbf{r} = \oint_{\partial V} u\cdot\hat{S}^{em}\cdot\mathbf{u}\,ds + \int_V \mathbf{w}^{em}\cdot\mathbf{u}\,d\mathbf{r} + \int_V \mathbf{w}^{e}\cdot\mathbf{u}\,d\mathbf{r} + \int_V \boldsymbol{\tau}^{v}\cdot\mathbf{u}\,d\mathbf{r} \tag{5.17.2}$$

When the quantities defined in block III are used, the internal torque $\mathbf{w}^e$ in Eqs. (5.17.1–5.17.2) vanishes in the one elastic constant approximation and the internal torque $\mathbf{w}^{em}$ vanishes in homogeneous and isotropic media. In deriving Eqs. (5.17.1–5.17.2), we assumed $\mathbf{v} = 0$ and appropriate anchoring conditions of $\mathbf{n}$ at the sample walls so to have no surface contribution from the elastic intrinsic and orbital angular momentum fluxes $\hat{S}^e$ and $\hat{L}^e$. An example of such anchoring conditions is a nematic film with homeotropic alignment at the walls as used in the experiments [26–28].

The flux densities $\hat{L}^{em}$ and $\hat{S}^{em}$ in Eqs. (5.17.1–5.17.2) are evaluated in the surrounding isotropic homogeneous medium, where both are divergence free. The closed surface ∂V is therefore essentially arbitrary and the surface integrals in Eqs. (5.17.1–5.17.2) can be well identified with the fluxes of L and S coming from the external optical field. The two fluxes are physically discriminated in Eqs. (5.17.1–5.17.2) on the grounds of the different mechanical effects they produce in the medium, so that $\hat{L}^{em}$ and $\hat{S}^{em}$ can be identified as the flux densities of the OAM and intrinsic angular momentum carried by the optical field through the surface ∂V, respectively. We emphasize that the gauge leading to Eqs. (5.17.1–5.17.2) [block III of Table 5.1] has been selected from the infinite possible ones, because it is the only one leading to angular momentum flux densities $\hat{L}^{em}$ and $\hat{S}^{em}$, both conservative in isotropic and homogeneous media. Moreover, they reduce to well-known expressions in the paraxial optics approximation. In the light of such an interpretation, Eqs. (5.17.1–5.17.2) show how the angular momentum of light naturally splits into two distinct parts in the transfer from the external optical field to the two distinct rotational degrees of freedom of LC. A pictorial representation of this concept is shown in Figure 5.2. It is worth noting that the separation of the total angular momentum flux of the electromagnetic field into an orbital and a spin part was performed on the grounds of the physical effects they produce inside matter rather than on mathematical grounds. It is well known, in fact, that in vacuum only the total electromagnetic angular momentum flux is meaningful and its decomposition into a spin and an orbital part is not gauge invariant [14, 17, 18].

In the physics of LCs, the inertial terms on the left-hand side of Eqs. (5.3)–(5.5) and (5.17.1–5.17.2) are usually neglected and the equations are solved with respect to the viscous torques and forces that are proportional to $\partial\mathbf{n}/\partial t$, to the gradients of $\mathbf{n}$, and to the fluid velocity $\mathbf{v}$. In most cases, the fluid motion can also be neglected. Then, setting $\mathbf{v} \approx 0$ in Eqs. (5.17.1–5.17.2) yields two integral relationships

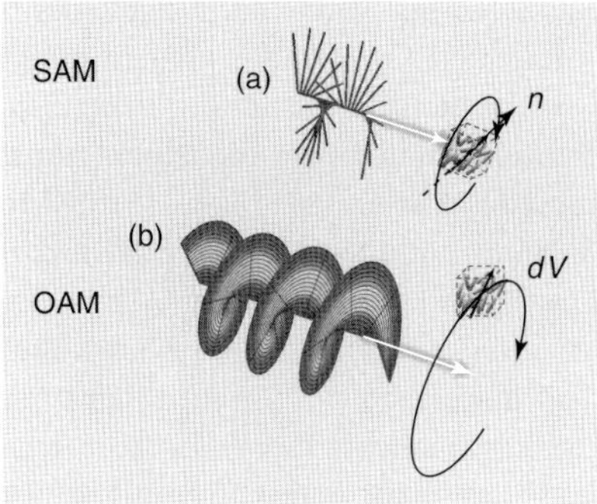

Figure 5.2 Pictorial representation of the separation of the orbital and spin parts of the total angular momentum of radiation on the grounds of the effects they produce inside matter. (a) Spin transfer induces a rotation of the director **n** (local optical axis). (b) Orbital transfer induces a rotational motion of the centers of mass of an elemental fluid volume. (Please find a color version of this figure on the color plates.)

involving only **n** and its time and space derivatives. A closer inspection shows that Eq. (5.17.1) couples $\hat{L}^{em}$ to the space derivatives of **n**, while Eq. (5.17.2) couples $\hat{S}^{em}$ to **n** itself. All these features reproduce what was claimed in previous works where the plane wave approximation was adopted [29] or where approximate models were proposed to describe the effects of the OAM of light in LCs [26–28].

5.2 Dynamical Effects Induced in Liquid Crystals by Photon SAM and OAM Transfer

In order to be specific and to get a practical insight into the mechanism of the optically induced molecular reorientation in the presence of OAM transfer from light to matter, we focus our attention on a number of actual experiments. In all the experiments reported here, the sample is a thin layer of NLC between two parallel walls, coated with an appropriate surfactant so as to have **n** everywhere perpendicular to the layer (homeotropic alignment). In the absence of external fields, the equilibrium molecular alignment is uniform with **n** independent of **r**. When the alignment is perturbed, elastic torques must be generated against the perturbation in order to restore the initial equilibrium. For small deformations, the elastic torques are proportional to the first spatial derivatives of **n**. The elastic torques are reported in block I of Table 5.1. If LC molecules are rotating, without any macroscopic displacement of the molecules ($\mathbf{v} = 0$), a viscous torque against rotation is expected. The complete analysis of molecular rotation in an LC involves five independent viscosity coefficients [25, 30–32]. For simplicity, a single phenomenological viscous constant γ_1 is often used. In this approximation, the viscous torque is given by

$$\tau^v = -\gamma_1 \mathbf{n} \times \frac{\partial \mathbf{n}}{\partial t} \tag{5.18}$$

The presence of an optical field yields an optical torque on the molecular director $\mathbf{n}$, on the one hand, and a rotation of the centers of mass of the fluid elements dV, on the other hand. In order to deduce the nonlinear optical response of the NLC as a whole, Eqs. (5.3)–(5.5) must be integrated over the film volume. Assuming $\mathbf{v} = 0$ and homeotropic anchoring at the sample walls, no elastic contribution results from the elastic intrinsic and orbital angular momentum flux densities $\hat{S}^e$ and $\hat{L}^e$, and Eqs. (5.17.1–5.17.2) result. Neglecting the inertial terms on the left-hand side of Eqs. (5.3)–(5.5) and (5.17.1–5.17.2), these transform into balance equations for the torques acting on the director $\mathbf{n}$ and for the torques acting on the center of mass of the elemental volume. The equations (5.17.1–5.17.2) are solved with respect to the viscous torques and forces that are proportional to $\partial\mathbf{n}/\partial t$, to the gradients of $\mathbf{n}$, and to the fluid velocity $\mathbf{v}$. In most cases, the fluid motion can also be neglected. Then, setting $\mathbf{v} \approx 0$ in Eqs. (5.17.1–5.17.2) yields two integral relationships involving only $\mathbf{n}$ and its time and space derivatives. A closer inspection shows that Eq. (5.17.1) couples $\hat{L}^{em}$ to the space derivatives of $\mathbf{n}$, while Eq. (5.17.2) couples $\hat{S}^{em}$ to $\mathbf{n}$ itself. In these conditions, where the motion of the fluid may be neglected, the OAM flux density $\hat{L}^{em}$ that the optical field releases in the medium behaves as an effective source of longitudinal torque affecting the orientation of the molecular director in the transverse plane.

In most part of the actual experiments with LCs, laser beams are used to excite the material, and small deformations of the molecular alignment are induced. For small deformations of the molecular director, in the paraxial optics approximations, the explicit expression for the overall SAM transferred from the electromagnetic field to the liquid crystalline medium, involved in Eq. (5.17.2), is [33]

$$\Delta S_z = -\frac{1}{\omega}\int dxdy\; I(x, y)\Delta s_3(x, y) \tag{5.19}$$

where $I(x, y)$ is the intensity profile of the beam, ω is the optical frequency, and Δs_3 is the change suffered by the reduced Stokes' parameter $s_3 = 2\mathrm{Im}(E_x E_y^*)/(|E_x|^2 + |E_y|^2)$ in traversing the medium ($s_3 = \mp 1$ for left-/right-handed polarization, respectively, and $s_3 = 0$ for linear polarization). The integral is carried out across the x, y-plane orthogonal to the beam.

In the same approximations, the explicit expression for the overall OAM transferred from the electromagnetic field to the medium, involved in Eq. (5.17.1), is [33]

$$\Delta L_z = \frac{1}{\omega}\int dxdy\; I_e(x, y)(\mathbf{r} \times \nabla)_z \Delta\Psi_e(x, y) \tag{5.20}$$

where $I_e(x, y)$ is the beam intensity transverse profile and $\Delta\Psi_e(x, y) = \Psi_e(x, y, L) - \Psi_e(x, y, 0)$ is the phase change of the extraordinary wave given by $\Delta\Psi_e(x, y) = 2\pi/\lambda \int_0^L [n_e(\theta) - n_o]\, dz \approx \tilde{L}(n_x^2 + n_y^2)$, where $\tilde{L}$ is the characteristic length of the sample. Consider that the ordinary wave suffers a phase change which is uniform in the transverse plane, so that it does not contribute to ΔL_z. If the medium has a regular refractive index distribution and if its surface has no dislocations, $\Delta\Psi_e(x, y)$

is a regular function of space coordinates. Then, if $I(x, y)$ is also a regular function of space, we may integrate Eq. (5.20) by parts, obtaining

$$\Delta L_z = -\frac{1}{\omega}\int dxdy\, \Delta\Psi_e(x, y)(\mathbf{r} \times \nabla)_z I_e(x, y) \tag{5.21}$$

From expressions (5.20) and (5.21) we see that no angular momentum is deposited in the medium if either $\Delta\Psi_e(x, y)$ or $I_e(x, y)$ is cylindrically symmetric around the propagation direction. In particular, Laguerre–Gauss beams (which are eigenstates of L_z) have an intensity profile that is cylindrically symmetric so that they cannot transfer their own OAM to transparent media such as LCs[1)].

The nonlinear process through which photon angular momentum is transferred to NLCs is known as self-induced stimulated light scattering (SISLS) [34, 35]. In the SISLS process, it is the *change* in the photon angular momentum that is transferred to matter. This is a distinctive difference between SISLS and the transfer of angular momentum by photon absorption. When the photon angular momentum is to be transferred to matter in an absorption process, the incident photons must be already prepared in a state with nonzero angular momentum. On the contrary, in the SISLS process, the angular momentum transfer can take place even when the incident light carries no angular momentum at all. The SISLS was first introduced, in connection to SAM transfer, to explain the collective rotation of LC molecules in the field of a normally incident circularly polarized laser beam [36]. During rotation, the viscous torque acting on the molecular director is balanced by a constant torque originating from the SAM that is constantly released into the medium by the portion of the incident photons transmitted with their spin reversed. SISLS was also recognized to be the physical grounds for complex dynamical regimes [37, 38] and for operating light-driven molecular motors [39]. Exploiting the same principle, manipulation of small transparent and birefringent particles trapped by optical tweezers was also achieved [40]. The first experiment realizing the orbital counterpart of the SISLS is much more recent [26] and put into practice an idea coming from a theoretical work by Allen *et al.* [15] according to which the measurement of the mechanical torque arising from the OAM has to be performed from Beth's experiment mould. The orbital SISLS was also exploited to control the transverse orientation of small transparent isotropic particles in optical tweezers [41, 42].

In typical experiments on laser-induced reorientation, the laser beam is focused into the LC sample to a spot of few hundreds microns or less. Above the threshold for the OFT, the optical reorientation is thus confined to a very small region in the focal zone, producing a spatially inhomogeneous distribution of the refractive index. The reoriented LC sample can be assimilated, therefore, to a thin microlens whose index profile may eventually change in time. If the intensity profile of the incident laser beam is elongated, the refractive index profile will be elongated as well, leading to an effective cylindrical laser-induced microlens. In general, the axis

1) This statement is no longer true when defects and/or dislocations are present in the LC texture.

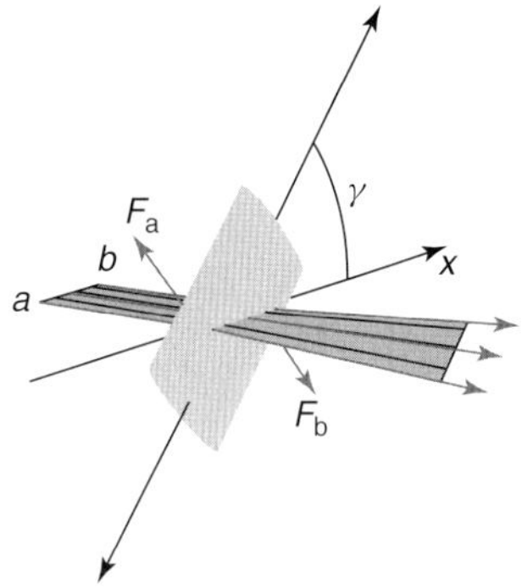

Figure 5.3 An NLC film optically distorted through an elliptically shaped laser beam behaves as a birefringent astigmatic microlens. Refraction of the incident beam by this lens results in a couple of forces $\mathbf{f}_a$ and $\mathbf{f}_b$ acting on the lens itself. (Please find a color version of this figure on the color plates.)

of the cylindrical microlens will not be aligned with the axis of the beam profile. In these conditions, the refraction of the elliptically shaped laser beam through the microlens may produce a torque on it, as shown in Figure 5.3. It is clear from the figure that the couple of forces acting on the lens is due to photon recoil during refraction, and, hence, this couple originates from the OAM of the light beam. It is expected, therefore, that the LCs are also sensitive to the orbital part of the light angular momentum. It is evident that the possibility of extracting the OAM from the light beam depends crucially on the presence of transverse gradients in the refractive index of the medium.

5.2.1 Experiments on OAM Transfer in Liquid Crystals

Several experiments have been performed so far that have been aimed at obtaining some insight into the mechanism of the OAM transfer in LCs and focusing on the reorientational effects induced by the longitudinal torque associated to it. In particular, the investigation of the effect on the molecular motion arising from the interplay of the alignment torque produced by the orbital momentum and the rotatory torque produced by the spin momentum has been aimed at. In all the experiments reported here, a thin layer of NLC with homeotropic alignment was irradiated by an elliptically rather than circularly shaped laser beam (Figure 5.4). The laser beam was made unpolarized [26], linearly polarized [33], or circularly polarized [28, 43–45]. In the second case, the angle between the optical field and the major axis of elliptical transverse cross section of the beam could be changed form 0 to $\pi/2$. Observations were performed varying the power incident on the sample and/or the beam ellipticity.

The sample was always chosen to be a nominally 50 μm thick E7 nematic film sandwiched between glass covers coated with *N,N*-Dimethyl-*N*-octadecyl-3-aminopropyltrimethoxysilyl chloride (DMOAP) for homeotropic alignment. The sample was pumped by a frequency-doubled cw Nd:YVO_4 laser source, working at $\lambda = 532$ nm. Two cylindrical lenses, with their axes orthogonal to each other, were used to obtain an elliptical beam waist at the sample position. The focal lengths of the lenses were $f_x = 500$ mm and $f_y = 30$ mm in the x and y directions,

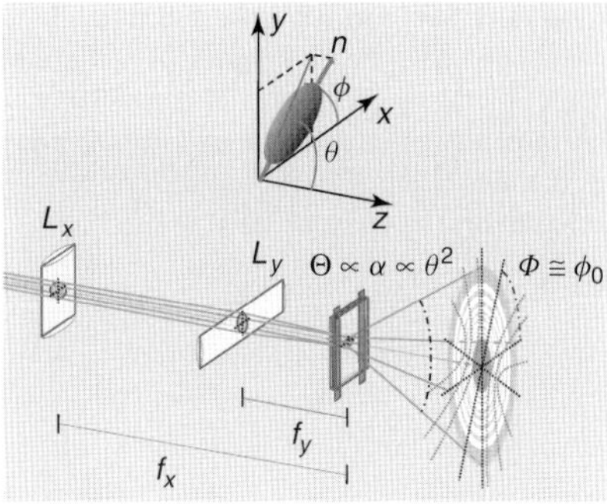

Figure 5.4 Schematic of the optical layout for photon OAM transfer in liquid crystals. Θ is the angular aperture of the far-field ring pattern and ϕ_0 is the inclination of its average polarization. The local direction of the optical field is tangent to the hyperbolic-like curves drown in the figure. (Please find a color version of this figure on the color plates.)

respectively. When the two cylindrical lenses were assembled so that their *second focal plane* were coincident (Figure 5.4), the beam radii in the common focal plane were found to be $w_x \approx 100$ μm and $w_y \approx 10$ μm, corresponding to a profile ellipticity $\mu \equiv w_x/w_y = 10$. The beam polarization was controlled by means of a suitable collection of half- and quarter-waveplates, besides electronically driven Pockels cells. The detection apparatus was designed to provide simultaneous and real-time measurements of the angular aperture Θ and of the average polarization direction angle Φ of the far-field self-diffraction pattern [6], which are formed beyond the LC sample when reorientation takes place [38]. For small LC distortion, one approximately has $\Phi \simeq \phi_0(t)$ and $\Theta(t) \propto \alpha(t) \propto \theta^2(t)$, where $\theta(t)$ is the zenithal polar angle of the molecular director $\mathbf{n}(\mathbf{r}, t)$ averaged over the sample, ϕ_0 is the azimuthal angle calculated at the exit face (Figure 5.4) and α is the phase difference between the extraordinary and ordinary wave over the sample [38].

The major difficulty in modeling this kind of experiment is that the dynamics of the NLC director is strongly affected by the finite size and shape of the incoming light beam so that the plane wave approach, where all the fields depend on only one coordinate, is inapplicable. All the three space coordinates and time must be retained. Only very recently we presented a self-consistent three-dimensional analytical – numerical model that is capable of describing all the reported experimental observations. The model accounts for the dependence on the incident beam intensity, polarization, and finite size and shape [45]. This model rests upon its capability to describe quantitatively the dynamics of, and beyond, the OFT under realistic experimental conditions almost three decades after its experimental discovery. The importance of the model is not restricted to the world of the dynamics of LCs, since it is the only "theoretical device" nowadays available to discern the role played by the finite-size effects in the dynamics with respect to shape effects. Only the latter are strictly connected to the OAM transfer according to all what has been said in the previous sections.

5.2.1.1 Orbital Photon Angular Momentum Transfer with Unpolarized Light

The first geometry referred to a fully unpolarized elliptically shaped laser beam. Such a laser beam carried neither SAM nor OAM, and only a transfer of OAM could take place, since, also in the medium, the beam remained unpolarized. To unpolarize the laser light, a Pockels cell was used. The cell was driven by a saw-tooth signal at λ-amplitude and a frequency of 1 kHz. This produced a 1 kHz periodic modulation of the light polarization through a sequence of states whose cycle-average returns zero. The NLC reorientational dynamics is much slower than the modulation period ($\approx$1s), so that it will only actually respond to the time-averaged Jones matrix of the polarization.

In the case of the unpolarized light, when the laser power P exceeded the threshold $P^u_{\mathrm{th}} = (312 \pm 3)$ mW for the OFT, the steady-state reorientation plane was found almost *independent* of P and parallel to the major axis of the beam intensity profile, in agreement with theoretical predictions [26]. This is shown in Figure 5.5a. The data were obtained by rotating the cylindrical lens gauge (and hence the intensity profile) to $\overline{\alpha} = 30^\circ$ with respect to the horizontal plane. Similar results were obtained at different angles $\overline{\alpha}$, thus proving the possibility of achieving angular control by transfer of the orbital photon angular momentum only.

Figure 5.5b,c refers to the measurements of the steady-state average value of the molecular director angle ϕ for different values of the beam power P, ellipticity μ, and orientation $\overline{\alpha}$ of the major axis of the beam profile. Data were taken at $\overline{\alpha} = 0$, $\overline{\alpha} = 30^\circ$, $\overline{\alpha} = 60^\circ$, and $\overline{\alpha} = 90^\circ$. In all cases we found that, for large enough ellipticity ($\mu \geq 2$) and large enough power ($P \geq 1.3P^u_{\mathrm{th}}$, where P^u_{th} is the OFT threshold power), the molecular director tends to move toward the plane containing the ellipse major axis.

For fixed power values, as the beam shape becomes more and more circular ($\mu \to 1$)[2)], the steady-state value of ϕ tends toward a well-defined but unpredictable value, changing from point to point in the sample. The behavior of the molecular director was observed to be likewise even with the beam power decreasing toward the OFT threshold for fixed values of μ. It is worth noting that, in an NLC film reoriented by unpolarized light, a similar attitude in the steady-state value of ϕ was observed in the previous experiments made with a circularly shaped laser beam [46, 47]. The unpredictable azimuthal reorientation may be ascribed to small uncontrolled and unavoidable factors breaking the perfect overall cylindrical symmetry of the system NLC + optical field, such as nonperfect parallelism of the sample walls, residual polarization in the incident light, small deviations from perfect normal incidence, and small pretilt at the sample surface.

5.2.1.2 Investigation of the Combined Effect of the Spin and Orbital Photon Angular Momentum Transfer with Linearly Polarized Light

In the second geometry investigated, a linearly polarized and elliptically shaped laser beam was normally incident on a homeotropically aligned nematic cell. In this

2) This was achieved changing the distances of both the lens L_x and L_y with respect to the sample.

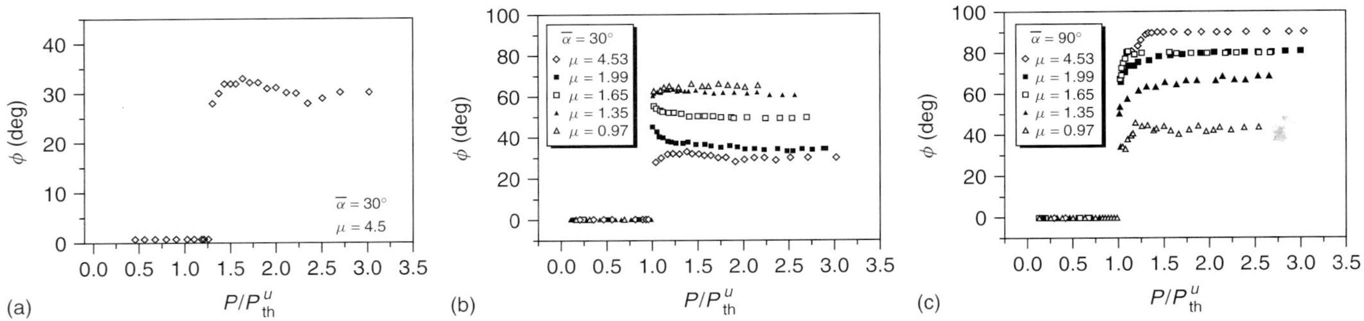

Figure 5.5 The steady-state azimuthal angle ϕ of the NLC reorientation plane as a function of the laser power. The laser was unpolarized and its intensity profile was elliptical. (a) The ellipticity of the beam profile was $\mu = 4.5$ and the major axis set at $\overline{\alpha} = 30°$ with respect to the horizontal reference plane. (b) The profile major axis was oriented at $\overline{\alpha} = 30°$ and the ellipticity was changed. (c) The profile major axis was oriented at $\overline{\alpha} = 90°$ and the ellipticity was changed.

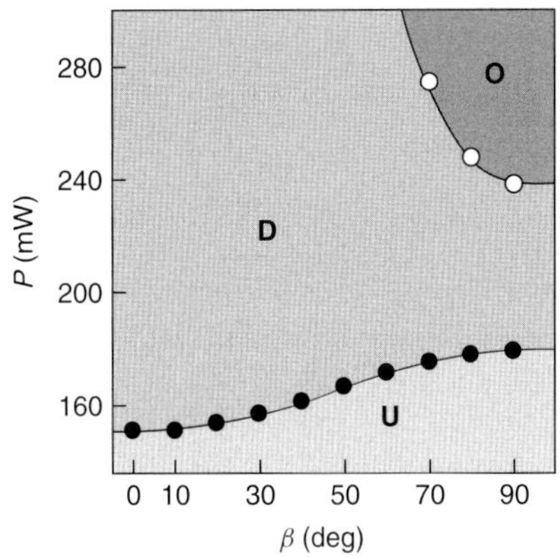

Figure 5.6 Map of the dynamical regimes in the parameter plane P, β. Three regions may be recognized: **U**, undistorted states; **D**, steady distorted states; **O**, oscillating states. The borderline between **U**- and **D**-regions represents the thresholds for the OFT and the borderline between **D**- and **O**-regions represents the thresholds for the oscillations start up as calculated from our model. Full circles on the first curve and open circle on the second are the experimental points. (Please find a color version of this figure on the color plates.)

case, the input beam carried zero-average OAM and SAM. Nevertheless, not only an OAM but also an SAM transfer could take place, thanks to the birefringence of LCs. In fact, when the molecular director is reoriented along a direction nonparallel to the major axis of the beam profile, the polarization inside the medium does not remain linear. The interest in this experimental geometry is due to the fact that deviations from the well-known behavior of the OFT can be ascribed to transverse effects connected to the elliptical shape and then to the OAM of light. Multistability and oscillatory behavior, both periodic and irregular, were observed corresponding to different values of the beam intensity and different polarization directions with respect to the major axis of the elliptical intensity profile [26, 33]. In Figure 5.6, the observed dynamical regimes have been reported in the plane of the control parameters P and β. The threshold power P_{th} for the optical reorientation is reported as a function of the angle β (continuous line). A second critical curve (dashed line) separating steady distorted states from nonlinear oscillations of $\mathbf{n}$ is also shown. The dots represent the experimental data and the lines represent the theoretical data. In Figure 5.7a, an example of multistability is reported when the angle β between the polarization direction and the ellipse major axis of the beam profile is $\beta = 50^{\circ}$. In Figure 5.7b, an instance of oscillatory regime for α and ϕ is shown, where the experimental data are superimposed on the theoretical curves.

5.2.1.3 **Investigation of the Combined Effect of the Spin and Orbital Photon Angular Momentum Transfer with Circularly Polarized Light**

The geometry of the experiment with circularly polarized light resembles in some ways the one faced with unpolarized light [26], since no favorite direction exists with the exception of the major axis of the laser beam transverse profile at the sample

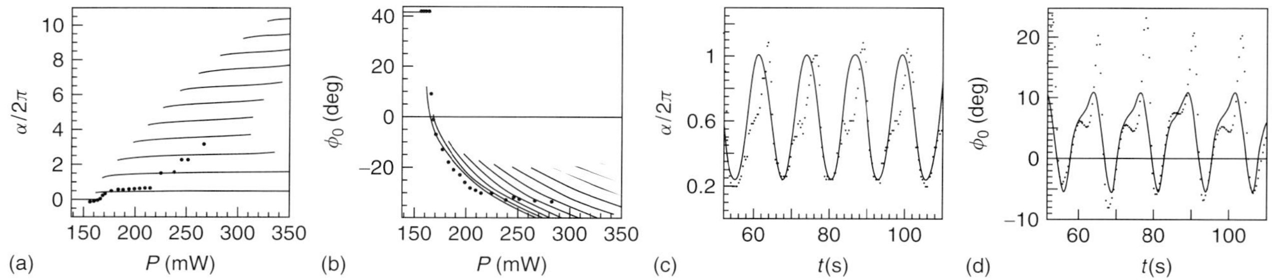

Figure 5.7 (a) The steady-state values of α and ϕ_0 as functions of the incident power P for $\beta = 50^\circ$. Full circles represent the experimental data and full lines the theoretical data. (b) Example of persistent time oscillation at $\beta = 90^\circ$ and $P = 260$ mW. The dots and the full line represent experimental and theoretical data, respectively.

position. However, different from the case of unpolarized light and from the case of linearly polarized light, now the pumping laser beam does carry an SAM of its own ($\pm\hbar$ per photon weather left- or right-circularly polarized), though it continues carrying zero-average orbital momentum. For a circularly shaped laser beam, our experimental conditions reproduce those for the well-known SISLS [34–36, 39] due to the photon SAM. Using an astigmatic laser beam, the azimuthal symmetry of the interaction scheme breaks and the photon OAM comes on the stage of the optically induced molecular reorientation. It is again SISLS, the mechanism through which the photon OAM may be transferred to LCs even from an incident light beam carrying a zero-average orbital momentum. Far from being the unique or the main character, OAM leaves a very specific and unambiguous mark on the molecular reorientation process, such as the appearance of steady distorted states above the OFT threshold for circularly polarized light, due to the SAM–OAM balance [43–45], which underlies a nonlinear process of self-induced spin-to-orbital conversion (SISTOC). In fact, it is well known that no state of equilibrium is possible when a circularly polarized, cylindrically symmetric laser beam is used. In that case, at the OFT threshold, the LC director jumps suddenly from the undistorted equilibrium state to a rotating regime [35, 36, 48]. In the case of an elliptically shaped laser beam, the symmetry breaking assists the formation, above threshold, of steady distorted states, which are forbidden by circular symmetry. In these states, SAM and OAM balance each other. As a consequence, the OFT turns to be second

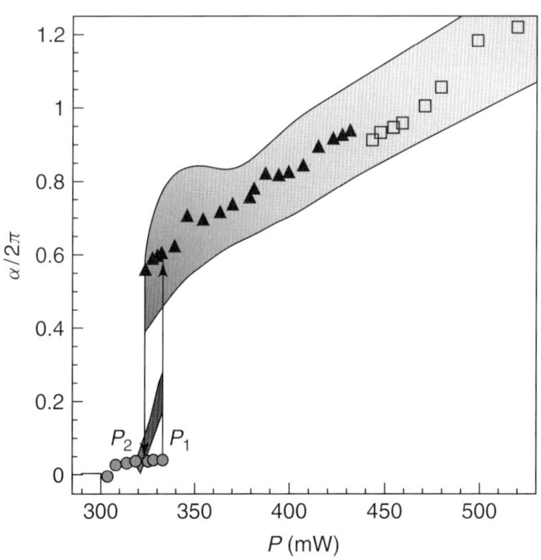

Figure 5.8 The optical phase difference α in unit of 2π as a function of the incident power P. (○) steady states; (▲) rotations; (□) intermittent states. The shadowed regions represent the oscillation amplitude of α and the experimental points mark the oscillation center. (Please find a color version of this figure on the color plates.)

rather than first order, as shown in Figure 5.8. The SISLS starts only after a second threshold via an inverse Hopf bifurcation owing to the presence of hysteresis. The stimulated scattering turns to be associated to a nutational rather than precessional regime and, above a critical value for the incident power, intermittent oscillations arise in the molecular nutation. To be more specific, an on–off intermittency in the chaotic rotation of the molecular director was observed [28]. The azimuthal angle $\phi(t)$ of the molecular director increased linearly in time on large time scales but, occasionally, it exhibited large fluctuations about its average value $\omega_0 t$, so that its angular velocity $\dot{\phi}(t)$ underwent an on–off intermittent motion. The intermittent signal $\omega(t) = \dot{\phi}(t) - \omega_0$ obeyed the scaling laws of on–off intermittency, including the symmetry between laminar and burst phases (Figure 5.9). The chaotic rotations were observed only when the SAM and the angular momentum of light were transferred simultaneously to the sample. Very recent theoretical investigations clearly indicated that the complex dynamical regimes observed with circularly polarized light cannot be accounted for by the simple interplay between SAM and OAM. In spite of the limitations in changing with ease the size of the beam elliptical

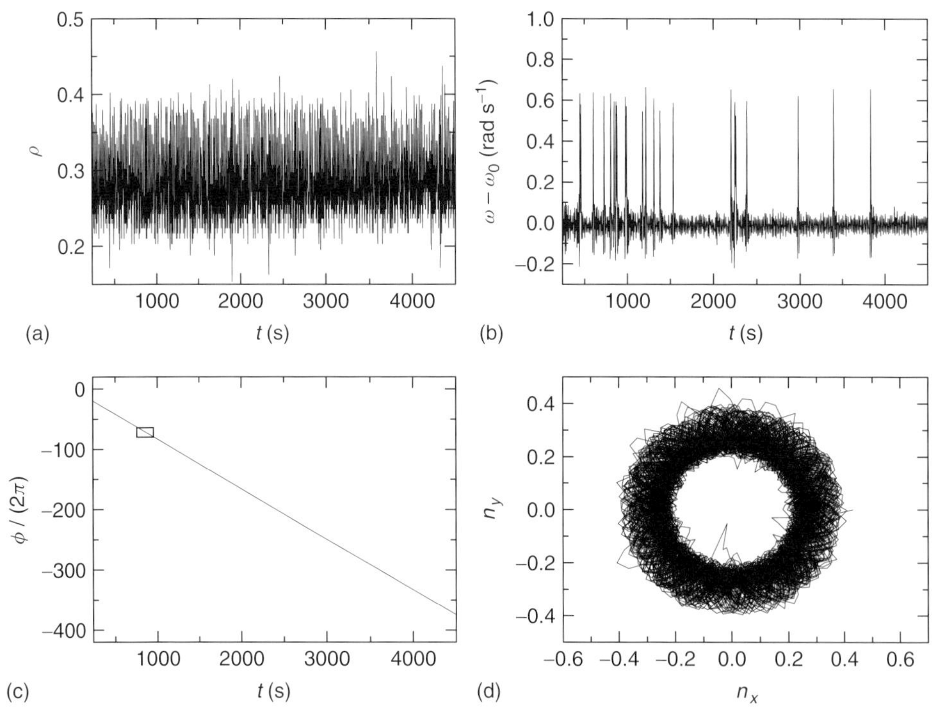

Figure 5.9 (a) radius $\rho(t)$ of the trajectory; (b) rotation angle $\phi(t)$; (c) instantaneous angular velocity $\omega(t)$; (d) trajectory in the x, y-plane of the chaotic rotation of the molecular director **n**. All time traces were taken at incident laser power $P = 445$ mW and in (a)–(c) the time scale is the same.

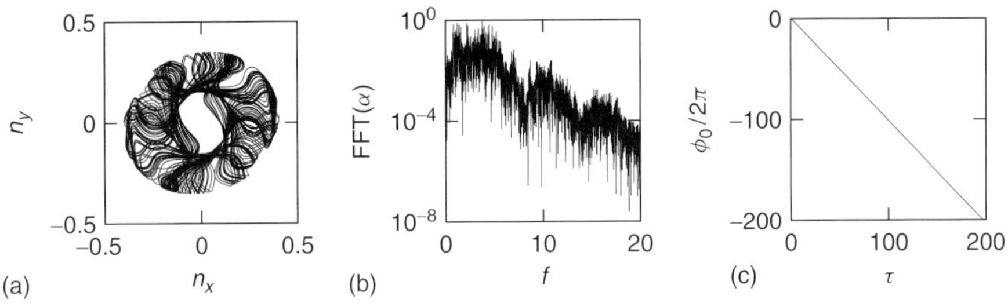

Figure 5.10 Theoretical prediction of complex rotation dynamics. (a) Director trajectory in the (n_x, n_y) plane. (b) Fourier spectrum of α where f is the normalized frequency. (c) $\phi_0(t)/2\pi$ illustrating the long-term uniform rotational motion.

profile in the experiment, complex dynamical regimes were actually observed only when the minor axis was significantly smaller than the film thickness. This induces to believe that finite beam size effects may be of decisive importance in this case [45], where a successful description of the complex sequence of the director rotations observed in [28] was obtained, adopting a full three-dimensional model able to describe quantitatively the dynamics of, and beyond, the OFT under the envisaged experimental conditions. The theoretical predictions for the trajectory of the chaotic rotation are shown in Figure 5.10.

As a matter of fact, several questions still remain open regarding the actual role played by *finite-size effects* – that is, effects due to the elastic response of the nematic when the beam width is significantly smaller than the film thickness – and the role played by *shape effects* – that is, effects due to the breaking of the usual cylindrical symmetry of the beam transverse profile and therefore due to OAM transfer. The model adopted in [45] will be used to make predictions of the dynamic behavior of the molecular director from the OFT threshold to highly nonlinear regimes under the action of both a circularly and a linearly polarized beam with elliptical cross section in order to discern the role played by OAM–SAM interplay from the finite beam size effects.

5.3 Conclusions

In this chapter we have discussed the problem of the photon angular momentum transfer to LCs, regarded as birefringent media endowed with internal orientational degrees of freedom. From the very beginning it has been clear that this problem intersected with the problem of the separation of the angular momentum of light in its intrinsic (or spin) and orbital parts. We have shown how it is possible to construct two electromagnetic fluxes $\hat{L}^{em}$ and $\hat{S}^{em}$ both conservative in vacuum (or in homogeneous isotropic media) which couple with the orbital and the intrinsic part of the angular momentum of matter, respectively. Therefore, we have been able to identify $\hat{L}^{em}$ and $\hat{S}^{em}$ on physical rather than mathematical grounds as the orbital and intrinsic angular momentum fluxes carried by the optical field. The fluxes $\hat{L}^{em}$ and $\hat{S}^{em}$ reduce to well-known expressions in the paraxial optics approximation, but may also be used beyond the paraxial approximation. When the inertial terms, in the dynamic equations of LCs, are neglected and appropriate approximations are made, simplified models can be worked out in order to predict the dynamics of the molecular director in the presence of both orbital and spin photon angular momentum transfer. In most situations, OAM behaves as an additional source of longitudinal torque acting on the molecular director in the same way as an SAM. A pictorial description of this torque can be obtained considering that an NLC film, nonlinearly distorted by means of an elliptically shaped laser beam, effectively behaves as an astigmatic birefringent microlens. The OAM transfer takes place during the alignment of the optically induced liquid crystalline microlens along the major axis of the laser elliptical cross section. The effects of the simultaneous transfer of photon SAM and OAM have been shown in several experimental geometries involving an elliptically shaped pumping beam carrying zero-average angular momentum. A variety of effects have been observed corresponding to different choices of the control parameters, such as the polarization state and the intensity of the incident beam, and the beam waist size along the directions of the major and minor axes of the intensity profile ellipse [26–28, 33, 43, 44, 49]. Using unpolarized light, steady distorted states were observed along the azimuthal directions closer to the ellipse major axis [26, 49, 50]. Using linearly polarized light, multistability and oscillatory behavior, both periodic and irregular were observed corresponding to different values of the beam intensity and different polarization directions with respect to the intensity ellipse major axis [26, 33]. A clear experimental evidence of on–off intermittency in the director rotation was also found using circularly polarized light [28]. Therefore, the breaking of the azimuthal symmetry due to the beam shape ellipticity has introduced a new actor in the process of the optically induced molecular reorientation, namely the photon OAM, which is transferred to LCs by SISLS. Though, on an average, an elliptically shaped light beam carries no OAM, this can be transferred to the molecular collective, optically reoriented by the beam itself within a noncylindrically symmetric profile, as shown through the comparatively simple models worked out in [26, 33, 44, 45].

References

1. de Gennes, P.G. (1974) *The Physics of Liquid Crystals*, Oxford University Press, Oxford.
2. Chandrasekhar, S. (1977) *Liquid Crystals*, Cambridge University Press, Cambridge.
3. Fréedericksz, V. and Zolina, V. (1933) Forces causing the orientation of an anisotropic liquid. *Trans. Faraday Soc.*, **29**, 919–930.
4. Zel'dovich, B.Y. and Tabiryan, N.V. (1980) Stimulated scattering of light in the mesophase of a smectic liquid crystal. *Sov. J. Quantum Electron.*, **10**, 440–442.

5. Zolot'ko, A.S., Kitaeva, V.F., Kroo, N., Sobolev, N.I. and Csillag, L. (1980) The effect of an optical field on the nematic phase of the liquid crystal OCBP. *Pis'ma Zh. Eksp. Teor. Fiz.*, **32**, 170 JETP Lett., 32, 158.
6. Durbin, S.D., Arakelian, S.M. and Shen, Y.R. (1981) Optical-field-induced birefringence and Fréedericksz transition in a nematic liquid crystal. *Phys. Rev. Lett.*, **47**, 1411.
7. Herman, R.M. and Serinko, R.J. (1979) Nonlinear-optical processes in nematic liquid crystals near freedericks transitions. *Phys. Rev. A*, **19**, 1757–1769.
8. Shelton, J.W. and Shen, Y.R. (1970) Phase-matched third-harmonic generation in cholesteric liquid crystals. *Phys. Rev. Lett.*, **25**, 24–26.
9. Shelton, J.W. and Shen, Y.R. (1971) Umklapp optical third-harmonic generation in cholesteric liquid crystals. *Phys. Rev. Lett.*, **26**, 538–541.
10. Shelton, J.W. and Shen, Y.R. (1972) Study of phase-matched normal and umklapp third-harmonic- generation processes in cholesteric liquid crystals. *Phys. Rev. A*, **5**, 1867–1882.
11. Marrucci, L., Manzo, C. and Paparo, D. (2006) Optical spin-to-orbital angular momentum conversion in inhomogeneous anisotropic media. *Phys. Rev. Lett.*, **96**, 163905.
12. Karimi, E., Piccirillo, B., Nagali, E., Marrucci, L. and Santamato, E. (2009) Efficient generation and sorting of orbital angular momentum eigenmodes of light by thermally tuned q-plates. *Appl. Phys. Lett.*, **94**, 231124.
13. Nagali, E., Sciarrino, F., De Martini, F., Marrucci, L., Piccirillo, B., Karimi, E., Santamato, E. (2009) Quantum information transfer from spin to orbital angular momentum of photons. *Phys. Rev. Lett.*, **103**, 013601.
14. Humblet, J. (1943) Sur le moment d'impulsion d'une onde electromagnetique. *Physica*, **10**, 585.
15. Allen, L., Beijersbergen, M.W., Spreeuw, R.J.C. and Woerdman, J.P. (1992) Orbital angular momentum of light and the transformation of Laguerre-Gaussian laser modes. *Phys. Rev. A*, **45**, 8185–8189.
16. van Enk, S.J. and Nienhuis, G. (1992) Eigenfunction description of laser beams and orbital angular momentum of light. *Opt. Commun.*, **94**, 147–158.
17. Cohen-Tannoudji, C.T., Dupont-Roc, J. and Grynberg, G. (1992) *Photons and Atoms: Introduction to Quantum Electrodynamics*, Wiley, New York.
18. Allen, L., Padgett, M.J. and Babiker, M. (1999) The orbital angular momentum of light. *Prog. Opt.*, **39**, 291–372.
19. Crichton, J.H. and Marston, P. (2000) The measurable distinction between the spin and the orbital angular momenta of electromagnetic radiation. *Electron. J. Diff. Eqns. Conf.*, **04**, 37–50.
20. Barnett, S.M. and Allen, L. (1994) Orbital angular momentum and nonparaxial beams. *Opt. Commun.*, **110**, 670–678.
21. Barnett, S.M. (2002) Optical angular-momentum flux. *J. Opt. B: Quantum Semicl. Opt.*, **4**, S7–S16.
22. Gordon, J.P. (1973) Radiation forces and momenta in dielecric media. *Phys. Rev. A*, **8**, 14–21.
23. Forster, D., Lubensky, T.C., Martin, P.C., Swift, J. and Pershan, P.S. (1971) Hydrodynamics of liquid crystals. *Phys. Rev. Lett.*, **26**, 1016–1019.
24. Landau, L.D. and Lifsitz, E.M. (1960) *Electrodynamics of continuous media*, Pergamon Press, New York.
25. Ericksen, J.L. (1960) Anisotropic fluids. *Arch. Ration. Mech. Anal.*, **4**, 231.
26. Piccirillo, B., Toscano, C., Vetrano, F. and Santamato, E. (2001) Orbital and spin photon angular momentum transfer in liquid crystals. *Phys. Rev. Lett.*, **86**, 2285.
27. Piccirillo, B., Vella, A. and Santamato, E. (2002) Competiton between spin and orbital photon angular momentum transfer in liquid crystals. *J. Opt. B: Quantum Semicl. Opt.*, **4**, S20–S24.
28. Vella, A., Setaro, A., Piccirillo, B. and Santamato, E. (2003) On-off intermittency in chaotic rotation induced in liquid crystals by competition between spin and orbital angular momentum of light. *Phys. Rev. E*, **67**, 051704.

29. Abbate, G., Maddalena, P., Marrucci, L., Saetta, L. and Santamato, E. (1991) Photodynamical effects induced by the angular momentum of light in liquid crystals. *Phys. Scr.*, **T39**, 389–393.
30. Ericksen, J.L. (1966) Inequalities in liquid crystal theory. *Phys. Fluids*, **9**, 1205.
31. Leslie, F.M. (1968) Some constitutive equations for liquid crystals. *Arch. Ration. Mech. Anal.*, **28**, 265.
32. Parodi, O. (1970) Stress tensor for a nematic liquid crystal. *J. Phys. (Paris)*, **31**, 581.
33. Piccirillo, B., Vella, A. and Santamato, E. (2004) Optical fréedericksz transition in liquid crystals and transfer of the orbital angular momentum of light. *Phys. Rev. E*, **69**, 021702.
34. Santamato, E., Romagnoli, M., Settembre, M., Daino, B. and Shen, Y.R. (1988) Self-induced stimulated light scattering. *Phys. Rev. Lett.*, **61**, 113–116.
35. Marrucci, L., Abbate, G., Ferraiuolo, S., Maddalena, P. and Santamato, E. (1992) Self-induced stimulated light scattering in nematic liquid crystals: theory and experiment. *Phys. Rev. A*, **46**, 4859.
36. Santamato, E., Daino, B., Romagnoli, M., Settembre, M. and Shen, Y.R. (1986) Collective rotation of molecules driven by the angular momentum of light in a nematic film. *Phys. Rev. Lett.*, **57**, 2423–2426.
37. Santamato, E., Abbate, G., Maddalena, P., Marrucci, L. and Shen, Y.R. (1990) Laser induced nonlinear dynamics in liquid crystal film. *Phys. Rev. Lett.*, **64**, 1377–1380.
38. Santamato, E., Maddalena, P., Marrucci, L. and Piccirillo, B. (1998) Experimental study of the molecular reorientation induced by the ordinary wave in a nematic liquid crystal film. *Liq. Cryst.*, **25**, 357–362.
39. Galstyan, T.V. and Drnoyan, V. (1997) Light-driven molecular motor. *Phys. Rev. Lett.*, **78**, 2760–2763.
40. Friese, M.E.J., Nieminen, T.A., Heckenberg, N.R. and Rubinsztein-Dunlop, H. (1998) Optical alignment and spinning of laser trapped microscopic particles. *Nature*, **394**, 348–350.
41. Santamato, E., Sasso, A., Piccirillo, B. and Vella, A. (2002) Optical angular momentum transfer to transparent isotropic particles using laser beam carrying zero average angular momentum. *Opt. Express*, **10**, 871–878.
42. Ambrosio, A., Piccirillo, B., Sasso, A. and Santamato, E. (2004) Experimental and theoretical study of the transient rotation of isotropic transparent microparticles in astigmatic optical tweezers. *Opt. Commun.*, **230**, 337–345.
43. Piccirillo, B. (2005) Optical angular momentum flux in liquid crystals. *Mol. Cryst. Liq. Cryst.*, **429**, 133–147.
44. Piccirillo, B., Vella, A., Setaro, A. and Santamato, E. (2006) Nonlinear dynamics induced in liquid crystals in the presence of the orbital and the spin angular momentum of light. *Phys. Rev. E*, **73**, 062701.
45. Brasselet, E., Piccirillo, B. and Santamato, E. (2008) Three-dimensional model for light-induced chaotic rotations in liquid crystals under spin and orbital angular momentum transfer processes. *Phys. Rev. E*, **78**, 031703.
46. Arnone, G., Sirleto, L., Marrucci, L., Maddalena, P. and Santamato, E. (1996) Optical Frédericksz transition induced by depolarized light. *Mol. Cryst. Liq. Cryst.*, **282**, 191–204.
47. Marrucci, L., Maddalena, P., Arnone, G., Sirleto, L. and Santamato, E. (1998) Liquid crystal reorientation induced by completely unpolarized light. *Phys. Rev. E*, **57**, 3033–3037.
48. Vella, A., Piccirillo, B. and Santamato, E. (2002) Coupled-mode approach to the nonlinear dynamics induced by an elliptically polarized laser field in liquid crystals at normal incidence. *Phys. Rev. E*, **65**, 31706.
49. Marrucci, L., Vetrano, F. and Santamato, E. (1999) Optical reorientation in nematic liquid crystals controlled by the laser beam shape. *Opt.Commun.*, **171**, 131–136.
50. Marrucci, L., Piccirillo, B., Vetrano, F. and Santamato, E. (2000) Longitudinal extra torque in liquid crystals due to the optical beam shape. *J. Opt. A: Pure Appl. Opt.*, **2**, 294–298.

6
Driving Optical Micromachines with Orbital Angular Momentum

Vincent L.Y. Loke, Theodor Asavei, Simon Parkin, Norman R. Heckenberg, Halina Rubinsztein-Dunlop, and Timo A. Nieminen

6.1
Introduction

The symmetry of an object such as an optically driven microrotor and the driving light beam is a key element in the generation of optical torque. The optical torque occurs when electromagnetic angular momentum is transferred from the driving beam to the microrotor by scattering. We discuss the effect of such symmetries on the generation of optical torque, and some consequent general principles for the design of optically driven micromachines. We describe the experimental test of a design based on these principles, including the quantitative evaluation of its performance through measurement of the orbital angular momentum (OAM) transfer. We describe the computational modeling of such rotors, and investigate the dependence of the performance on the geometry of the microrotor.

6.2
Symmetry, Scattering, and Optically Driven Micromachines

Holographic techniques are widely used to produce laser beams carrying OAM about their axes [1]. The off-axis "fork" hologram is deservedly popular, since the angular separation of the diffracted orders allows a beam of a particular OAM to be selected. The on-axis hologram (Figure 6.1) – often spiral as shown in Figure 6.2 – is, however, simpler, and provides a clear illustration of the connection between discrete rotational symmetry and angular momentum.

The symmetry of an object plays a central role in its interaction with light. If we are interested in the exchange of angular momentum – that is, the generation of optical torque – then it is the rotational symmetry that is of most interest. Consideration of the effects of rotational symmetry allows some quite general aspects of optical torque to be understood [2, 3]. Essentially, if an object has pth order discrete rotational symmetry, incident light with $m_0\hbar$ angular momentum per photon about the symmetry axis of the particle will be scattered into modes with

Twisted Photons: Applications of Light with Orbital Angular Momentum.
Edited by Juan P. Torres and Lluis Torner

ISBN: 978-3-527-40907-5

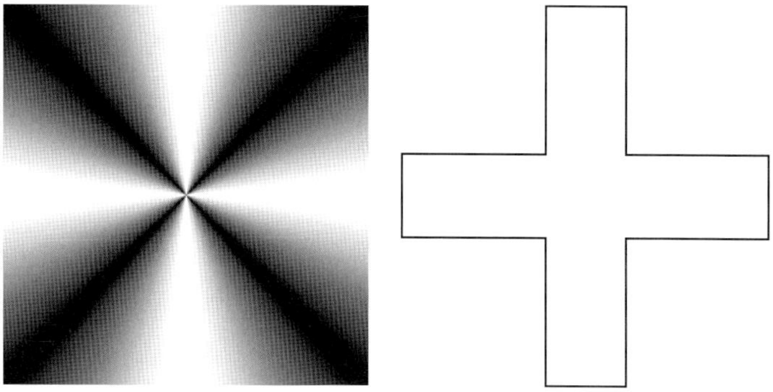

(a) (b)

Figure 6.1 On-axis hologram for conversion of Gaussian beam to LG_{04}. (a) An amplitude hologram – the normalized interference pattern between a plane wave and an LG_{04} beam. (b) A structure that will act as a binary approximation of the corresponding phase hologram, suitable for use as an optically driven microrotor.

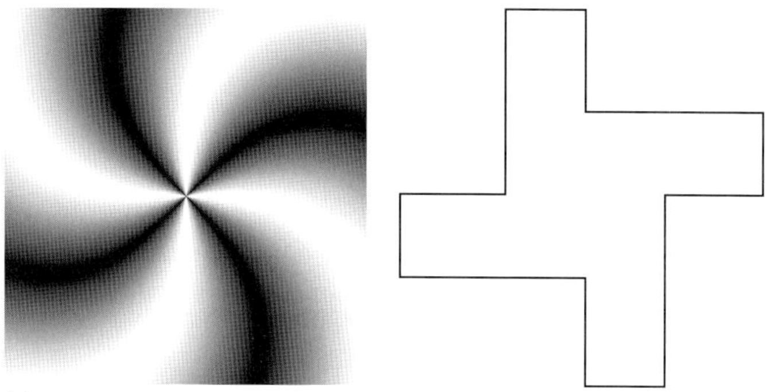

(a) (b)

Figure 6.2 Spiral on-axis hologram for conversion of Gaussian beam to LG_{04}. (a) An amplitude hologram – the normalized interference pattern between a spherical wave and an LG_{04} beam. (b) A structure that will act as a binary approximation of the corresponding phase hologram, suitable for use as an optically driven microrotor.

$m_i\hbar$ angular momentum per photon, where $m_i = m_0 - ip$, where i is an integer. This is the key element in the function of holograms such as those shown in Figures 6.1 and 6.2. This is the rotational analog of the scattering of a plane wave by a periodic structure into a discrete plane wave spectrum, which produces a similar relationship for the transverse linear momentum of the modes as we have here for angular momentum.

For paraxial optical vortex beams, we have a well-defined OAM per photon (zero for a Gaussian beam, $\ell\hbar$ per photon for a Laguerre–Gauss mode of order ℓ [4]).

There is a simple separation between spin and orbital angular momentum, and it is possible to account for the spin in terms of left- and right-circularly polarized component, where the total angular momentum per photon is $\ell \pm 1$ for each of the two components. This is sufficient for dealing with the optical angular momentum of paraxial beams and macroscopic holograms.

However, with the interaction of nonparaxial beams with microscopic objects, it is necessary to consider the total angular momentum, rather than separate spin and orbital components. Since the driving beam incident on an optically driven microrotor is typically produced by tightly focusing a paraxial beam (such as a Laguerre–Gauss beam carrying $\ell\hbar$ OAM per photon), and focusing the optical system will not alter the angular momentum per photon [5, 6], the nonparaxial beam has the same total angular momentum content as the initial paraxial beam. A convenient way to represent such a nonparaxial beam is as a superposition of vector spherical wavefunctions (VSWFs) [7], which have an azimuthal phase variation of $\exp(im\phi)$, where m is the azimuthal mode index, with the z component of the angular momentum being equal to $m\hbar$ per photon. Thus, the nonparaxial VSWF representation of the beam will consist of modes with $m = \ell \pm 1$.

Thus, for an object with pth order rotational symmetry, incident modes will scatter into modes with $m = \ell \pm 1, \ell \pm 1 \pm p, \ell \pm 1 \pm 2p, \ell \pm 1 \pm 3p$, and so on. For particles with $p = 2$, such as elongated or flattened particles, the left- and right-circular incident modes will scatter into the same set of modes, resulting in interference, and therefore affecting the polarization of the scattered light – this gives rise to the shape birefringence of such particles. On the other hand, if $p > 2$, the scattered modes corresponding to the incident circular polarizations are distinct, and this interference will not occur, and the incident polarization will be affected only weakly. Therefore, the torque will be a consequence of the transfer of OAM.

We can assume that the coupling of incident modes with small $|m|$ will usually be stronger than for high $|m|$ modes. Thus, an object illuminated by light carrying the OAM will usually experience a torque (except in special cases, including when opposing spin reduces the total angular momentum to zero). If the object is mirror symmetric (i.e., achiral), then the scattering is independent of the handedness of the angular momentum, that is, the coupling of, for example, m_1 to m_2 is the same as from $-m_1$ to $-m_2$. Therefore, an achiral rotor with $p > 2$ will experience no torque in a plane-polarized Gaussian beam. Such a rotor will be ideal for illumination by Laguerre–Gauss beams, and will be equally rotatable in both directions. A chiral rotor, on the other hand, can rotate in a Gaussian beam, but will generally rotate at different speeds when illuminated by Laguerre–Gauss beams of opposite handedness.

A majority of optically driven microrotors produced and tested so far are operated immersed in water, typically in an optical tweezers apparatus. Owing to the relatively small difference between the refractive index of such rotors and the surrounding water, they reflect only weakly, and the majority of the light is transmitted. The coupling to the different scattered modes will depend strongly on phase and intensity variations in the transmitted light. Therefore, we can productively view optically driven rotors as microscopic holograms – the thickness of the structure should ensure a large phase difference between light that passes through the structure and light that does not.

On the basis of this idea, we can straight away suggest the possible designs for optically driven microrotors, as shown in Figures 6.1 and 6.2. The key difference is the solid central region, which will be required for the structural integrity of a microrotor. For a chiral structure, as in Figure 6.2, the symmetry of scattering to positive and negative orders of scattering (i.e., to $m < m_0$ and $m > m_0$) will differ, and a torque will result from the generation of OAM by scattering by the structure. Therefore, a structure of this type can be rotated by a Gaussian beam.

However, if we wish to have a structure that can be rotated with equal torque in either direction, we need to avoid a chiral shape, and have particles that have mirror symmetry about a plane containing the axis of rotational symmetry (Figure 6.1), in which case the coupling from $m_0 = 0$ to $\pm m$ will be identical, since these modes are mirror images of each other. This type of structure is ideal for driving with a beam carrying the OAM. If, in this case, the incident beam has $m_0 = +2$, there will be significant scattering to VSWF modes with $m = -2$, and a torque will result. The direction of the applied torque can be reversed by changing the handedness of the driving beam.

6.3 Experimental Demonstration

The above principles can be demonstrated by using synthetic micro-objects, fabricated using two-photon photopolymerization [8, 9]. The production, trapping, and rotation of such objects have been demonstrated by a number of groups around the world [10, 16]. This method provides an ideal opportunity for the development of initial designs and their evaluation, prior to further engineering and optimization. It is possible to calculate the optical forces and torques acting on such structures, but it requires a major computational effort. Therefore, it is useful to apply such computational techniques to the optimization and improvement of an existing design, rather than through the creation of the initial design *ex nihilo*. Such an initial test of a design, and the general design principles outlined above, was our major goal here, to be followed by computational optimization of the design discussed in Section 6.4.

6.3.1 A Preliminary Design

Since the achiral rotor with $p > 2$ offers an interesting and complex range of behavior, we use such a structure to demonstrate the action of OAM, and also demonstrate zero torque when trapping using a plane-polarized Gaussian beam. From the above principles, we can do this with a four-armed, mirror-symmetric rotor. A practical issue is that the symmetry axis of the rotor should coincide with the beam axis. This can be ensured by using a central stalk which will align along the beam axis. For a typical photopolymerized object in water, we have refractive indices of approximately $n_{\mathrm{obj}} = 1.50$–1.55 for the object and $n_{\mathrm{med}} = 1.33$ for the surrounding water. To obtain a phase difference of close to a half-wave, we want a thickness of $\lambda_{\mathrm{med}}/(2(n_{\mathrm{obj}} - n_{\mathrm{med}}))$, where λ_{med} is the wavelength in the medium. This is a thickness of approximately double the free-space wavelength of the trapping beam.

6.3.2 Fabrication

The two-photon photopolymerization technique was pioneered by Strickler and Webb in 1991 [17], following the application of two-photon excitation in laser scanning fluorescence microscopy [18]. The first 3D microfabricated structures with two-photon photopolymerization were reported in 1997 [19]. Since then, various micromachines have been produced (micropumps, microgears, microneedles) with resolution on the order of 100 nm [10, 20, 21]. The method has been recently reviewed [8, 9]. Our own photopolymerization setup and its performance have been described in [22].

We produce the microrotors using NOA63 resin from Norland Products. The 3D object is represented by 2D layers (bitmaps) corresponding to the areas that need to be polymerized. The stage is raster-scanned, with the beam being turned on and off as required. The stage is moved in the z direction after each x–y scan. The bitmap size is 100×100 pixels, which corresponds to $10\,\mu\mathrm{m} \times 10\,\mu\mathrm{m}$ travel in the x and y directions hence each individual pixel is $100 \times 100\,\mathrm{nm}^2$ in size giving a lateral resolution of 100 nm. The steps in the z direction are 200 nm.

After the polymerization, the unexposed resin is washed off with acetone, leaving the 3D structure attached to the cover slip. An SEM image of a typical rotor is shown in Figure 6.3.

6.3.3 Optical Trapping and Rotation

Optical trapping was performed in an in-house-built inverted microscope. The trapping laser is a 5 W, 1070 nm Nd:YAG fiber laser (IPG Photonics, Oxford, MA, USA), focused on the sample by a 100× Olympus oil immersion objective lens with high numerical aperture 1.3. The output power is controlled by a half-wave plate and a polarizing beam splitter. The sample is imaged onto a CCD camera by the same objective lens.

The angular momentum of the incident trapping beam was controlled by using a quarter wave plate for spin angular momentum and a computer-generated hologram [23] creating Laguerre–Gauss modes (designated as $\mathrm{LG}_{p\ell}$, where p is the radial mode index and ℓ is the azimuthal mode index) for OAM. In our experiments we used a hologram that generates LG_{02} modes in the first order when the incident beam is the TEM_{00} Gaussian beam of the laser. These modes have an OAM of $2\hbar$

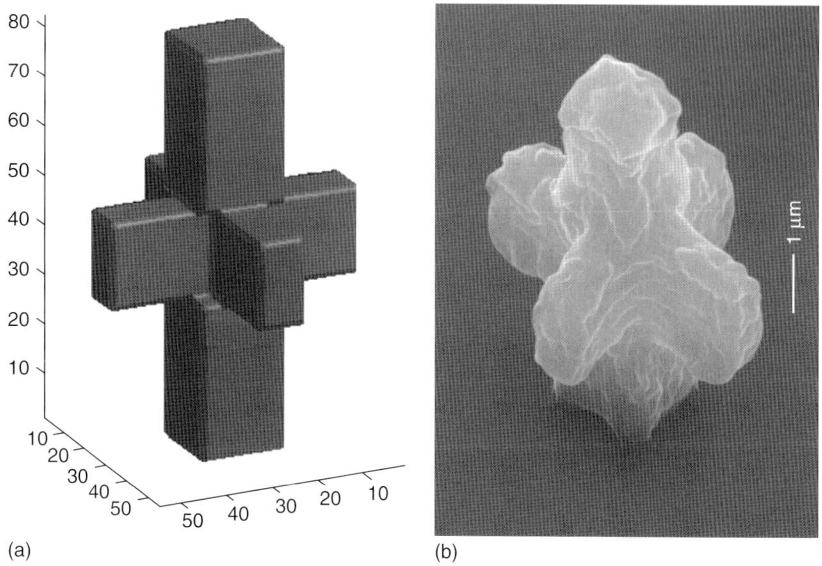

Figure 6.3 Microrotor: design and realization. (a) The design with a scale of 100 nm wide voxels and (b) a scanning electron microscope (SEM) image of the microfabricated structure attached to the cover slip. (Please find a color version of this figure on the color plates.)

per photon. The hologram we used here was produced photographically [24], but a spatial light modulator (SLM) or other means of generating beams with OAM could be used instead.

Demineralized water was added to immerse the rotor, which was still attached to the coverslip. This was then placed on the sample stage of the trap. The trapping system uses a water-immersion condenser, which allows the top of the sample to remain open (recall that the trap is based on an inverted microscope); this allows mechanical access to the sample. The microstructure was detached using the tip of a needle mounted on a translation stage.

The rotor was easily trapped and rotated around the axis of the beam carrying angular momentum. The rotor was trapped stably with its long axis (i.e., the central stalk) aligned along the beam axis. The rotation rate was on the order of 1 Hz for a trapping beam power of 20 mW at the sample. The structure did not rotate when trapped in a Gaussian beam.

6.3.4
Optical Measurement of Torque

The torque exerted on the microstructure is the sum of the two contributions from spin and orbital angular momentum transfer.

The spin torque measurement is based on the fact that any coherent beam can be represented as a sum of two circularly polarized components with opposite handedness with a coefficient of circular polarization σ_S given by $\sigma_S = (P_L - P_R)/P$, where P_L and P_R are the powers of the left and right-circularly polarized components, respectively, and P is the total power of the beam [25]. Therefore, the spin torque τ_s can be written as $\tau_s = \Delta\sigma_S P/\omega$, with $\Delta\sigma_S$ being the change in the coefficient of circular polarization due to the spin angular momentum transfer, P is the incident beam power, and ω is the optical angular frequency of the beam. Hence, by measuring $\Delta\sigma_S$ and knowing P and ω, the torque can be found. The change in the coefficient of circular polarization is measured by two photodetectors (PD1 and PD2), which are placed after a polarizing beam splitter cube. The outgoing beam is collimated by the condenser and then split into two orthogonal linearly polarized components by the quarter wave plate ($\lambda/4$) and the cube. The two linearly polarized components correspond to the left- and right-circularly polarized components of the outgoing beam. The two detectors measure the power of each beam, and hence the coefficient $\Delta\sigma_S$ that is directly related to the spin torque per photon can be found. From the above-mentioned equation $\tau_s = \Delta\sigma_S P/\omega$, writing $P = N\hbar\omega$ with N being the number of photons per time, one can find out that the spin torque per photon has the value of $\Delta\sigma_S\hbar$.

Similar to the spin component of the torque, one can write the orbital torque τ_o as being $\tau_o = \Delta\sigma_o P/\omega$, with $\Delta\sigma_o$ being a coefficient related to the orbital torque per photon in the same way as $\Delta\sigma_S$ is related to the spin torque per photon. Thus the orbital torque per photon has the value of $\Delta\sigma_o\hbar$.

In order to measure the orbital torque τ_o, we used a method previously described in [26]. It is based on the steady rotation of the microstructure, which means that the total optical torque is equal to the drag torque due to rotation in the liquid. In our case, the surrounding medium behaves as a Newtonian fluid and hence the drag torque is proportional to the angular speed of rotation Ω. The rotor was trapped well clear of any surfaces so that no corrections for frictional forces or wall effects were necessary.

The total torque is

$$\tau = \tau_s + \tau_o = D\Omega \tag{6.1}$$

where D is the rotational drag coefficient for the microstructure in the fluid. Using this, we can calculate τ_o and D by measuring τ_s and Ω for three different degrees of polarization of the incident light (left handed, right handed, and linearly polarized light). Just as the torque efficiency provides a power-independent description of the optical torque, it can be useful to introduce a spin torque efficiency Q_s, such that $\tau_s = Q_s P/\omega$, and an orbital torque efficiency Q_o such that $\tau_o = Q_o P/\omega$.

The intensity variation at an off-axis point in the transmitted beam was also measured. Owing to the symmetry of the structure, the rotation rate was one quarter of the signal frequency. With a laser power of $P = 40$ mW at the focus, the rotation frequency varied from 2.25 Hz (right circular) to 3 Hz (left circular), with an uncertainty of 5%.

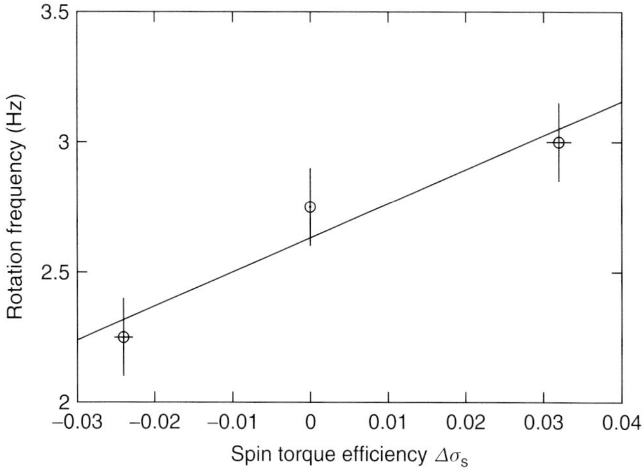

Figure 6.4 The rotation frequency of the trapped rotor as a function of the spin torque per photon for three different polarizations (left circular, right circular, and linear). From the fit to the data, the orbital torque per photon is found.

The relationship between rotation rate and the spin component of the torque is shown in Figure 6.4. Since the spatial structure of the beam is the same for all three polarizations (left circular, right circular, and linear), we can assume that the orbital torque is the same for all three, and the only difference in torque is due to the different spin torques, which are measured optically. The difference in rotation rates due to the different spin torques allows us to find the rotational drag coefficient, and hence the total torque from the rotation rates. We find that the orbital torque efficiency is 0.20 ± 0.03, and the orbital torque is 4.8 ± 0.7 pN μm, which is 10 times higher than the spin component. For a plane-polarized incident beam, this is also the total torque.

The total torque can also be simply found from the rotation rate if the viscous drag torque coefficient is known. The drag coefficient can be found by computational modeling of the fluid flow around the rotor; we performed this calculation (in the Stokes/creeping-flow limit) using a finite-element package [27]. At the rotation rate measured with the plane-polarized beam, 2.75 Hz, this gave a torque of 5.4 pN μm, which was in close agreement with the optically measured torque. These are compared in Figure 6.5.

6.3.5
Discussion

Qualitatively, the rotor performs as expected – the structure rotates when trapped in a beam carrying the OAM, and does not rotate in a Gaussian beam. The torque efficiency of 0.2 is much higher than that typical for shape-birefringent objects (e.g., 0.02–0.05). If the geometry of the structure is known, and the properties

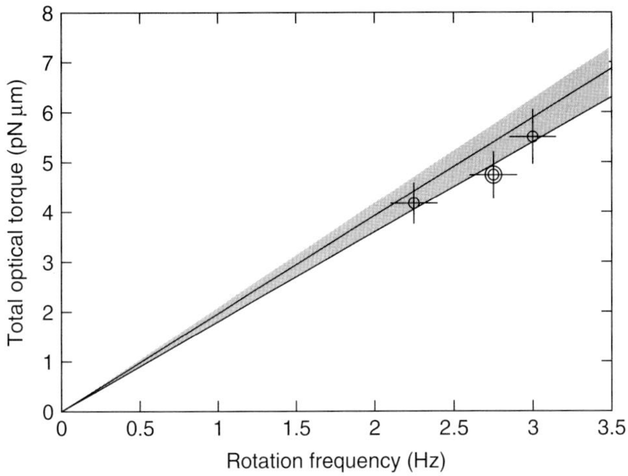

Figure 6.5 The total optical torque as a function of the rotation frequency (thick line). The torque determined from the measured rotation rate and the calculated viscous drag torque coefficient is also shown (thin line), with the error bound due to the experimental error in the frequency measurement (gray region).

of the fluid are known, the viscous drag torque coefficient can be calculated, and the torque can be simply found from the rotation rate. If these are not known, Parkin's method for the all-optical measurement of torque [26] provides an accurate measurement of the torque, including the dominant contribution due to OAM.

We have demonstrated that consideration of the effects of symmetry on the scattering of light provides a sound theoretical basis for the design of optically driven microrotors. In particular, it provides a simple set of qualitative guidelines for the development of initial designs that can then be experimentally or computationally evaluated. Rotors exploiting the OAM of light can make use of angular momentum fluxes of over $\hbar$ per photon in the driving beam, and can have higher efficiencies as a result; in the example case in this paper, the orbital component of the torque was 10 times higher than the spin torque.

The total optical torque exerted on the micrometer-sized objects rotating in an optical trap can be measured accurately by optical means, that is, by polarimetric measurement of the spin component of the optical torque. Even when the orbital component of the torque is much greater than the spin torque, as it was in the test case of the method here, the results are accurate.

We have also demonstrated an efficient method for calculating the viscous drag torque acting on such a microrotor, by computationally solving the 3D Laplace equation. This can be used for finding the optical torque acting on a microrotor from the rotation rate, as we did here, or for the prediction of performance, including near a surface.

The two-photon photopolymerization technique has proved to be a powerful tool for fabricating microstructures with potentially any arbitrary shape, which can be useful in studies dealing with transfer of linear and/or angular optical momentum. While it might not be the method of choice for the large-scale manufacture of devices, it is perfect for the fabrication of prototypes and test devices.

6.4 Computational Optimization of Design

In the preceding section, we showed that considering microrotors as microholograms altering the angular momentum content of the driving beam can be used for the conception of designs. One design – a symmetric cross-shaped rotor – was fabricated and tested, and performed qualitatively as expected. The tested rotor was designed to be three-dimensionally trapped, and included a long central stalk for maintaining proper alignment within the trap.

However, suggested applications of optically driven rotors often involve the rotor being mounted on an axle within a microfluidic device, as opposed to a "free-swimming" rotor like the test design. In this case, since the rotor is not free to move along the beam axis, it can lie in the focal plane of the driving beam. This simplifies the design procedure, since there are fewer variables to consider – no stalk is needed, and there is no need to be concerned with where the equilibrium trapping position is along the beam axis.

A free rotor, on the other hand, will be trapped past the focal plane due to the action of the scattering force, as is seen for all particles in single-beam traps. One result of this is that the rotor is typically expected to be trapped along the beam axis where the beam begins to strongly diverge. This will result in a reduction in the torque efficiency [27]. This is shown in Figure 6.6.

Therefore, we can expect improved performance from a rotor integrated within a device, mounted on an axle. However, this presents significantly greater difficulty in fabrication, and is thus an ideal candidate for a more extensive computational exploration.

6.4.1 Computational Modeling of Microrotors

The calculation of optical forces and torques is essentially an electromagnetic scattering problem – the incident field carries energy, momentum, and angular momentum toward the particle in the trap, and the superposition of the scattered and incident fields carries these away. The difference between the inward and outward fluxes gives the absorbed power, and optical force and torque. In principle, a wide variety of methods can be used for the scattering calculation, and methods such as ray optics and the Rayleigh approximation remain popular due to their simplicity. However, the particles typically trapped and manipulated using optical tweezers are too small for short wavelength approximations such as geometric optics and too

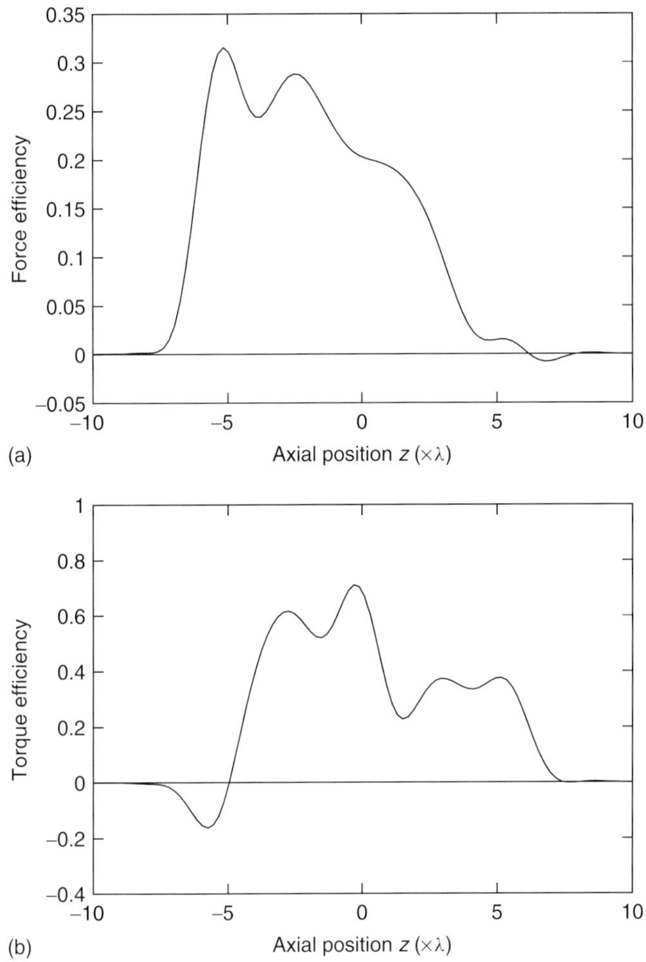

Figure 6.6 Force and torque as a function of axial position of cross-rotor trapped in plane-polarized LG_{02} beam. The equilibrium position is past the focal plane (a), and the torque efficiency has dropped substantially below the maximum value at this position (b).

large for long wavelength approximations such as Rayleigh scattering. For particles in this intermediate size range, a resort to computational electromagnetics is usual. General methods such as the finite-difference time-domain (FDTD) method and the finite-element method (FEM) can be used, but tend to be excessively slow, especially in view of the repeated calculations needed to model an optical trap.

One method that provides efficient repeated calculations is the T-matrix method; this also has the considerable advantage of allowing the force and torque to be calculated without needing to calculate the actual fields or numerically integrate the Maxwell stress tensor. Fundamentally, the T-matrix method makes use of a discrete basis set of functions $\psi_n^{(\text{inc})}$, where n is a mode index labeling the functions, each of which is a divergence-free solution of the Helmholtz equation, to represent the incident field

$$U_{\text{inc}} = \sum_n^{\infty} a_n \psi_n^{(\text{inc})} \tag{6.2}$$

and $\psi_k^{(\text{scat})}$ to represent the scattered wave, so that the scattered field can be written as

$$U_{\text{scat}} = \sum_k^{\infty} p_k \psi_k^{(\text{scat})} \tag{6.3}$$

The expansion coefficients a_n and p_k together specify the total field external to the particle.

When the electromagnetic response of the scatterer is linear, the relationship between the incident and scattered fields must be linear, and can be written as the matrix equation

$$p_k = \sum_n^{\infty} T_{kn} a_n \tag{6.4}$$

or

$$\mathbf{P} = \mathbf{T}\mathbf{A} \tag{6.5}$$

The T_{kn}, which are the elements of the transition matrix, or system transfer matrix, often simply called the T-matrix, are a complete description of the scattering properties of the particle at the wavelength of interest. The T-matrix is independent of the incident illumination, which is why the method is efficient for repeated calculation – it is only necessary to find the expansion coefficients of the incident light, and repeat the matrix–vector product (Eq. (6.5)), using the same T-matrix, to complete the scattering calculation.

When the scatterer is finite and compact, the most useful set of basis functions is VSWFs [5, 7, 28, 29]. In particular, the convergence of the VSWFs is well behaved and known [30], and this allows the sums given above to be truncated at some finite $n_{\max}$ without significant loss of accuracy.

The T-matrix method does not prescribe any particular method for the calculation of the T-matrix [29, 31], although the extended boundary condition method (EBCM) is usual [5, 28]. For a spherical particle, the T-matrix is given analytically by the Lorenz–Mie solution [32, 33]; this extension of the Lorenz–Mie solution to arbitrary illumination is usually called *generalized Lorenz–Mie theory (GLMT)* [34]. For a complex particle, especially one for which the EBCM fails, a more general method can be used. One such method is the discrete dipole approximation (DDA) [35–38], where the scatterer is represented as a collection of coupled dipole scatterers (hence the alternative name, the coupled dipole method [39]). DDA is a very attractive method for modeling scattering by microrotors, since it can be used for geometrically complex objects, and allows optimizations based on the symmetry of the scatterer, reducing the computational requirements by orders of magnitude.

In particular, we exploit the discrete rotational symmetry and mirror symmetry of typical microrotors [38].

The basic procedure for calculating the optical force and torque is as follows:

1) calculate the T-matrix,
2) calculate the incident field expansion coefficients a_n,
3) find the scattered field expansion coefficients p_k using $\mathbf{P} = \mathbf{TA}$, and
4) calculate the inflow and outflow of energy, momentum, and angular momentum from the total field.

For microrotors, step 1 is the bulk of the computational task [38]. Steps 2–4 can be carried out using our T-matrix-based optical tweezers computational toolbox, the *optical tweezers toolbox* [40].

6.4.2
Performance of a Four-Armed Rotor

A simple type of rotor was chosen in order to computationally explore the optimization of design. Again, fourfold symmetry was used. The arms were sectorial rather than straight, joined by a simple cylindrical hub. Since we considered illumination by optical vortex beams, the central region lies in the dark central region of the beam, and similar performance would result if the rotor is secured in a device by an axle passing through the hub or is fabricated with a stalk for use as a free-floating rotor. Therefore, the variable parameters are thickness of the arms, angle occupied by the arms, outer radius of the rotor, and radius of the inner hub (Figure 6.7).

The symmetry optimizations discussed above were used in this project. The fourfold discrete rotational symmetry of the microrotor was exploited to reduce the memory footprint and to speed up the DDA calculations. The discrete rotation symmetry and mirror symmetry were used in the near-field point matching, and the mode redundancy associated with the discrete rotation symmetry was used when solving the scattering coefficients, which both reduce calculation time by orders of magnitude.

We considered LG_{02} and LG_{04} incident beams, coaxial with the microrotor, so incident azimuthal modes were limited to $m = 1, 3, 5$. Therefore, it was not necessary to calculate the entire T-matrix.

We assumed that the rotors had the same refractive index as those we produced by two-photon photopolymerization, $n = 1.54$, and were surrounded by water. The

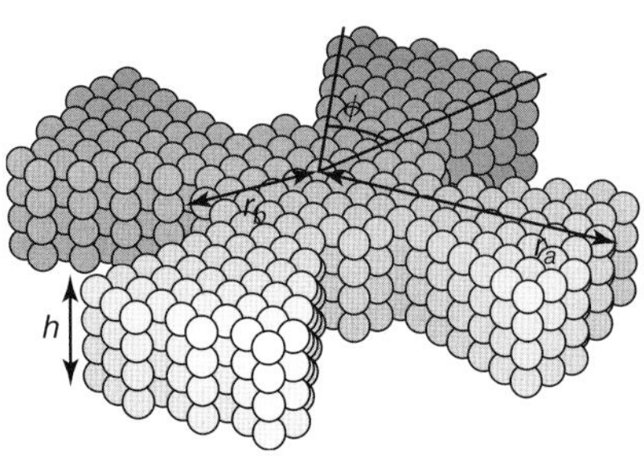

Figure 6.7 Dipole model of the four-armed microrotor.

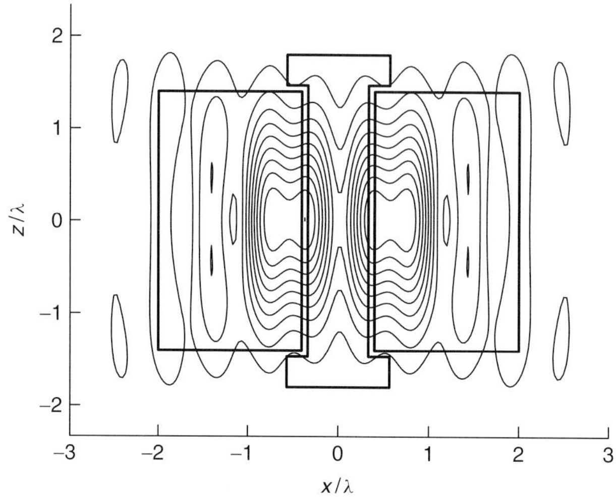

Figure 6.8 Intensity contour of the LG_{02} incident beam and quatrefoil rotor profile.

driving beams were assumed to be tightly focused by an objective with numerical aperture of 1.3, giving a beam convergence angle of 78° in water. The beams we used were assumed to have the back aperture of the objective illuminated by a plane-polarized LG_{02} or LG_{04} beam. Figures 6.8 and 6.9 show cross sections of the focal regions of the incident LG_{02} and LG_{04} beams, respectively, superimposed on the rotor.

Figures 6.8 and 6.9 also show an important feature of tightly focussed optical vortex beams that even those produced by focusing zero-radial-order beams ($LG_{0\ell}$, with radial mode index $p = 0$) do not consist of a single bright ring

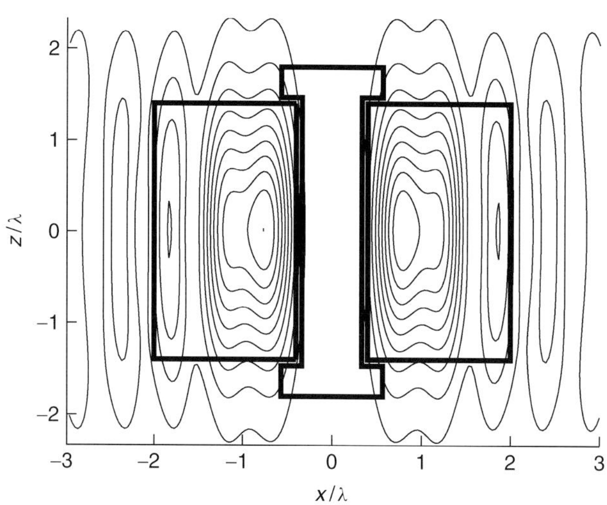

Figure 6.9 Intensity contour of the LG_{04} incident beam and quatrefoil rotor profile.

in the focal region, but of a series of concentric rings. The focal region of such a beam approximates a truncated Bessel beam. While an infinitely wide Bessel beam (which also requires infinite power) is nondiverging, a truncated Bessel beam diverges as it propagates through the spreading of the outer ring, followed by the next ring, and so on, until it spreads to the innermost ring. This can also be seen in tightly focused vortices, such as that we have here [41].

In the hologram picture of microrotors, the thickness of the arms is an essential parameter, since it determines the phase retardation imparted on light passing through the structure. However, the simple ray model used for the preliminary design can be criticized on many grounds – the features of the structure are too small for ray optics to be reliable, and the rotor is located in the focal region of the beam that is not described properly in the ray optics approximation. However, one would need to be careful with a ray picture of this type even if the objections above do not apply. For example, if we consider a paraxial beam passing through a phase retarder printed on a holographic plate, there is a very clear difference between a beam with no retardation, and one with a full-wave retardation, as shown in Figure 6.10.

With these considerations in mind, the dependence of the torque efficiency on the thickness of the structure is especially interesting. This is shown, for both LG_{02} and LG_{04} beams, in Figures 6.11 (for outer radius $r_a = 1\lambda$) and 12 (for outer radius $r_a = 2\lambda$).

While, for the reasons advanced above, the torque efficiency is not expected to fall to zero as suggested by the ray picture, the plateau seen in Figures 6.11 and 6.12 is remarkable. This effect will result from two major causes. First, as seen in Figures 6.8 and 6.9, the focal region of the beam is approximately three wavelengths in length, and a rotor thicker than the length of the focal region will be illuminated partly from the side, not just from the end. In this case, additional thickness will make little difference to the torque efficiency.

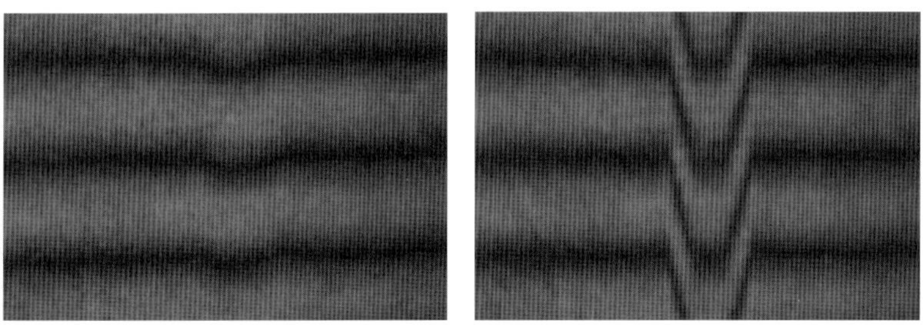

(a) (b)

Figure 6.10 Phase structure of a beam that has passed through a holographic plate with a phase-retarding stripe made visible through interference with a plane wave. (a) A stripe with a very small retardation has very little effect on the beam, as expected, while in (b), a full-wave retardation is very noticeable. In a simple ray picture, the effect of both would be the same, but the continuity of the electric field of the beam as it propagates through the holographic plate prevents this.

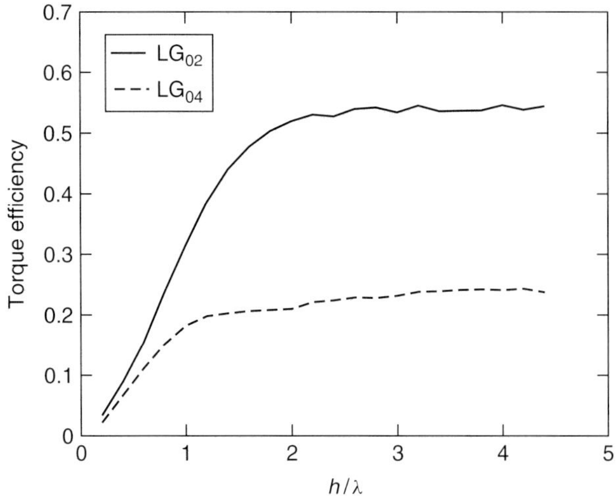

Figure 6.11 The torque efficiency of the microrotor versus thickness with outer radius $r_a = 1\lambda$ and hub radius $r_b = 0$.

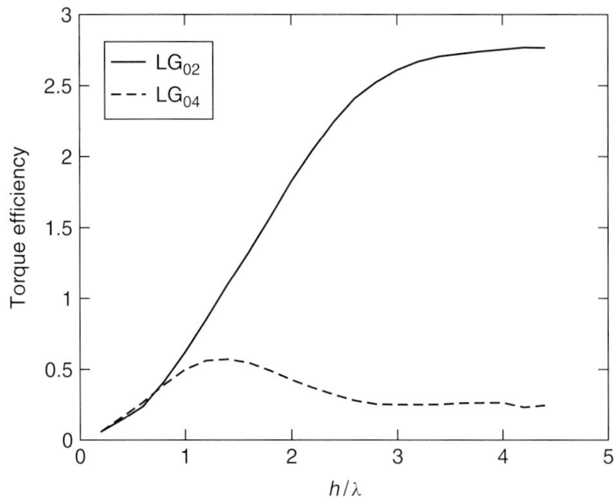

Figure 6.12 The torque efficiency of the microrotor versus thickness with outer radius $r_a = 2\lambda$ and hub radius $r_b = 0$.

A second cause is that the arms of the rotor, being of higher index than the surrounding medium, will act as waveguides. Light will tend to concentrate within the arms, forming a guided mode. When these guided modes are established, the phase of the wave in the surrounding medium immediately outside the arm will be determined by the phase within the arm, and thicker arms will not result in greater relative phase shifts. Since the space between the arms is comparable to

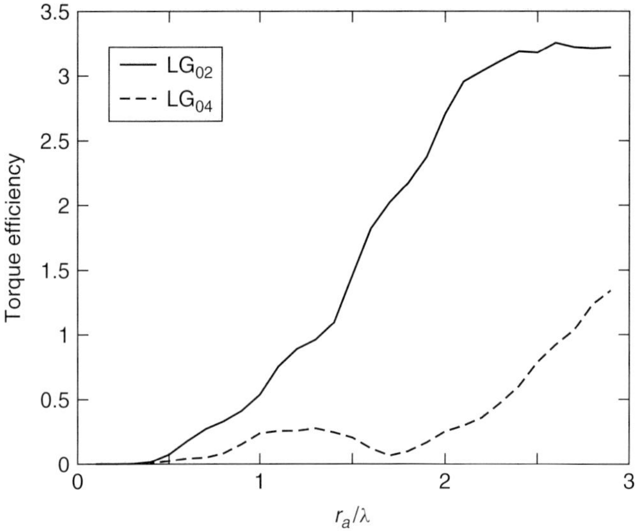

Figure 6.13 The torque efficiency of the microrotor versus blade radius r_a for thickness of $h = 3.4\lambda$.

the wavelength, essentially all of the beam will be close enough to the arm to be affected by this.

We can also see that the radius of the rotor is very important; the maximum torque efficiency for a rotor of radius $r_a = 2\lambda$ is much greater than that for a radius of $r_a = 1\lambda$. This occurs despite the smaller rotor still being large enough to intercept the innermost (and brightest) bright ring. However, Bessel beams have equal power in each bright ring, and we can expect the power in the second and further successive bright rings to be a large fraction of the total power. Therefore, it is important for the rotor to intercept more of the beam than just the innermost ring. The dependence of the torque efficiency on the radius of the rotor is shown in Figure 6.13. For the LG_{02} beam, a radius of $r_a = 2.6\lambda$ is sufficient for the maximum torque to be achieved, and little improvement results from radii over $r_a = 2.2\lambda$. Since the bright ring of an LG_{04} beam has a larger radius than that of an LG_{02} beam, we expect that a larger rotor radius would be required to reach the same torque efficiency when using an LG_{04} beam. However, the difference in widths of the LG_{02} and LG_{04} beams, as shown in Figures 6.8 and 6.9, is insufficient to explain the lower torque efficiencies. This is discussed below, after exploration of the dependence of torque efficiency on the angle occupied by the rotor blades.

For subsequent calculations, we used a thickness of $h = 2.2$ and rotor radius of $r_a = 2.2$; these are sufficient to give performance close to the optimum, while the increased size required to achieve the actual optimum performance would greatly increase the required computational time and resources.

The remaining parameters to be explored are the hub radius r_b and the rotor blade angle. The effect of the hub radius is shown in Figure 6.14. It can be seen

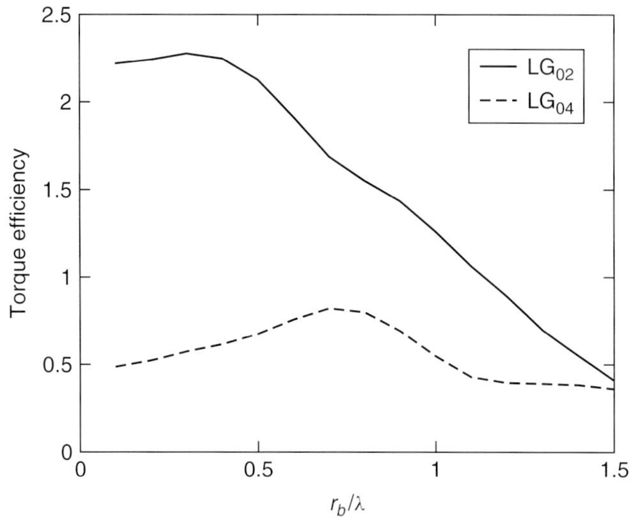

Figure 6.14 The torque efficiency of the microrotor versus hub radius r_b, for thickness $h = 2.2\lambda$ and rotor radius $r_a = 2.2\lambda$.

that the torque efficiency drops rapidly after $r_b = 0.4$, as the hub begins to intercept a significant portion of the beam power. In the extreme case, $r_b = r_a$, the rotor becomes a cylinder without arms, and no torque can be generated.

The effect of the blade angle is shown in Figure 6.15. It appears that the blades should occupy an angle between 35° and 40° to achieve optimal torque (for the LG_{02} beam). Viewing the rotor as a hologram, it would be best if half of the power passes through the arms or the rotor, with the other half of the power passing between them. With a ray model, this is achieved by having the arms occupy half of the area, that is, a blade angle of 45°. However, since the beam will be concentrated within the arms due to the waveguiding effect discussed above, half of the power will correspond to less than half of the area.

For the LG_{04} beam, the dependence on angle is interesting and complex. Noting that the blade of the rotor is comparable to the wavelength in width, we can expect the "waveguide" mode, which is best supported, to be relatively uniform within the blade, which would be the lowest-order TEM mode in an optical fiber. As such, we can expect poor coupling of the incident light to this mode when the blade angle is 45°, since the incident beam is a half-wave out of phase on opposite sides of the blade. The phase of the incident light across the blade is an odd function, and will not couple strongly to a low-order mode where the phase is uniform. Stronger coupling should occur where the phase variation across the blade is a 1/4 or 3/4 wave – an even function such as this will have a significant zero-frequency Fourier component, as opposed to the vanishing zero-frequency component for an odd function.

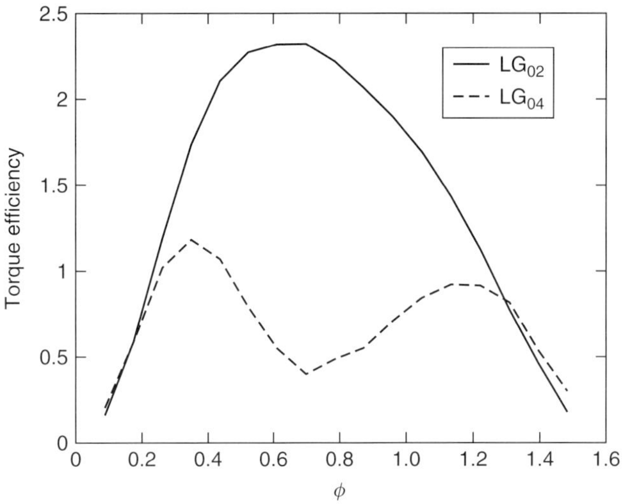

Figure 6.15 The torque efficiency of the microrotor versus blade angle, for thickness $h = 2.2\lambda$ and rotor radius $r_a = 2.2\lambda$.

6.4.3 Discussion

First, the above discussed computational results show that the hologram picture of microrotors is useful, and gives qualitatively correct predictions; these predictions agree with and are supported by the qualitative expectations.

Second, we have shown that, given a preliminary design, computational modeling of a microrotor allows optimization of the design. The parameters investigated here were thickness, outer radius, hub radius of the rotor, and angle occupied by the blades. Other parameters that might be of interest are the (relative) refractive index of the structure, the number of arms, the details of the driving illumination, and so on. The calculations above were carried out on desktop PCs (from 2008 to 2009), with the larger structures needing 64-bit computers due to memory requirements. Therefore, such computational optimization is possible with readily available hardware. The computational resources required to carry out the scattering calculation to determine the optical force and torque increase rapidly with increasing size, and would, at the time of writing, require high-performance computing resources using the methods described. An increase in computational power of typical desktop computers, which can be expected over time, will also increase the usefulness of these methods. Most of the software used for the above results are already available in our optical tweezers toolbox [40], and the DDA–T-matrix software [38] will be included in the future. The symmetry optimizations were of great practical value, since they allowed the calculations to be carried out in a reasonable time on readily available hardware.

Third, some general conclusions can be drawn from these results. It is clear that the exact values of parameters such as the rotor thickness and radius are not important, as long as they are above particular values. Thus, a particular microrotor can be expected to perform well under a variety of conditions; for example, being driven by beams of various wavelengths or focused to varying degrees.

The relative performance when driven by an LG_{02} beam versus an LG_{04} beam also merits further discussion. From a qualitative consideration of the hologram picture of microrotors, we might expect similar performance from both. However, the LG_{02} beam produced much higher torques. Since the action of the rotor arms as "waveguides" appears to be important, we can expect the coupling of an incident LG_{02} to a transmitted $LG_{0,-2}$ beam to be strong. If the waveguide effect is dominant, the output will closely resemble the superposition of LG_{02} and $LG_{0,-2}$ beams, as shown in Figure 6.16. In this case, the left- and right-helical output modes would be of equal power, and a torque efficiency of 2 would be achieved, due to half of the power being in modes with an angular momentum change of $4\hbar$ per photon. This is close to the calculated torque efficiency.

The maximum torque efficiencies here are notably greater than the observed orbital torque efficiency of 0.2 in Section 6.3. In Figure 6.6, it can be seen that the torque efficiency is greatest when the rotor lies in the focal plane, rather than lying past the focal plane in the axial trapping equilibrium position. Thus, a rotor mounted on an axle can produce greater torque compared with a free rotor trapped in a single beam. Alternatively, a free rotor could be trapped in the focal plane using counterpropagating beams to eliminate the net axial scattering force. However, the maximum torque efficiency for the original design is still much lower than the

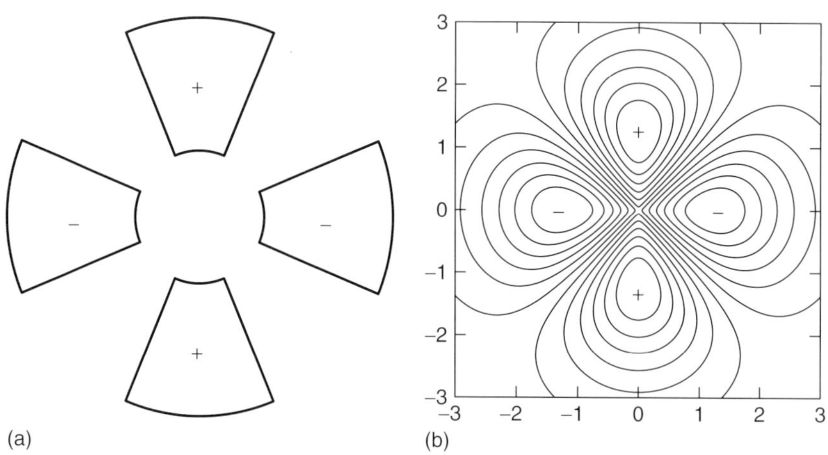

Figure 6.16 Light transmitted through rotor arms illuminated by LG_{02} beam, due to "waveguiding" effect. The field in each rotor arm resembles that of a lowest-order mode in a waveguide, and will be approximately uniform in intensity and phase. (a) Owing to the phase variation of the incident light, the light in each successive rotor arm is a half-wave out of phase. (b) The superposition of $LG_{0,2}$ and $LG_{0,-2}$ beams.

torque efficiency for the optimized design here, with only approximately one third of the torque efficiency. The potential benefit of optimization is clear.

6.5 Conclusion

Consideration of the effects of symmetry on the scattering of light provides a sound theoretical basis for the design of optically driven microrotors. The symmetry principles are the same as seen in the interaction between light and holograms; the analogy between the holographic generation of optical vortex beams carrying the OAM and the application of OAM to the generation of optical torque on microrotors is strong, and provides a convenient and familiar picture for the conception of designs. This provides a simple set of qualitative guidelines for the development of initial designs that can then be experimentally tested or computationally optimized.

Two-photon photopolymerization is an excellent method for the experimental evaluation of designs. Coupled with the optical measurement of the torque – Parkin's method [27, 42] provides an accurate measurement of both the total optical torque and the orbital component – or computational modeling of the fluid flow about the rotor [27], the performance of the rotor can be both qualitatively and quantitatively evaluated. The two-photon photopolymerization technique has proved to be a powerful tool for fabricating microstructures with potentially any shape. While it might not be the method of choice for the large-scale manufacture of devices, it is perfect for the fabrication of prototypes and test devices and the study of the transfer of linear and/or angular optical momentum to them.

Systematic computational modeling can be used to optimize a preliminary design. We demonstrated computational modeling of this nature using a hybrid DDA–T-matrix method [38] coupled with our optical tweezers toolbox [3] for modeling optical forces and torques in arbitrary beams. Symmetry optimizations exploiting discrete rotational symmetry and mirror symmetry of the rotor allow the calculations to be carried out on a desktop PC. We showed that it is possible to construct rotors that will give close-to-optimum performance under a variety of conditions. This type of modeling, ideally including the modeling of nonelectromagnetic effects such as heating and convective flow, can be used to carry out the engineering and design of such in a systematic and guided manner.

References

1. Heckenberg, N.R., McDuff, R., Smith, C.P., and White, A.G. (1992) Generation of optical phase singularities by computer-generated holograms. *Opt. Letters*, **17** (3), 221–223.
2. Nieminen, T.A., Parkin, S.J., Heckenberg, N.R., and Rubinsztein-Dunlop, H. (2004) Optical torque and symmetry. *Proc. SPIE*, **5514**, 254–263.

3. Nieminen, T.A., Asavei, T., Loke, V.L.Y., Heckenberg, N.R., and Rubinsztein-Dunlop, H. (2009) Symmetry and the generation and measurement of optical torque. *J. Quant. Spectrosc. Radiat. Transf.*, **110** (14–16), 1472–1482.
4. Allen, L., Beijersbergen, M.W., Spreeuw, R.J.C., and Woerdman, J.P. (1992) Orbital angular momentum and the transformation of Laguerre-Gaussian laser modes. *Phys. Rev. A*, **45**, 8185–8189.
5. Waterman, P.C. (1971) Symmetry, unitarity, and geometry in electromagnetic scattering. *Phys. Rev. D*, **3**, 825–839.
6. Nieminen, T.A., Stilgoe, A.B., Heckenberg, N.R., and Rubinsztein-Dunlop, H. (2008) Angular momentum of a strongly focused Gaussian beam. *J. Opt. A*, **10** (11), 115005.
7. Nieminen, T.A., Rubinsztein-Dunlop, H., and Heckenberg, N.R. (2003) Multipole expansion of strongly focussed laser beams. *J. Quant. Spectrosc. Radiat. Transf.*, **79–80**, 1005–1017.
8. Maruo, S. and Fourkas, J.T. (2008) Recent progress in multiphoton microfabrication. *Laser Photon. Rev.*, **2** (1–2), 100–111.
9. Park, S.-H., Yang, D.-Y., and Lee, K.-S. (2009) Two-photon stereolithography for realizing ultraprecise three-dimensional nano/microdevices. *Laser Photon. Rev.*, **3** (1–2), 1–11.
10. Galajda, P. and Ormos, P. (2001) Complex micromachines produced and driven by light. *Appl. Phys. Lett.*, **78**, 249–251.
11. Galajda, P. and Ormos, P. (2002) Rotors produced and driven in laser tweezers with reversed direction of rotation. *Appl. Phys. Lett.*, **80** (24), 4653–4655.
12. Gauthier, R.C. (2001) Optical levitation and trapping of a micro-optic inclined end-surface cylindrical spinner. *Appl. Opt.*, **40**, 1961–1973.
13. Knöner, G., Parkin, S., Nieminen, T.A., Loke, V.L.Y., Heckenberg, N.R., and Rubinsztein-Dunlop, H. (2007) Integrated optomechanical microelements. *Opt. Express*, **15** (9), 5521–5530.
14. Higurashi, E., Ukita, H., Tanaka, H., and Ohguchi, O. (1994) Optically induced rotation of anisotropic micro-objects fabricated by surface micromachining. *Appl. Phys. Lett.*, **64** (17), 2209–2210.
15. Ukita, H. and Kanehira, M. (2002) A shuttlecock optical rotator – its design, fabrication and evaluation for a microfluidic mixer. *IEEE J. Sel. Top. Quantum Electron.*, **8** (1), 111–117.
16. Ukita, H. and Nagatomi, K. (2003) Optical tweezers and fluid characteristics of an optical rotator with slopes on the surface upon which light is incident and a cylindrical body. *Appl. Opt.*, **42** (15), 2708–2715.
17. Strickler, J.H. and Webb, W.W. (1991) Three-dimensional optical data storage in refractive media by two-photon point excitation. *Opt. Lett.*, **16** (22), 1780–1782.
18. Denk, W., Strickler, J.H., and Webb, W.W. (1990) Two-photon laser scanning fluorescence microscopy. *Science*, **248**, 73–76.
19. Maruo, S., Nakamura, O., and Kawata, S. (1997) Three-dimensional microfabrication with two-photon-absorbed photopolymerization. *Opt. Lett.*, **22** (2), 132–134.
20. Maruo, S., Ikuta, K., and Korogi, H. (2003) Submicron manipulation tools driven by light in a liquid. *Appl. Phys. Lett.*, **82** (1), 133–135.
21. Maruo, S. and Inoue, H. (2006) Optically driven micropump produced by three-dimensional two-photon microfabrication. *Appl. Phys. Lett.*, **89**, 144101.
22. Asavei, T., Nieminen, T.A., Heckenberg, N.R., and Rubinsztein-Dunlop, H. (2009) Fabrication of microstructures for optically driven micromachines using two-photon photopolymerization of UV curing resins. *J. Opt. A: Pure Appl. Opt.*, **11**, 034001.
23. He, H., Heckenberg, N.R., and Rubinsztein-Dunlop, H. (1995) Optical particle trapping with higher-order doughnut beams produced using high efficiency computer generated holograms. *J. Mod. Opt.*, **42** (1), 217–223.
24. Parkin, S.J., Nieminen, T.A., Heckenberg, N.R., and

Rubinsztein-Dunlop, H. (2004) Optical measurement of torque exerted on an elongated object by a noncircular laser beam. *Phys. Rev. A*, **70**, 023816.

25. Nieminen, T.A., Heckenberg, N.R., and Rubinsztein-Dunlop, H. (2001) Optical measurement of microscopic torques. *J. Mod. Opt.*, **48**, 405–413.

26. Parkin, S., Knöner, G., Nieminen, T.A., Heckenberg, N.R., and Rubinsztein-Dunlop, H. (2006) Measurement of the total optical angular momentum transfer in optical tweezers. *Opt. Express*, **14** (15), 6963–6970.

27. Asavei, T., Loke, V.L.Y., Barbieri, M., Nieminen, T.A., Heckenberg, N.R., and Rubinsztein-Dunlop, H. (2009) Optical angular momentum transfer to microrotors fabricated by two-photon photopolymerization. *New J. Phys.*, **11**, 093021.

28. Mishchenko, M.I. (1991) Light scattering by randomly oriented axially symmetric particles. *J. Opt. Soc. Am. A*, **8**, 871–882.

29. Nieminen, T.A., Rubinsztein-Dunlop, H., and Heckenberg, N.R. (2003) Calculation of the *t*-matrix: general considerations and application of the point-matching method. *J. Quant. Spectrosc. Radiat. Transf.*, **79–80**, 1019–1029.

30. Brock, B.C. (2001) *Using Vector Spherical Harmonics to Compute Antenna Mutual Impedance from Measured or Computed Fields*. Sandia report SAND2000-2217-Revised. Sandia National Laboratories, Albuquerque, New Mexico.

31. Michael Kahnert, F. (2003) Numerical methods in electromagnetic scattering theory. *J. Quant. Spectrosc. Radiat. Transf.*, **79–80**, 775–824.

32. Lorenz, L. (1890) Lysbevægelsen i og uden for en af plane Lysbølger belyst Kugle. *Vidensk. Selsk. Skrifter*, **6**, 2–62.

33. Mie, G. (1908) Beiträge zur Optik trüber Medien, speziell kolloidaler Metallösungen. *Ann. Phys.*, **25**, (3), 377–445.

34. Gouesbet, G. and Grehan, G. (1982) Sur la généralisation de la théorie de Lorenz-Mie. *J. Opt. (Paris)*, **13** (2), 97–103. On the scattering of light by a Mie scatter center located on the axis of an axisymmetric light profile.

35. Draine, B.T. and Flatau, P.J. (1994) Discrete-dipole approximation for scattering calculations. *J. Opt. Soc. Am. A*, **11** (4), 1491–1499.

36. Purcell, E.M. and Pennypacker, C.R. (1973) Scattering and absorption of light by nonspherical dielectric grains. *Astrophys. J.*, **186**, 705–714.

37. Mackowski, D.W. (2002) Discrete dipole moment method for calculation of the *t* matrix for nonspherical particles. *J. Opt. Soc. Am. A*, **19** (5), 881–893.

38. Loke, V.L.Y., Nieminen, T.A., Heckenberg, N.R., and Rubinsztein-Dunlop, H. (2009) T-matrix calculation via discrete-dipole approximation, point matching and exploiting symmetry. *J. Quant. Spectrosc. Radiat. Transf.*, **110** (14–16), 1460–1471.

39. Chaumet, P.C. and Nieto-Vesperinas, M. (2000) Coupled dipole method determination of the electromagnetic force on a particle over a flat dielectric surface. *Phys. Rev. B*, **61**, 14119–14127.

40. Nieminen, T.A., Loke, V.L.Y., Stilgoe, A.B., Knöner, G., Brańczyk, A.M., Heckenberg, N.R., and Rubinsztein-Dunlop, H. (2007) Optical tweezers computational toolbox. *J. Opt. A: Pure Appl. Opt.*, **9**, S196–S203.

41. Nieminen, T.A., Parkin, S., Asavei, T., Loke, V.L.Y., Heckenberg, N.R., and Rubinsztein-Dunlop, H. (2008) Optical vortex trapping and the dynamics of particle rotation, in *Structured Light and Its Applications: An Introduction to Phase-Structured Beams and Nanoscale Optical Forces*, (ed. D.L. Andrews), Academic Press, San Diego, pp. 195–236, chapter 8.

42. Parkin, S.J., Knöner, G., Nieminen, T.A., Heckenberg, N.R., and Rubinsztein-Dunlop, H. (2006) Torque transfer in optical tweezers due to orbital angular momentum. *Proc. SPIE*, **6326**, 63261B.

7
Rotational Optical Micromanipulation with Specific Shapes Built by Photopolymerization

Péter Galaja, Lóránd Kelemen, László Oroszi, and Pál Ormos

7.1
Introduction

Photons carry spin; consequently, a light beam can have angular momentum and while interacting with matter, this angular momentum can be transferred to objects. This theoretical concept became a practical tool with the development of optical micromanipulation.

With the emergence of lasers, where high light intensities could be generated in small areas, it was demonstrated – primarily through the original works of Arthur Ashkin [1, 2] – that light can effectively manipulate microscopic particles. Optical tweezers were introduced, where particles could be trapped in the focus of a high numerical aperture laser beam. The method developed to become an innovative and useful noncontact manipulation tool, with countless applications, notably in biology. In the basic configuration, optical tweezers can grab a spherical particle that has an index of refraction larger than that of the surrounding medium. The particle is held in the focus that forms an elastic trap. The position of the particle is monitored and controlled. This system already has great potential. The noncontact manipulation method itself allows previously unfeasible experiments: trapping of cells and intracellular objects, manipulation of macromolecules through attached plastic beads, stretching single molecules (DNA [3, 4], titin [5] etc.), mechanical characterization of molecular motors [6, 7], and so on, just to name a few.

It is clear that if the optically trapped object is a sphere made of an isotropic material, the only relevant coordinate is its position. There is no additional means of manipulation in this respect. On the other hand, when we manipulate nonspherical objects in general, additional degrees of freedom such as rotation/orientation are also important.

To extend the traditional, purely translational trapping, the obvious next degree of freedom to be manipulated is the rotation of the trapped object around the propagation direction of the light as an axis. Here, the torque exerted by the trapping light plays the dominant role. The particle can be orientated, rotated, and this gives major new potential for manipulation. In recent years, a number of techniques have been developed that are based on different optical phenomena.

Twisted Photons: Applications of Light with Orbital Angular Momentum.
Edited by Juan P. Torres and Lluis Torner

ISBN: 978-3-527-40907-5

Two basic principles exist to achieve rotational manipulation. In the first and distinctly elegant scheme, the incoming light beam itself carries the angular momentum, and its interaction with the trapped object generates the torque for rotational manipulation. The classic example is the interaction of circularly polarized light with a birefringent material, first demonstrated by the Rubinsztein-Dunlop group [8]. Subsequently, different schemes were introduced along this concept.

In another basic procedure, the incoming manipulating beam carries no angular momentum, but it interacts with an object having helical shape. In this case, the light gains angular momentum following the scattering. This interaction also exerts torque on the manipulated object. This rotation mechanism is roughly analogous to a propeller driven by blowing wind. The shape of the object is crucial in this approach. By using various shapes, the interaction of light and particle can be controlled to a large extent, and different and surprising possibilities can be achieved. Key to the practical realization of this system is the production of shapes with the appropriate specificity. We explored the possibilities of this approach.

7.2
Microfabrication by Photopolymerization

7.2.1
Fabrication by Scanning a Single Focused Laser Beam

Building of micrometer-sized structures to perform different tasks in microfluidics is an emerging field and different methods have been developed. A straightforward approach uses silicon-based microlithographic methods [9–11]. In this procedure, 3D objects are built in several steps of layer-by-layer repetitive production processes such as photolithography, etching, doping, and so on. Laser ablation methods are also suitable for fabrication [12].

Photopolymerization is an alternative approach to producing micrometer-sized 3D structures. In this procedure, laser light excitation hardens an appropriate photopolymer. A laser beam is focused on the material and with a carefully chosen parameter set crosslinking of monomers occurs in the close vicinity of the focal spot. By scanning the focal point in the photopolymer along a predefined path, entire objects can be built in one (or just a few) step(s). The procedure is schematically shown in Figure 7.1. One can scan either the sample relative to the laser focal spot or vice versa. The spatial resolution of the method can be improved by the use of femtosecond lasers, where the intensity in the pulses is sufficiently high to induce two-photon absorption (TPA). Owing to the quadratic intensity dependence of the two-photon process, the size of the polymerized region can be reduced, thereby increasing resolution [13–15]. Recent papers reported successful two-photon polymerization of organic–inorganic materials with submicron resolution [16]. The method is capable of generating different microscopic mechanical devices such as springs [15, 17], gears [14, 18, 19], and micromanipulator arms [20]. The actuation of photopolymerized objects mainly

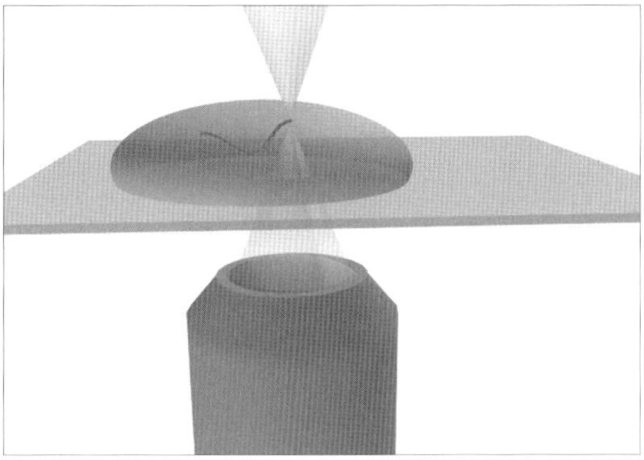

Figure 7.1 Illustration of the laser-induced photopolymerization technique. Laser light is focused onto a light-sensitive polymer. Using two-photon excitation, the polymer hardens only in the laser focus, and the object is drawn by the moving focus.

involves optical tweezers that either simply hold or relocate objects [20] or, in case of special-shaped objects, even rotate them [10, 14, 21]. In the experiments discussed here, we used particles built by this procedure.

We worked with several photoresists – depending on the desired properties of the final structures, different materials may be advantageous. The two basic types were versions of Norland optical adhesives (Norland, Cranbury, NJ, USA) and SU8 photoresist (Michrochem, Newton, MA, USA). The Norland material is a composite of acrylate monomer with mercapto-ester. Structure building takes place in a single step. In addition, the polymer is of optical quality. The disadvantage of this material is its mechanical properties: during the polymerization process the material shrinks, resulting in a certain degree of distortion in the final structure. In contrast, SU8 is an epoxy-based negative photoresist. It has excellent mechanical qualities: it does not shrink during the process and the final material is significantly harder than the Norland [22]. The disadvantage is that it is more complicated to process (e.g., a careful pre- and postexposure bake is needed), and the optical quality of the hardened material is not good: it scatters a significant amount of light. Depending on the particular application, we used both materials.

Both materials are light-sensitive for wavelengths of less than about 400 nm, so the light from a mode-locked Ti:sapphire laser (FemtoRose 100 TUN, R&D Ultrafast Lasers Ltd, Budapest, Hungary) is appropriate for two-photon excitation with the following parameters: 150 fs pulses with 80 MHz repetition rate and up to 6.3 nJ pulse energy (500 mW average power) in mode-locked operation at 790 nm wavelength. In general, an average power of 1–8 mW is sufficient for the polymerization. A 100x magnification, 1.25 NA oil immersion objective (Zeiss Achroplan, Carl Zeiss, Germany) was used to focus the beam into the photopolymer.

Scanning the focus is achieved by moving the microscope stage along a predetermined trajectory: we use piezoelectric devices, primarily an X–Y piezo-translator

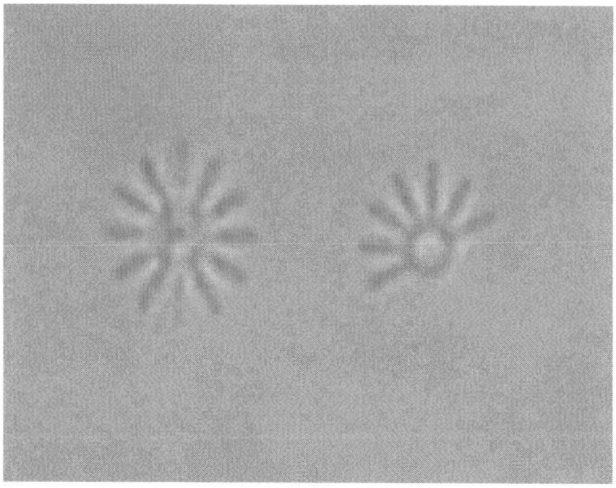

Figure 7.2 Snapshot of the structure building process. A cogwheel is being produced – after crosslinking, the refractive index of the polymer changes, making the process visible. A completed wheel is on the left.

(Physik Instrumente, Germany) of 100 × 100 μm travel range and nanometer precision displacement perpendicular to the optical axis, while the positioning along the Z-direction is realized by mounting the microscope objective also on a piezo-translator (PIFOC, Physik Instrumente, Germany) with 80-μm travel range along the optical axis.

Figure 7.2 shows an image of the fabrication process in the case of Norland photoresists – a cogwheel is being produced. The polymerized structure is visible because the refractive index of the material increases following excitation. This is not the case for the SU8 photoresist: the illuminated region is not visible, and the result can be seen and evaluated only after the structure is developed. After completion of the photopolymerization, the nonhardened material is removed and the object is ready to be used.

7.2.2
Parallel Photopolymerization using Diffractive Optics

Following the structure building process, the free-floating micrometer-sized particles are quite difficult to collect. Consequently, a large number of identical particles have to be built to cope with the limited recovery yield. The rate-limiting step here is the scanning because the laser beam has to go through each particle one by one, and it becomes necessary to speed up the process.

The rate of the production can be increased by parallelizing the procedure. The most trivial way to achieve this is to multiplex the laser beam by appropriate diffractive optical elements. In the literature, one can find examples for using passive transparent-type devices such as microlens arrays [23] and kinoforms [21]. We investigated the latter option becausethis way, the process can be multiplexed and identical objects can be built – the level of multiplexing is limited by only the available laser power [21].

Higher level of complexity and control can be achieved by the use of active beam-shaping devices such as spatial light modulators (SLMs). We also studied the possibilities of this approach using computer-controlled reflective SLM (LC-R 2500, Holoeye Photonics AG, Berlin-Adlershof, Germany). In such systems, simple multiplexing can also be achieved in a controlled manner, but in addition, complex illumination patterns can be generated enabling highly accelerated or even single-shot photopolymerization [21].

Structure building by photopolymerization has developed into a most useful and highly capable rapid prototyping technique. In general, the complexity of the structure to be built is not a limiting factor, and when particles with well-defined, specific three-dimensional shapes have to be fabricated in moderate copy numbers (i.e., $<10\,000$), it is the most advisable procedure [24, 25].

7.3 Light-Driven Rotors, Micromachines

7.3.1 Propeller

The basic topic of this section is light-induced rotation, where the origin of the torque is the scattering of light on a helical object: a propeller. It often happens when experimenting with optical tweezers that microscopic particles trapped in the focus of a laser beam tend to rotate. The phenomenon may first appear surprising, but after all, an irregular shape may have some degree of helicity with a fairly high probability, and this is sufficient to induce rotation. The phenomenon is interesting from a purely intellectual standpoint, but at the same time it has far-reaching implications as it forms the basis for prospective nanotechnology applications.

For the systematic study of this phenomenon, we built well-defined propeller shapes that can be rotated in optical tweezers. While the working principles of a propeller are trivial, the details here are not really obvious for several reasons. There have been attempts to estimate the light forces by ray optics by simply assuming reflection and refraction on the surfaces [26]. However, since the wavelength of the trapping/rotating light is comparable to the particle size, the use of ray optics is an oversimplification; an exact calculation of the force and torque on the particle with complex shape is not trivial. In addition, even the position of the particles within the trap is most difficult to determine. Obviously, the rotating torque due to the scattering of the light crucially depends on the position of the body in the focus. Instead of using unreliable approximate calculations, we have tested numerous shapes that we expected to assume a regular position in the trap and rotate efficiently. We have tried many classes of forms, the variations of sprinkler shapes with a central axis were best both with respect to stability of position in the

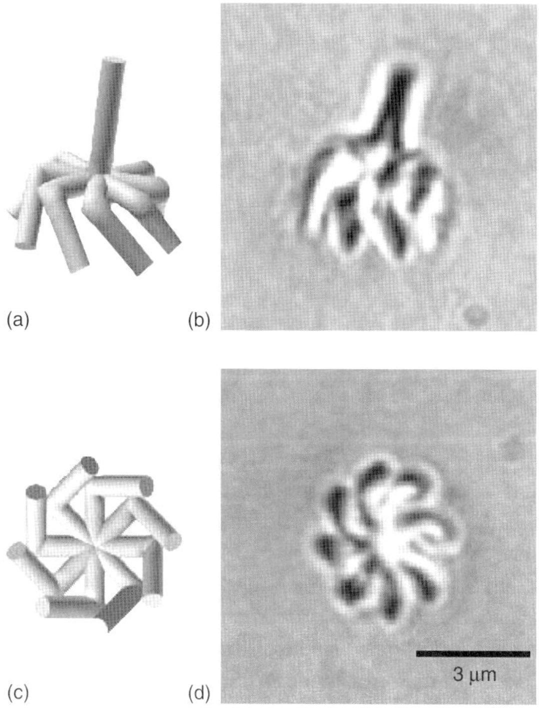

Figure 7.3 Geometry and picture of an efficient microscopic light-driven rotor. The rotor can take up two stable positions in the laser trap. (a,c) Drawing of the design. (b,d) Image of the actual rotor in the microscope in equivalent positions.

trap and efficiency of rotation [27]. The shape shown in Figure 7.3 has proved to be very effective [14]: It was both very stable in the focus and showed efficient rotation: at 20-mW power, frequencies of several hertz were achieved.

We have observed that the rotor can assume two stable positions in the trap: with the axis pointing toward the microscope objective or away from it. In both positions the propeller rotates, with somewhat different efficiencies (Figure 7.4). This observation indicates the complexity of the situation. At first glance the two opposite orientations should be equivalent. However, if we consider that due to the scattering force the equilibrium position is not exactly in the focus, it becomes clear that the two cases are not equivalent. This also points to the difficulty of an exact description of the system. However, we can make reasonable estimations to characterize the dynamics. In stationary rotation, the torque exerted by the scattered light is in equilibrium with the viscous drag. A rough but reasonable estimate can be made for the driving torque by assuming that on an average, light is deflected by 10° as a result of scattering and the average distance from the axis is 3 μm. From the momentum change for light of 995-nm wavelength and power of 10 mW, the torque is about $M \approx 2 \times 10^{-17}$ Nm. The viscous drag of the propeller can be estimated as follows: the estimation of the drag torque

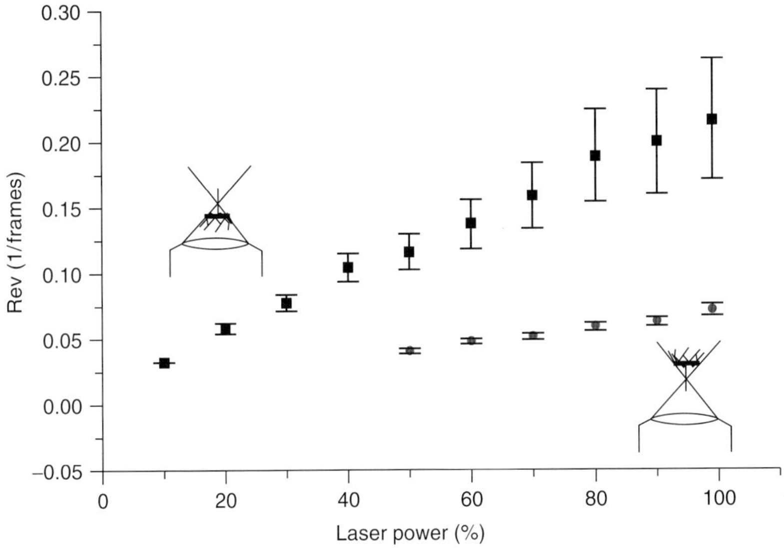

Figure 7.4 Dependence of the rate of revolution upon laser power. The two plots correspond to the two indicated positions the rotor can assume in the laser trap. (Please find a color version of this figure on the color plates.)

is easy for regular three-dimensional shapes, for example, for a sphere of radius r, $D = 8\pi\eta r^3$ at viscosity η. Our case is more complicated: The structures are built up from cylinders, and here the drag force is not simply linearly proportional to the velocity. The viscous drag force on a cylinder of length l moving in a direction perpendicular to its axis with velocity v is

$$F = \frac{4\pi vl}{\frac{1}{2} - C - \ln\frac{Rv\rho}{4\eta}} \tag{7.1}$$

where $\eta = 3.2 \times 10^{-4}$ kg (m s)$^{-1}$, ρ is the density of the liquid medium ($\rho =$ 790 kg m^{-3} for acetone), R is the diameter of the cylinder, and C is the Euler constant ($C = 0.577$). The drag torque on a complex structure can be calculated numerically. For the shape of our propeller (Figure 7.3) at the observed $\omega = 7$ s^{-1} at 10 mW power, the viscous drag torque is 3.6×10^{-17} Nm, which is in good agreement with the former estimate, especially if we remember that only a rough agreement can be expected because of the ambiguous position of the propeller in the trap. In the size range of the light-driven rotor, the linear correlation between speed and drag torque is a good approximation across a wide range of velocities, hence an effective viscous drag torque coefficient can be calculated. For the rotor in Figure 7.3 with a diameter of 10 µm and 10 s^{-1} rotation, we get a torque of 5.7×10^{-17} Nm. The equivalent sphere has a radius of 4.5 µm. This is a remarkable result in itself: the equivalent radius has just the size of the propeller: it seems that in the low Reynolds-number environment, the actual shape of the particle is less important for the determination of the viscous drag; it is the overall size that matters.

7.3.2
Propeller with Reversed Direction of Rotation

The propeller discussed before is analogous to a windmill rotated by the blowing wind. In this case, the direction of rotation is given by the shape of the propeller for a given wind direction, and there is no way to reverse it. Still, for optical tweezers, there are some strategies we can use. If we look closer, we realize that the analogy is limited; there is a fundamental difference between wind and our optical tweezers. Wind is a moving mass of air with parallel streamlines (more or less, even if the flow is turbulent), so a closer analogy would be a collimated light beam. In contrast, our propeller is held and driven by a focused beam of high numerical aperture. One basic difference between the two scenarios is that in the focused case, light carries momentum in the direction perpendicular to the axis of propagation. In fact, for a high numerical aperture beam, a large part of the total momentum falls into this category. In addition, for a particular light ray, the component perpendicular to the propagation may even have an opposite sign in front of and behind the focus: in front of the focus this perpendicular component points toward the axis, while behind the focus it points away from the axis. This beam geometry makes it possible to realize light-induced bidirectional rotation with appropriate rotor design.

We built rotors that utilized this "radial" component of light [28]. The shape of the rotor has to be such that the in-axis component of the light does not transfer momentum, while in the perpendicular direction, the transfer should be maximal. A logarithmic spiral in the plane perpendicular to the axis is such a shape. The characteristics of this spiral is that at any point of the curve, the tangent makes an angle identical to the radius drawn to this point and this angle characterizes the curve. It is easy to see that if light propagating in or out in the radial direction is reflected away from a structure with this shape, torque transfer will be maximum if the above angle is 45°. Figure 7.5 illustrates the concept of how the planar rotor uses the radial component of the light in the optical tweezers, and how it can rotate in two directions if positioned on either side of the focus. We included several features in the design to fulfill the requirements of the practical realization of this rotor. To ensure a stable axial position of the rotation center of the rotor in the laser tweezers, an axis was added to the rotor with asymmetry in the in-axis direction (Figure 7.6). This rotor can assume two stable positions in the trap: we select the one where the shortest part of the axis is pointing toward the microscope objective. Here, the equilibrium position of the rotor is between the focus and the objective. The axial position of the rotor is changed by moving the objective in the axial direction (focusing the microscope): moving the objective lens away from the sample pulls the rotor toward the microscope coverslip. When it reaches the glass surface and the objective is pulled further, the trapped body is pushed toward the focus and then beyond. Thus, the axial position of the rotor can be easily changed, relative to the focus.

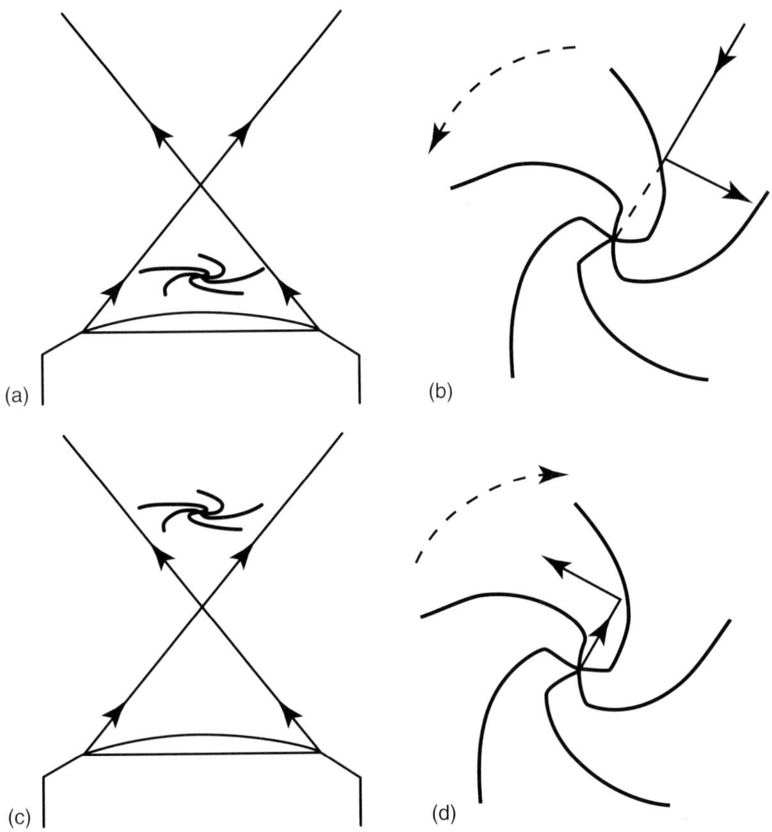

Figure 7.5 The concept of the rotor that can reverse the rotational direction depending on its relative position in the focus. (a) Position of the rotor in the focused laser beam, (b) characteristic reflections to drive rotation in one direction, and (c,d) respective equivalent figures for rotation in the opposite direction.

The rotor rotates in the equilibrium positions. The rate of rotation is linearly proportional to the laser power, for 20 mW the frequency is 2 s^{-1}, and the direction is counterclockwise as expected for the shape in Figure 7.6 and as depicted in Figure 7.5a,b (i.e., direction of momentum pointing toward the axis). On the other side of the focus, the rotation changes direction as shown in Figure 7.5c,d (the whole effect can be best observed in a real-time movie, provided on the web page *www.brc.hu/ormosgroup*). We modeled the behavior of the rotor with a ray-tracing simulation. The driving light is represented by a large number of individual rays and the momentum transfer for each of them is calculated for the event when the ray first hits the surface of the rotor. The observed effect is a sum of the impacts of all these rays. This approach gives a good description of the system; that is, the dynamics of the rotation is well explained and understood.

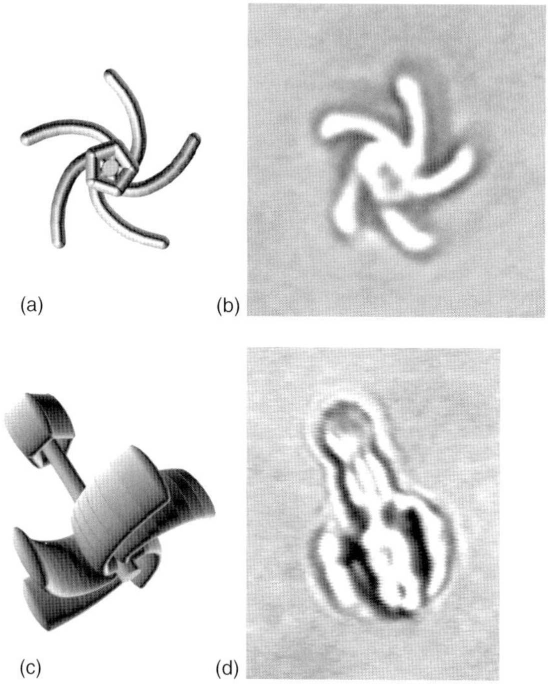

Figure 7.6 The rotor used in the experiments. (a) Schematic diagram, (b) photomicrograph of the rotor in an on-axis view, and (c,d) the same from an arbitrary viewpoint.

The described system demonstrates that the very high numerical aperture used in laser tweezers can produce interesting effects and by appropriate optical arrangement, the light-driven rotors can utilize these characteristics to make the optomechanical system significantly more flexible. This is most important for future applications. The above experiments also illustrate that by appropriately selecting the shape of rotors and system geometry, a great degree of control can be achieved over the motion of the trapped particle. Modifications can be made to make the application more practical in view of the particular task. These may include addition of a site where an object to be rotated can be fixed. Likewise, one can easily add an extension to convert the microrotor into a tool with a specific task (drill, mill, mixer, etc.) – the possibilities are numerous.

7.3.3 Complex Micromachines

The two-photon polymerization method and the light-driven rotors discussed here also offer the possibility of constructing more complicated micromechanical systems. We explored and tested the concept by building complex micromechanical systems that are built and driven by light. As a test and illustration of the general idea, we built systems where light-driven rotors (as seen before) drive cogwheels. Both the rotors and the cogwheels are produced simultaneously. The cogwheels rotate on axes. They are fabricated together with the axis on which they rotate. The fabrication process is straightforward and based on photopolymerization: first the axis is drawn such that it is fixed to the glass coverslip of the sample compartment (it grows out of the glass surface). Subsequently, the cogwheel is built onto the axis, so that it cannot get detached. The light-driven rotor is freely floating, that is, it is held by the laser tweezers. As the concept was to show the basic elements of complex micromechanical devices, we assembled systems where several cogwheels are engaged to each other and are rotated by the light-driven rotor. Several layouts were built and tested and these examples are shown in Figure 7.7.

The rotor and cogwheel are very easy to engage (by appropriately positioning the rotor with the laser tweezers) and the rotor turns the cogwheels: the machine

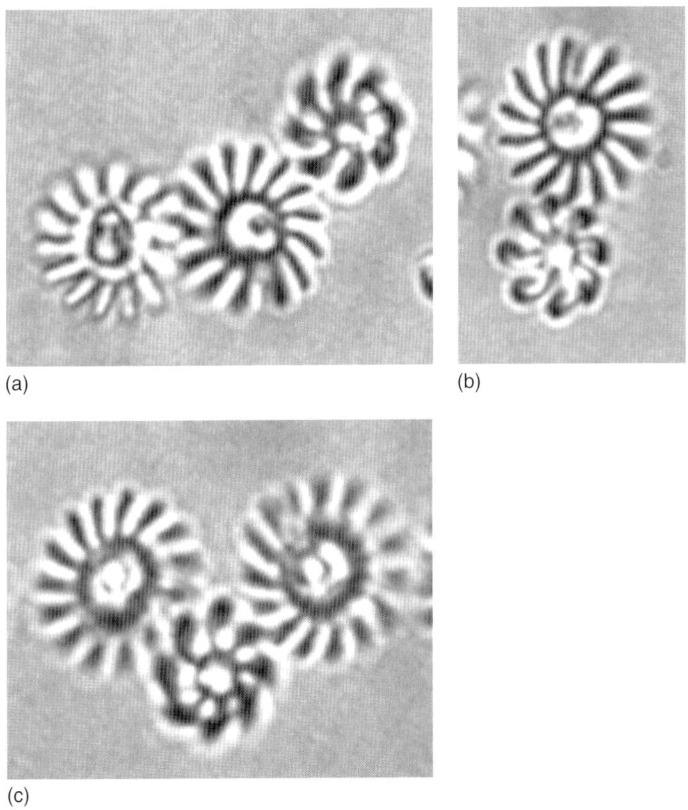

Figure 7.7 Complex micromachines built by the two-photon technique: cogwheels are rotated by a light-driven rotor as shown in Figure 7.3. The rotor is held and rotated by the laser tweezers and the rotating propeller drives the system. Several arrangements are shown: one, two, and two-engaged cogwheels are rotated by the light-driven rotor.

"works." Of course, the rotor can also be fixed to an axis – in this case, the light provides only the driving force and rotation in both directions can be achieved. More complicated machines consisting of both static and mobile parts can be constructed (forming micropumps, switches, etc.) in a straightforward manner. The example shows that in the same compartment, driving rotors and mechanisms of significant complexity can be put together.

The above rotors and machines demonstrate the power of two-photon photopolymerization, and if combined with light-induced rotation, this offers a highly promising method to construct light-driven machines on a micrometer scale.

7.4
Integrated Optical Motor

In the case of the devices introduced so far, the light-driven rotors were all such that light not only provided the driving force, but also held them in position: the optical tweezers were a fundamental part of the system. While this arrangement is practical when the added manipulation potentials of the optical tweezers are used – for example, we can move around the rotating propeller freely – in a number of cases, it may be a drawback that a complicated bulky device, such as a microscope, is needed for the system to work. In an alternative concept, all components (moving and actuating) are integrated on a single chip and no additional actuating device is needed. Simple driving without the need for complicated microscopes and additional beam-shaping equipment would also offer advantages and would expand the application area of such "self-contained," optically controlled microfluidic systems significantly. If a possibly simple stand-alone microscopic device, for example, a microfluidic lab-on-a-chip system is to be constructed, this latter approach may be of advantage. In the process of exploring the possibilities of light-actuated complex micromechanical systems, we have developed the prototype of a fully integrated optical motor, where all components are built on a glass substrate surface [18]. Instead of the focused beam of an optical trap, the light that drives the motor is guided in a surface-supported optical waveguide that is also an integral part of the system.

Two-photon excited photopolymerization was used here too, to create all components. Two types of photopolymers were used: the mechanical machine was built of SU8 (index of refraction: 1.63), because it has excellent mechanical qualities after polymerization. But on the other hand, it absorbs considerably in the visible region (and we used green (532 nm) light to drive the system) and therefore, was not appropriate for the several millimeter-long optical waveguide. To build the waveguide, Norland 81 optical adhesive was used (refractive index: 1.56 after curing) with excellent transmission in the visible range.

The design of the motor is shown in Figure 7.8. The whole system is constructed on the surface of a glass: a microscope coverslip of 170 μm thickness. A cogwheel-shaped rotor is held by a fixed, polymerized axis structure. The light to drive the rotor is carried by the optical waveguide and hits the rotor in a tangential

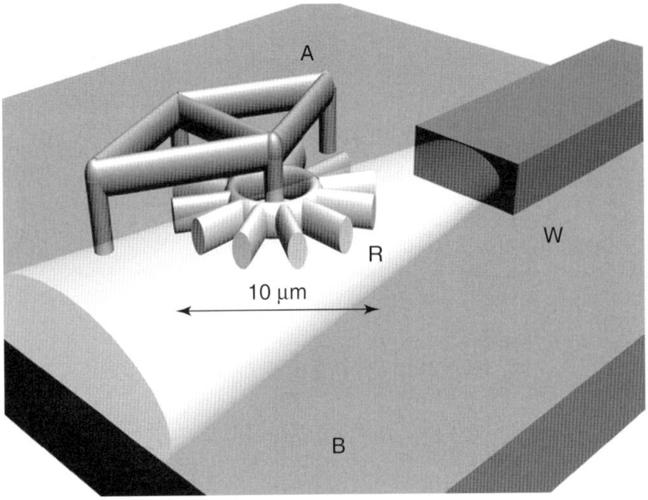

Figure 7.8 Design of the integrated optical motor. B, cover glass baseplate; W, optical waveguide; A, axis assembly; R, rotor. The diameter of the rotor is 10 μm.

direction. The rotor was driven by the 532-nm light of a diode-pumped solid-state laser (Verdi, Coherent, USA).

The construction process consisted of two steps. First, the waveguide was built using the following procedure. The coverslip was spin-coated with the Norland adhesive to give a film thickness of 10 μm. This is somewhat thicker than the final height of the waveguide. Then, the waveguide was drawn into the photopolymer layer. The non crosslinked resist was dissolved. This 10 μm wide waveguide carried the driving light with minimal loss over a distance of several millimeters. In the next step, the SU-8 photoresist was spin-coated over the substrate with the light guide already in place. The rotor assembly was built from the SU-8 photoresist by the three-dimensional drawing method introduced earlier. Special attention was paid to the precise positioning of the cogwheel structure relative to the tip of the light guide.

Figures 7.9 and 7.10 show the final structure with both the rotor assembly and the optical waveguide. Figure 7.9 is a bright-field transmission image taken with an optical microscope, whereas Figure 7.10 is a scanning electron micrograph that shows the structure in much more detail.

The motor, immersed in water, is driven by light carried to the rotor by the waveguide shown in the figures. A single-mode optical fiber was used to deliver the light to the system. The fiber was coupled to the device by placing the fiber at the end of the waveguide using a micromanipulator. The fiber position was optimized visually for maximum intensity at the output of the waveguide. A satisfactory coupling could be achieved: typically about 25 ± 8 mW ($50 \pm 15\%$) of the light from the fiber was coupled to the waveguide.

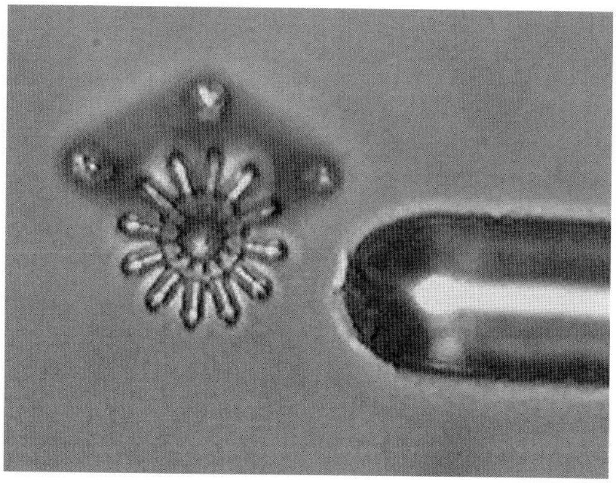

Figure 7.9 The completed integrated motor imaged in the light microscope.

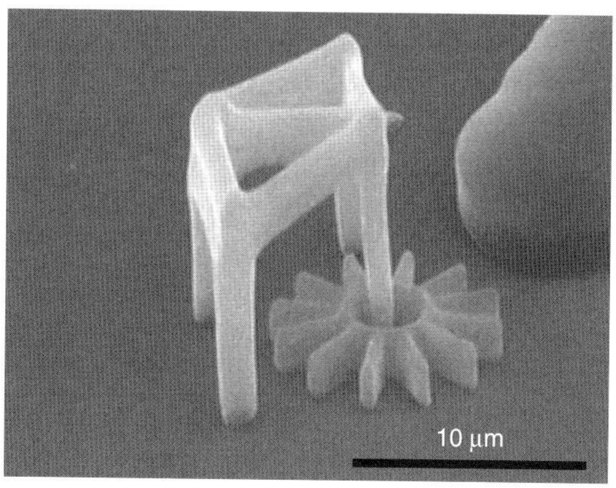

Figure 7.10 Scanning electron microscope image of the integrated optical motor.

The motor rotated even at fairly low light intensities: it started to rotate at about 5–10 mW power at the rotor. Above this threshold, the rotation rate increased roughly linearly with the power; at the maximum of 25 mW the rate was 2 Hz. We determined the efficiency of the motor by estimating the torque originating from the light hitting the rotor blades, and also by estimating the viscous drag of water. First, the forces due to reflection were estimated. The shape of the output beam from the waveguide was determined in independent experiments on identical waveguides where the output beam was visualized using fluorescent dyes. The output light has a numerical aperture of $\sim$0.07. The reflection forces were estimated by using the difference between the indices of refraction of water and the solidified resin of 1.33 and 1.63, respectively. Assuming a planar rotor blade surface and taking into account the reflections on five blades of different angles with respect to the direction of the driving light, the torque is estimated to be $6.2 \pm 2.1 \times 10^{-18}$ Nm at the maximum intensity. The viscous drag was calculated by approximating the rotor with a disc with identical height and diameter. Using the observed rotation rate of 2 s^{-1} at the maximum driving intensity, we estimate it to be $1.9 \times 10^{-17} \pm 3.8 \times 10^{-18}$ Nm.

If we compare these torque values with those measured on other propeller systems driven in optical tweezers or flat particles oriented by tweezers formed by linearly polarized light (discussed later), we see that in terms of the torque achieved they are comparable.

7.5
Angular Trapping of Flat Objects in Optical Tweezers Formed by Linearly Polarized Light

In addition to rotation of objects of specific shapes, orientation is also an interesting manipulation possibility. Orientation of a trapped object in optical tweezers is possible based on the interaction of the focused laser beam with the anisotropic optical character of the object. Several methods have been reported previously. For example, if the shape of the object is elongated (e.g., it is a rotational ellipsoid), it will be grabbed with the long axis aligned along the optical axis [29–31]. Orientation around the optical axis of the trapping beam has been described for birefringent particles: if held in the focus formed by linearly polarized light, they can be orientated [8, 31]. If such birefringent particles are trapped by laser tweezers formed by circularly polarized light, they will rotate and the direction of rotation is determined by the direction of the circular polarization [8, 31]. Several additional rotation mechanisms were also reported [14, 27, 28, 32–34].

It was observed earlier that flat objects can be oriented in the optical tweezers if the trap is formed by an anisotropic beam. This can be realized by various beam-shaping methods, for example, simply placing a rectangular, elongated aperture in the light path [35].

Experience shows that there is a fairly simple additional way to align particles around the optical axis. A particle with a flat shape will be oriented in a trap formed by linearly polarized light, even if its material is not birefringent. This is easy to understand in general: consider the ray optics regime, just remembering that the reflection and transmission at the interface of two materials depend on the polarization of the incident light. There were earlier observations that pointed to this angular trapping effect. The anisotropic character of the trap formed by linearly polarized light was investigated before [36]. It was calculated that the trapping force depends slightly upon the direction: the lateral restoring force on a spherical bead with a diameter almost equal to the wavelength is about 10% larger in the direction of polarization. It was also reported that small rod-shaped particles held in optical tweezers pushed against the glass coverslip surface (so that they are perpendicular to the direction of the laser beam) can be orientated by linearly polarized light [37].

We studied the orientating torque of linearly polarized light acting on flat particles. In this process, we created microscopic objects that were designed and built to be convenient to manipulate by optical tweezers utilizing the orienting effect of a linearly polarized beam [38].

We modified our optical tweezers system to enable such studies. We controlled the polarization state of the beam by positioning optical retarder plates in the light path before the infinity-corrected objective of the microscope. A quarter-waveplate in a rotation stage was used to change the polarization state from linear to circular. In addition, a half-waveplate was also added to rotate the direction of the linear polarization. The half-waveplate was rotated by an electric motor and the instantaneous position of the plate was known at all times.

We found that linearly polarized light indeed orients flat objects. What is required for this effect is simply for the trapped particle to be flat as observed in the direction of light propagation. For the quantitative analysis of the phenomenon, we produced test particles with a well-defined shape. We have chosen a cross consisting of two perpendicular axes of different lengths as shown in Figure 7.11. This particle is grabbed in the optical tweezers, preferably with the longest part pointing along the optical axis. The short axis is orientated by the polarized light: in equilibrium, it is parallel to the direction of polarization.

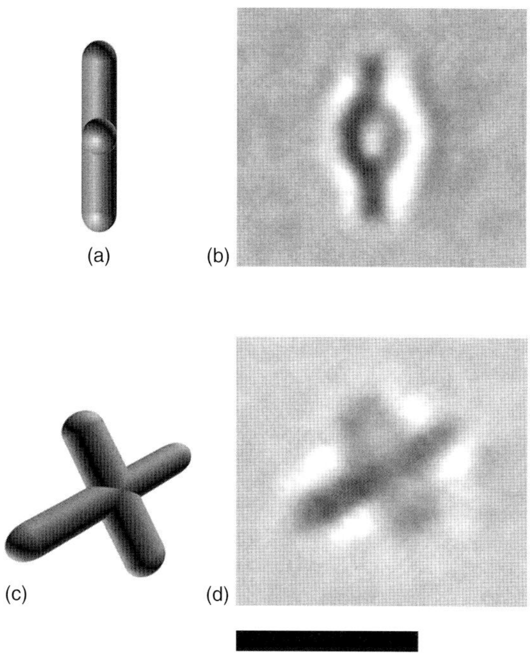

Figure 7.11 Cross-shaped test object to study angular trapping in optical tweezers formed by linearly polarized light. (a,c) Schematic diagram of the object from two different views. (b,d) Photomicrographs of the fabricated objects viewed from respective directions. The scale bar is 3 μm long.

We measured and characterized the torque exerted on the test object. By rotating the half-waveplate with a rate ω, the plane of polarization rotates with 2ω. Owing to the viscous drag caused by the rotation, the particle follows the polarization with a phase delay, and this delay is increasing with increasing rotation rate. The drag torque on the rotating cross was calculated as described earlier, using formulae given for cylinders moving in viscous liquids. The orientating torque of the polarized light was calculated based on the assumption that the origin of the torque is the polarization dependence of the refraction and reflection of light on the surface of the particle, as described by the Fresnel formulae. These equations give the reflected and transmitted electromagnetic power for polarization directions parallel and perpendicular to the plane of incidence. We applied ray optics using these formulae to numerically calculate the torque acting on the test object; that is, we did not take into account the fact that the particle size is in the order of the wavelength of light.

The result of the experiments and the calculations is shown in Figure 7.12. The experimental and calculated values agree well with each other when using realistic parameters (the following parameters were used: 20 mW laser power, index of refraction of 1.36 for the medium and 1.56 for the particle with a length of 2.6 μm, elliptical cross section with axes of 1 μm in the direction of the light propagation, and 0.7 μm perpendicular to it). The calculation shows that the torque increases to a phase delay of about 45°. Beyond this threshold the system becomes unstable: further increase in the phase delay causes a decrease in the torque, so the particle cannot follow the rotation of polarization and therefore, it slips. This phenomenon is also observed in the experiments.

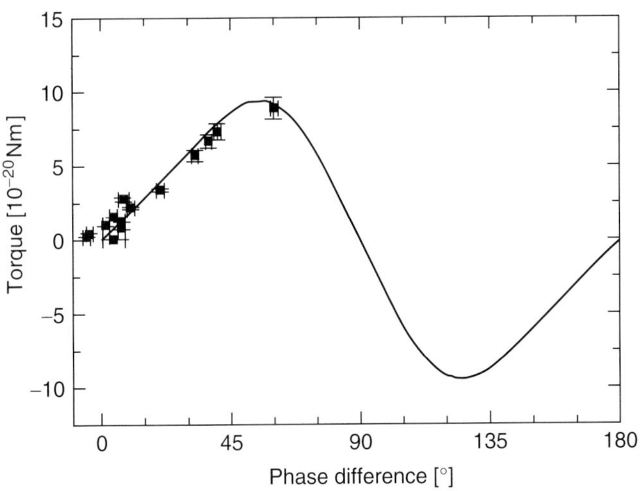

Figure 7.12 Torque exerted by the polarized light upon the trapped body as a function of the angle between the polarization plane and the long axis of the trapped flat object. ■: measured data, – model calculation.

The results show that this orientation mechanism yields a torque in the order of 10^{-19} Nm for realistic system parameters. This torque is suitable for torsional manipulation of biological objects such as ATPase motors [39] or DNA molecules [40].

The angular trapping effect provides the means to align objects in the optical tweezers. By changing the plane of polarization with the half-waveplate, the anisotropic object can be precisely oriented and/or rotated in the focus. This effect can be used to align nonspherical objects – and since real biological objects are generally not perfectly spherical, it offers an additional manipulation tool for biology. We have explored this potential for typical microscopic biological objects, for example, chromosomes and chloroplasts [41].

The presented cross-shaped microscopic objects make the application of the phenomenon, as a manipulation tool, practical. They are held in stable axial position in the focus, consequently moving the particle to arbitrary locations and rotation around the axis can be controlled separately. Since the torque exerted by the laser beam on the trapped object is known, this system can also be used to exert and measure torque on biological objects. And since the torque observed for realistic laser powers and particle shapes and sizes falls exactly within the range predicted for and observed in biological systems, we are dealing with a method that holds great promise for biology.

7.6
Torsional Manipulation of DNA

The effect described before, the angular trapping of flat particles in polarized light, can be used as an optical torque wrench. As an illustration of the method, we show how we used it to measure the torsional stiffness of double-stranded DNA.

In general, the rotational manipulation of biopolymers (like DNA and proteins) is a most interesting topic; for numerous biological problems, torsional manipulation is an indispensable tool in crucial experiments. A number of groups have developed various experimental techniques for this purpose. Different methods were developed to twist, for example DNA molecules, with indirect and direct ways to measure the torque. Magnetic beads [42, 43], multiple beads attached to different segments of the molecule and held in optical tweezers [44], birefringent objects manipulated in circularly and linearly polarized light [45], just to name the most characteristic approaches. Our method is an easy-to-implement alternative for the direct measurement of torque based on flat test particles.

In our system, the DNA molecule (λ-DNA of 15.6 μm contour length) is attached to a flat disc-shaped polystyrene particle, approximately 2 μm in diameter, at one end and to a fixed plastic surface at the other end. The flat particles that were to act as orientating tools were produced from plastic beads. Polystyrene microspheres of 1-μm diameter have been squeezed mechanically to form discs.

The linearly polarized trapping light has an angular trapping effect on the anisotropic test particle [29, 38, 46, 47]. By rotating the plane of polarization, the orientation of the disc can be changed, thus the attached molecule can be twisted and torsionally stressed. As turns are added, the orientation of the disc drops behind the plane of polarization, because the torsional strain of the molecule acts against the orientating power of the light. By measuring the relevant angles (using video analysis), the molecular torsional stiffness can be directly compared to the angular trapping power of the light. After calibrating the angular trap, the torsional modulus can be determined.

We used the pH-induced nonspecific binding of DNA to polystyrene to fix the DNA to the coverslip surface as well as to the flat discs. An MES puffer of 50 mM, pH 5.0 was used in the experiments. Two microliters of the λ-DNA buffer and 5 μl of the "disc suspension" was added to 100 μl MES. This mixture was dispensed between two coverslips separated by a 150-μm thick spacer and incubated for several hours. The lower coverslip had a thin plastic layer on it, created prior to incubation by spin-coating with polystyrene–toluene (50 mg ml^{-1}) solution at 2500 rpm for 30 s [48]. During incubation, the DNA molecules attached to the plastic layer and/or to the discs with a certain probability [48]. Ultimately, the sample contained a significant number (approximately 10%) of discs that were connected to the surface via a single DNA molecule. The probability that a disc is connected to the surface by two or more DNA molecules is negligible at these concentrations.

7.6.1 Direct Measurement of Torque

The anisotropic object is trapped in the laser focus in a translational and angular sense simultaneously. The object has an equilibrium position in the translational trap, around which Brownian fluctuations can be observed. Similarly, the object has an equilibrium orientation in the trap (defined by the polarization plane of the trapping light and the anisotropic properties of the object) around which rotational Brownian motion can be observed.

In our case, the translational trap is strong compared to the angular trap, so the translational Brownian fluctuations can be neglected. The characterization of the angular trapping is fairly straightforward. If the trap is linear, the torque τ exerted by the light beam is proportional to the angle α between the equilibrium (no external torque) and the actual orientation of the object. The proportionality factor is the angular trapping constant k.

$$\tau = -k\alpha \tag{7.2}$$

In this case, the potential energy E of the trapped object is harmonic:

$$E = \frac{1}{2}k\alpha^2 \tag{7.3}$$

The object fluctuates around its equilibrium orientation due to Brownian motion. According to the Boltzmann energy distribution, the probability density function of α is Gaussian:

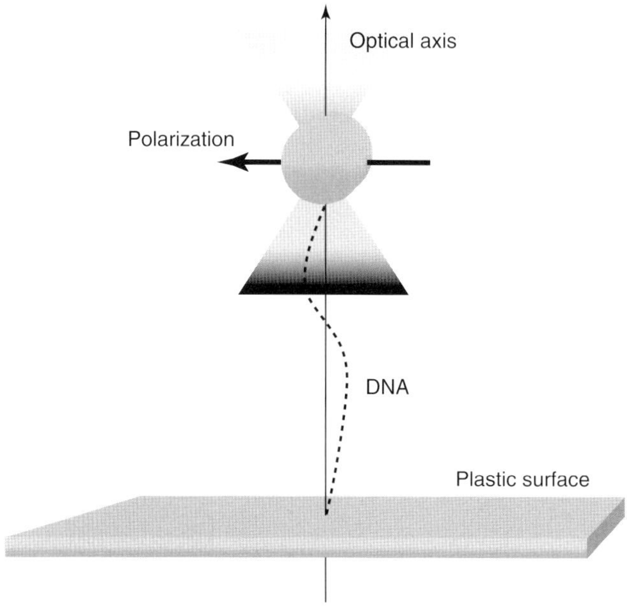

Figure 7.13 Arrangement of the DNA twisting experiment. The trapped disc is orientated by the linearly polarized light. By rotating the polarization plane the disc can be rotated; thus, the attached DNA molecule can be twisted. The orientation of the disc is determined from images with a standard view along the optical axis.

$$\rho(\alpha) \propto e^{-\frac{E}{k_B T}} = e^{-\frac{k\alpha^2}{2k_B T}} \tag{7.4}$$

where T is the temperature and k_B is Boltzmann's constant. By monitoring α over a sufficiently long period of time, the density function of α can be derived and fitted by a Gaussian, and thus k can be determined. Hence, the angular trap can be calibrated.

If a molecular strand links the disc to a fixed surface (i.e., a coverslip) with nonrotating bonds (Figure 7.13), an additional torque is generated by the torsional strain of the molecule. In this case, the orientation of the trapped object will fluctuate around a new equilibrium orientation (EQ) determined by both the direction corresponding to the zero torsional strain of the molecule (M) and the polarization plane of the trapping light (P) (Figure 7.14). In the equilibrium state (EQ) the torques generated by the molecule and the light are equal: $\tau_M = \tau_L$. Each torque can be expressed as the product of its respective angular trapping constant and the angle between its equilibrium state (M or P) and the EQ state:

$$k_M \alpha_M = k_L \alpha_L \tag{7.5}$$

where k_M is the trapping constant (torsional spring constant) of the molecule and k_L is the trapping constant of the polarized light. This means that by knowing the

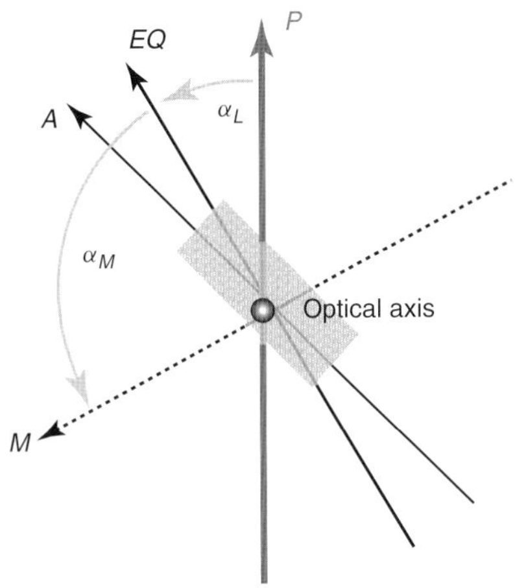

Figure 7.14 Explanation of the measurement of torque and torsional elasticity. M denotes the orientation where the molecular torsional strain vanishes, P indicates the plane of polarization of the trapping light, EQ is the equilibrium state where the two orienting torques (molecular and optical) cancel. The orientation of the trapped object fluctuates around EQ due to Brownian motion, A represents the actual orientation at a certain time. (Please find a color version of this figure on the color plates.)

relevant angles, the trapping constant of the molecule can be compared to that of the light directly.

The orientation of the trapped object fluctuates around EQ due to rotational Brownian motion. If both torques are linear to the angle, the effective angular trapping constant k_{eff} is the sum of k_M and k_L:

$$k_{\text{eff}} = k_M + k_L \tag{7.6}$$

The potential energy remains harmonic and k_{eff} can be obtained by analyzing the rotational Brownian motion of the trapped object (using Eq. (7.4)).

By combining Eqs. (7.5) and (7.6) k_M could be obtained if we were able to measure α_M and α_L angles. However α_M is unknown and it is technically easier to measure the change of angles rather than absolute values in our system. If we rotate the plane of polarization (P) by Δ_P, the equilibrium state of the trapped object (EQ) shifts by Δ_{EQ}. Using Eqs. (7.5) and (7.6) again, the following expression can be derived for k_M:

$$k_M = k_{\text{eff}} \left(1 - \frac{\Delta_{\text{EQ}}}{\Delta_P} \right) \tag{7.7}$$

Finally, the torsional modulus G of the molecule is (assuming no supercoiling takes place):

$$G = k_M l \tag{7.8}$$

where l is the (contour) length of the molecule.

In the case of dsDNA, the torsional spring constant of the molecule is three orders of magnitude smaller than the angular trapping constant of light in our system. In order to measure this small effect, we twisted the DNA molecule several times in positive and negative directions symmetrically while staying in the linear torsional regime of the polymer [44]. The measurement proceeded as follows: The plane of polarization and thus the disc itself was rotated by several turns in one direction with a relatively high speed ($180^\circ\ s^{-1}$) at maximum laser power. After this, the power of the trapping laser was reduced to a level where the Brownian fluctuations of the disc became visible (and detectable by image analysis). Here, we started to rotate the plane of polarization at a much lower speed ($3.6^\circ\ s^{-1}$) and added two more turns (one turn for the $\lambda/2$ plate). From the data (orientation of the disc) recorded during these two turns (200 s), we calculated the equilibrium orientation and analyzed the Brownian fluctuations in order to determine the effective angular trapping constant k_{eff}. Next, the laser power was set again to maximum and the polarization was rotated back to its initial state, and the same procedure was repeated in the opposite direction. Knowing the shift of the equilibrium angle measured at the two twist extrema Δ_{EQ}, the change of the polarization Δ_P and k_{eff}, the torsional modulus of the molecule could be calculated using Eqs. (7.7) and (7.8).

We note that k_M can be determined in principle without any additional rotation at the twist extremities by simply analyzing the angular fluctuations of the disc at fixed polarization planes. However, rotating the polarization during the averaging period is an efficient way to reduce slight systematic anisotropic errors present in the experimental apparatus (the detection of the very small additional torque originating from the DNA requires extreme accuracy).

We measured the torsional stiffness at different relative extensions of the molecule. At medium extension, the torsion applied to the polymer is distributed in bending and torsional deformations. Using the statistical mechanics model of twist-storing polymers [49, 50], the local torsional modulus of dsDNA (the case when no bending is allowed) could be calculated. The local torsional modulus was determined to be 420 ± 44 pN nm^2 (corresponding to a value of 102 ± 10 nm for the twist persistence length of the polymer).

7.7
Conclusion

In the introduced examples, we have shown how objects of special shape can enhance the scope of optical micromanipulation. We have developed three-dimensional fabrications by laser-induced photopolymerization to a level where structures of arbitrary complexity can be built with submicron spatial resolution. These structures offer exciting applications to achieve new types of manipulations, and control more degrees of freedom. They also allow building of complex structures and micromechanical machines, where light can be used as a driving/controlling agent. We believe that this approach opens up new avenues for scientific studies, as well as technological development.

Acknowledgment

This work was supported by grant Országos Tudományos Kutatási Alap **NK72375**.

References

1. Ashkin, A. (1970) Acceleration and trapping of particles by radiation pressure. *Phys. Rev. Lett.*, **24**, 156–159.
2. Ashkin, A., Dziedzic, J.M., Bjorkholm, J.E., and Chu, S. (1986) Observation of a single-beam gradient force optical trap for dielectric particles. *Opt. Lett.*, **11**, 288–290.
3. Yin, H., Wang, M.D., Svoboda, K., Landick, R., Block, S.M., and Gelles, J. (1995) Transcription against an applied force. *Science*, **270**, 1653–1657.
4. Smith, S.B., Cui, Y., and Bustamante, C. (1996) Overstretching B-DNA: the elastic response of individual double-stranded and single-stranded DNA molecules. *Science*, **271**, 795–799.
5. Kellermayer, M.S.Z., Smith, S.B., Granzier, H.L., and Bustamante, C. (1997) Folding–unfolding transitions in single titin molecules characterized with laser tweezers. *Science*, **276**, 1112–1116.
6. Svoboda, K., Schmidt, C.F., Schnapp, B.J., and Block, S.M. (1993) Direct observation of kinesin stepping by optical trapping interferometry. *Nature*, **365**, 721–727.
7. Finer, J.T., Simmons, R.M., and Spudich, J.A. (1994) Single myosin molecule mechanics: piconewton forces and nanometre steps. *Nature*, **368**, 113–119.
8. Friese, M.E.J., Nieminen, T.A., Heckenberg, N.R., and Rubinsztein-Dunlop, H. (1998) Optical alignment and spinning of laser-trapped microscopic particles. *Nature*, **394**, 348–350.
9. Gauthier, R.C., Tait, R.N., Mende, H., and Pawlowicz, C. (2001) Optical selection, manipulation, trapping, and activation of a microgear structure for applications in micro-optical–electromechanical systems. *Appl. Opt.*, **40**, 930–937.
10. Higurashi, E., Ukita, H., Tanaka, H., and Ohguchi, O. (1994) Optically induced rotation of anisotropic micro-objects fabricated by surface micromachining. *Appl. Phys. Lett.*, **64**, 2209–2210.
11. Voldman, J., Gray, M.L., and Schmidt, M.A. (1999) Microfabrication in biology and medicine. *Annu. Rev. Biomed. Eng.*, **1**, 401–425.
12. Gower, M.C. (2000) Industrial applications of laser micromachining. *Opt. Express*, **7**, 56–67.
13. Maruo, S. and Kawata, S. (1998) Two-photon-absorbed near-infrared photopolymerisation for three-dimensional microfabrication. *J. Microelectromech. Syst.*, **7**, 411–415.
14. Galajda, P. and Ormos, P. (2001) Complex micromachines produced and driven by light. *Appl. Phys. Lett.*, **78**, 249–251.
15. Kawata, S., Sun, H.B., Tanaka, T., and Takada, K. (2001) Finer features for functional microdevices. *Nature*, **412**, 697–697.
16. Straub, M., Nguyen, L.H., Fazlic, A., and Gu, M. (2004) Complex-shaped three-dimensional microstructures and photonic crystals generated in a polysiloxane polymer by two-photon stereolithography. *Opt. Mater.*, **27**, 359–364.
17. Maruo, S., Nakamura, O., and Kawata, S. (1997) Three-dimensional microfabrication with two-photon-absorbed photopolymerisation. *Opt. Lett.*, **22**, 132–134.
18. Kelemen, L., Valkai, S., and Ormos, P. (2006) Integrated optical motor. *Appl. Opt.*, **45**, 2777–2780.
19. Sun, H. (2000) Real three-dimensional microstructures fabricated by photopolymerization of resins through two-photon absorption. *Opt. Lett.*, **25**, 1110–1113.
20. Maruo, S., Ikuta, K., and Korogi, H. (2003) Force-controllable, optically driven micromachines fabricated by single-step

two-photon microstereolithography. *J. Microelectromech. Syst.*, **12**, 533–539.

21. Kelemen, L., Valkai, S., and Ormos, P. (2007) Parallel photopolymerisation with complex light patterns generated by diffractive optical elements. *Opt. Express*, **15**, 14488–14497.
22. Feng, R. and Farris, R.J. (2003) Influence of processing conditions on the thermal and mechanical properties of SU8 negative photoresist coatings. *J. Micromech. Microeng.*, **13**, 80–87.
23. Matsuo, S., Juodkazis, S., and Misawa, H. (2005) Femtosecond laser microfabrication of periodic structures using a microlens array. *Appl. Phys.*, **A 80**, 683–685.
24. Rodrigo, P.J., Kelemen, L., Alonzo, C.A., Perch-Nielsen, I.R., Dam, J.S., Ormos, P., and Glückstad, J. (2007) 2D optical manipulation and assembly of shape-complementary planar microstructures. *Opt. Express*, **15**, 9009–9014.
25. Rodrigo, P.J., Kelemen, L., Palima, D., Alonzo, C.A., Ormos, P., and Glückstad, J. (2009) Optical microassembly platform for constructing reconfigurable microenvironments for biomedical studies. *Opt. Express*, **17**, 6578–6583.
26. Gauthier, R.C. (1995) Ray optics model and numerical computations for the radiation pressure micromotor. *Appl. Phys. Lett.*, **67**, 2269–2271.
27. Galajda, P. and Ormos, P. (2002a) Rotation of microscopic propellers in laser tweezers. *J. Opt. B.*, **4**, S78–S81.
28. Galajda, P. and Ormos, P. (2002b) Rotors produced and driven in laser tweezers with reversed direction of rotation. *Appl. Phys. Lett.*, **80**, 4653–4655.
29. Gauthier, R.C. (1997) Theoretical investigation of the optical trapping force and torque on cylindrical micro-objects. *J. Opt. Soc. Am.*, **B14**, 3323–3333.
30. Gauthier, R.C., Ashman, M., and Grover, C.P. (1999) Experimental confirmation of the optical-trapping properties of cylindrical objects. *Appl. Opt.*, **38**, 4861–4869.
31. Cheng, Z., Chaikin, P.M., and Mason, T.G. (2002) Light streak tracking of optically trapped thin microdisks. *Phys. Rev. Lett.*, **89**, 108303.
32. He, H., Friese, M.E.J., Heckenberg, N.R., and Rubinsztein-Dunlop, H. (1995) Direct observation of transfer of angular-momentum to absorptive particles from a laser-beam with a phase singularity. *Phys. Rev. Lett.*, **75**, 826–829.
33. Yamamoto, A. and Yamaguchi, I. (1995) Measurement and control of optically induced rotation of anisotropic shaped particles. *Jpn. J. Appl. Phys.*, **34**, 3104–3107.
34. Simpson, N.B., Dholakia, K., Allen, L., and Padgett, M.J. (1997) Mechanical equivalence of spin and orbital angular momentum of light: an optical spanner. *Opt. Lett.*, **22**, 52–54.
35. O'Neil, A.T. and Padgett, M.J. (2002) Rotational control within optical tweezers by use of a rotating aperture. *Opt. Lett.*, **27**, 743–745.
36. Wohland, T., Rosin, A., and Stelzer, E.H.K. (1996) Theoretical determination of the influence of the polarization on forces exerted by optical tweezers. *Optik*, **102**, 181–190.
37. Bonin, D.B., Kourmanov, B., and Walker, T.G. (2002) Light torque nanocontrol, nanomotors and nanorockers. *Opt. Express*, **10**, 984–989.
38. Galajda, P. and Ormos, P. (2003) Orientation of flat particles in optical tweezers by linearly polarized light. *Opt. Express*, **11**, 446–451.
39. Yasuda, R., Noji, H., Kinoshita, K. Jr., and Yoshida, M. (1998) F1-ATPase is a highly efficient molecular motor that rotates with discrete 120 degrees steps. *Cell*, **93**, 1117–1124.
40. Allemand, J.F., Bensimon, D., Lavery, R., and Croquette, V. (1998) Stretched and overwound DNA forms a Pauling-like structure with exposed bases. *Proc. Natl. Acad. Sci. U.S.A.*, **95**, 1452–1457.
41. Garab, G., Galajda, P., Pomozi, I., Finzi, L., Praznovsky, T., Ormos, P., and Amerongen, H. (2005) Alignment of biological microparticles by polarized laser beam. *Eur. Biophys. J.*, **34**, 335–343.
42. Smith, S.B., Finzi, L., and Bustamante, C. (1992) Direct mechanical

measurement of the elasticity of single DNA molecules by using magnetic beads. *Science*, **258**, 1122–1126.

43. Strick, T.R., Bensimon, D., and Croquette, V. (1999) Micro-mechanical measurement of the torsional modulus of DNA. *Genetica*, **106**, 57.
44. Bryant, Z., Stone, M.D., Gore, J., Smith, S.B., Cozzarelli, N.R., and Bustamante, C. (2003) Structural transitions and elasticity from torque measurements on DNA. *Nature*, **424**, 337.
45. La Porta, A. and Wang, M.D. (2004) Optical torque wrench: angular trapping, rotation and torque detection using quartz microparticles. *Phys. Rev. Lett.*, **92**, 190801.
46. Bayoudh, S., Nieminen, T.A., Heckenberg, N.R., and Rubinsztein-Dunlop, H. (2003) Orientation of biological cells using plane polarized Gaussian beam optical tweezers. *J. Mod. Opt.*, **50**, 1581–1590.
47. Bishop, A.I., Nieminen, T.A., Heckenberg, N.R., and Rubinsztein-Dunlop, H. (2003) Optical application and measurement of torque on microparticles of isotropic nonabsorbing material. *Phys. Rev. A.*, **68**, 033802.
48. Allemand, J.F., Bensimon, D., Jullien, L., Bensimon, A., and Croquette, V. (1997) pH-dependent specific binding and combing of DNA. *Biophys. J.*, **73**, 2064.
49. Moroz, J.D. and Nelson, P. (1998a) Torsional directed walks, entropic elasticity, and DNA twist stiffness. *Proc. Natl. Acad. Sci. U.S.A.*, **94**, 14418–14422.
50. Moroz, J.D. and Nelson, P. (1998b) Entropic elasticity of twist-storing polymers. *Macromolecules*, **31**, 6333–6347.

8 Spiral Phase Contrast Microscopy

Christian Maurer, Stefan Bernet, and Monika Ritsch-Marte

8.1 Phase Contrast Methods in Light Microscopy

In optical imaging, owing to the lack of endogenous pigments that can absorb in the visible region, biological samples typically have no strong intrinsic contrast; that is, the majority of samples represents thin phase objects with only small refractive index variations. Thus, a variety of staining techniques have been developed for microscopy. Fluorescence microscopy [1], for instance, has become an indispensable tool in biomedical research. Here, one often utilizes fluorescence markers that are designed to bind to specific structures in a living cell, thus making cell components or binding sites of selected molecules visible. A disadvantage is the fact that the preparation of the sample with fluorescence markers can also introduce undesired effects; for example, some markers are phototoxic and most of them have problems with photostability, that is, they bleach.

The basic idea behind phase contrast (PC) methods is to convert phase variations into intensity modulations by means of optical manipulations [2], allowing one to image native, unstained samples with good contrast. The first purely optical method of PC (apart from Schlieren imaging) was introduced by Zernike [3] in the early twentieth century, which brought him the Nobel Prize in Physics in 1953. In short, the elegant idea behind "Zernike PC" [4] with the famous Zernike phase ring is to shift the phase of the transmitted light by a quarter-wavelength with respect to the diffracted light.

For thin phase objects, one can expand the phase function $\Phi(x, y)$ of the transmission function $T(x, y)$ of the sample to the lowest order in a Taylor series

$$T(x, y) = e^{i\Phi(x,y)} \approx 1 + i\Phi(x, y) \tag{8.1}$$

This shows that (for thin phase objects where the approximation above is valid) light that is scattered from inhomogeneities in the optical thickness of the sample is a quarter-period out of phase with the unscattered light. To achieve good visibility for small refractive index variations on a homogeneous background, one can induce a relative phase shift of $\pm\,\pi/2$, which for positive or negative phase contrast

Twisted Photons: Applications of Light with Orbital Angular Momentum.
Edited by Juan P. Torres and Lluis Torner

ISBN: 978-3-527-40907-5

respectively, compensates the quarter period phase difference between higher and zeroth Fourier components.

Differential interference contrast (DIC) is another widespread optical method designed to enhance the contrast of phase samples [5, 6]. In the original scheme suggested by Normanski [5], two Wollaston prisms are implemented in the optical path of the microscope, which split the image wave into two orthogonal linear polarizations traveling at a slightly different angle (the "shearing angle") to the image plane. Since the shearing distances are very small, the method does not require coherent light.

DIC images are easily recognized by their typical "relief-like" appearance, giving the impression of a three-dimensional object that is illuminated from a particular direction (which is determined by the shearing vector induced by the Wollaston prism). Similar images can also be created by using, for example, oblique illumination [7] or amplitude gradient illumination [8] or by knife-edge or slit aperture techniques as in Schlieren contrast microscopy [9].

8.2 Fourier Filtering in Optical Imaging

In a wider sense, PC can be seen as spatial filtering in *k*-space, where one can "tailor" the transmission of various spatial frequencies. Using an illumination source with sufficient spatial coherence, the spatial frequencies are separated in any plane in the far-field, with the low frequency components (pertinent to the general outline of the structures in the sample) being in the center and the high frequency components (pertinent to the fine details in the sample) in the periphery. This provides one with a means to selectively manipulate specific spatial frequencies, by phase-shifting, attenuating, or blocking them. Normally this Fourier filtering, as it is called, is not really carried out in the far-field, but a suitably placed optical lens performs a Fourier transform [10] "pulling *k*-space into its focal plane."

The generic arrangement for Fourier filtering is indicated in Figure 8.1. The Fourier components are arranged around the focused spot of the illumination light representing the spatial carrier wave, also called *zero-order Fourier component* or *DC-component*, of the image, and thus can individually be modulated with a filter mask. Blocking of the zero-order Fourier component, for instance, results in dark field microscopy, where scattering structures such as edges appear bright on a dark background. Shifting the phase of the zero-order Fourier component by $\pi/2$ with respect to the remaining wave emulates (central) PC. It is also possible to emulate Zernike PC by phase-shifting an annular area representing the Zernike phase ring [11], which has to be matched to an illumination ring. We have recently suggested the use of spatial light modulators (SLMs) for emulating various classic and novel Fourier filters in microscopy [12–15]. The term SLM is often used to encompass several types of systems, for example, adaptive deformable mirror and digital micromirror devices, where actuators or membranes are moved by electric or magnetic fields, or liquid crystal displays (LCDs). The latter are miniaturized LCDs

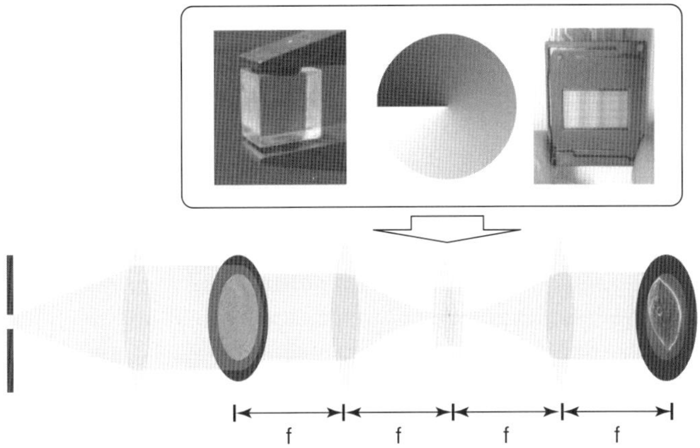

Figure 8.1 Generic setup for Fourier filtering in optical imaging: An object illuminated with a plane wave diffracts the light to outer regions in the Fourier plane, that is, the focal plane of the objective lens. One can access and specifically manipulate the various spatial frequencies there with a spatial light modulator or a phase plate. As an example, spiral phase filtering, which includes isotropic edge enhancement is shown in the upper part of the figure. (Please find a color version of this figure on the color plates.)

which can dynamically influence the amplitude and/or phase of light going through or being reflected from the panel. The active area is typically on the order of 2 cm^2 with a resolution of up to 2 million pixels. Both amplitude and phase modulations arise from the birefringence of the liquid crystal: Amplitude modulations originate from polarization modulations in combination with a subsequent polarizer. For state-of-the-art LCD–SLMs, phase modulations in the range of 2π are typically achievable. Wavefronts with much higher phase retardation cannot be generated; only their correspondingly "wrapped" versions can be generated, as in diffractive optical elements.

Often the Fourier filtering is not carried out "on-axis," as depicted in Figure 8.1, but one modifies the setup to an "off-axis" configuration, where – in order to get a clear image–only the first-order diffracted beam from a blazed grating is used for imaging. When using an SLM, the blazed grating can directly be implemented by simply superposing onto the hologram used for Fourier filtering each structure, where the phase rises linearly from 0 to 2π within one grating period and which diffracts the incoming wave to the CCD camera [16]. For a digitally displayed blazed grating approximated by eight modulation steps, the theoretical diffraction efficiency is above 95%. A minor disadvantage of the "off-axis configuration" of Fourier filtering is that one has to restrict the field of view to some extent, a consequence of the need for separating the zeroth from the first diffraction order of the grating.

In the off-axis configuration, it is possible to modulate both the phase (for phase contrast microscopy) and the amplitude (for dark field microscopy) of the SLM transmission function. The phase can be coded in the spatial position of the displayed grating structures, since this converts into a phase offset of the first-order diffracted wave. And, the amplitude can be spatially modulated by locally varying the *contrast* of the grating fringes. Dark field imaging is an extreme example, where the contrast in the center is set to zero.

8.3 Spiral Phase Fourier Filtering

Optical vortices have a helical phase profile and thus carry orbital angular momentum [17–19]. They have been investigated in various contexts [20–22]. In optical micromanipulation, phase holograms with a helical phase profile are used to create Laguerre–Gauss beams, which carry orbital angular momentum and can trap dielectric particles depending on the relative refractive index, either in the center or in the intensity ring [23–25].

In image processing, the spiral phase signature of an optical vortex has been used as a filter mask for edge enhancement [26]. This is in the spirit of a two-dimensional generalization of the Hilbert transform [27], which has also been used for numerical processing of images [28]. First attempts to implement a spiral phase filter in an optical setup had already been made earlier [29, 30]. In 2005, we realized that it is possible to use the Fourier filtering techniques to implement spiral phase filtering in optical microscopy, either by means of an SLM or a built-in phase plate [31, 32], which can *directly emulate* various contrast enhancing techniques without the need of numerical processing of the images [12, 33]. Finally, we would like to mention that the concept of spiral phase Fourier filtering, is also entering other areas of imaging, such as soft X-ray diffraction imaging [34, 35] or optoacoustics [36].

The transmission function of a spiral phase filter is given by $T(x, y) = \exp(i\varphi)$ with φ being the azimuthal angle (Figure 8.2). If such a filter is introduced into the Fourier plane of an image-carrying wave, the resulting field amplitude in the camera plane is given by the convolution of the image's Fourier transform with the complex transmission function of the sample:

$$E(x, y) = O(x, y) * \mathcal{F}\{e^{i\varphi}\} = O(x, y) * i\frac{e^{i\varphi}}{2\pi r^2} \tag{8.2}$$

Here, $O(x, y)$ is the complex object transmission function, $\mathcal{F}$ denotes the Fourier transform, and $*$ the convolution operation. $r = \sqrt{x^2 + y^2}$ and $\varphi = \arctan(y/x)$

Figure 8.2 Spiral phase filter: The phase shift produced by an SLM or a phase plate corresponds to the helical phase profile typical for a Laguerre–Gauss beam. Note that opposite points are out of phase by π. (Please find a color version of this figure on the color plates.)

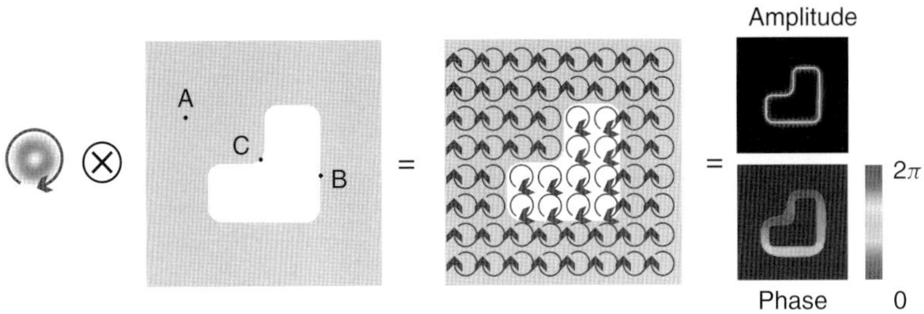

Figure 8.3 Graphical representation of the convolution of an object with a spiral phase filter: When the integration is carried out, the helical phase profile of the PSF of the spiral phase filter in unstructured regions leads to perfect cancellation of the signal by destructive interference, except at edges where either the phase or the amplitude of neighboring points differ. Note: convolution kernel not to scale. (For an unambiguous assignment of the phase values please consult the color version of this figure on the color plates.)

are the polar coordinates in a plane orthogonal to the optical axis. The point spread function (PSF) of the spiral phase filter is a doughnut-shaped intensity ring with phase between 0 and 2π around the ring. In the convolution process, the complex object amplitude is weighted with the PSF at each point of the sample and then integrated over the whole area. In "flat" regions, neighboring points have doughnut rings that are identical in phase and amplitude, which leads to destructive interference because of the π phase step across the doughnut. Thus, unstructured regions (e.g., around points of type A in Figure 8.3) of the sample appear dark. At the edges of the object (e.g., around points of type B), the PSF of neighboring points differs either in amplitude or in phase retardation, which is indicated in Figure 8.3 by a different color and a different rotational offset of the circles, respectively, both of which result in brightening.

For large apertures and homogeneous plane wave illumination, the image field of thin and nonabsorbing phase objects is the product of the input field and the phase gradient $\vec{G}_{\text{ph}}(x, y) = \nabla_T \Phi(x, y)$ of the sample [37]

$$E_{\text{out}}(x, y) \propto E_{\text{in}}(x, y)\, g_{\text{ph}}(x, y) e^{i\delta_{\text{ph}}(x,y)} \tag{8.3}$$

Here, $g_{\text{ph}}(x, y)$ and $\delta_{\text{ph}}(x, y)$ denote amplitude and phase of $\vec{G}_{\text{ph}}$, respectively. In the assumed ideal case, the phase along the contour of a phase step sample, as depicted in Figure 8.3, is given by the geometrical orientation of the edge of the sample.

Spiral phase contrast (SPC) uses the helicity $\ell = \pm 1$ for Fourier filtering to induce large enhancement of the brightness of edges. Filtering with $\ell = \pm 2$ highlights areas with curved edges (Figure 8.4c). For such "simplistic" samples, this can easily be made plausible by considering that the convolution involves an integral of the form

$$I_\ell(\Phi(x, y)) = \int_0^{2\pi} d\varphi\, e^{i\Phi(x,y)}\, e^{i\ell\varphi} \tag{8.4}$$

The above integral vanishes for constant Φ as around point A. Around point B, it takes on a nonzero value for $\ell = 1$ because the phase of Φ changes *within* the

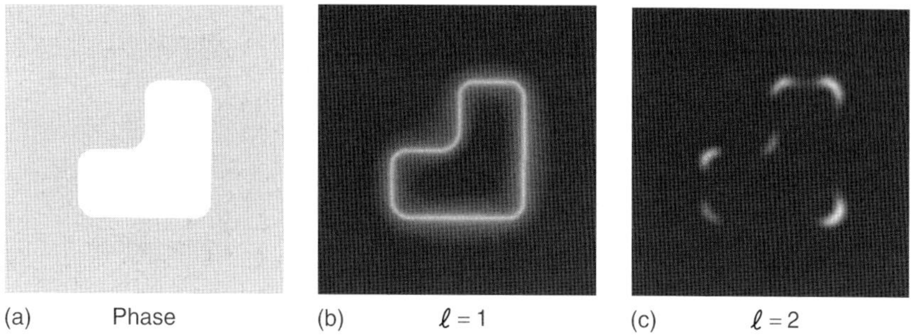

Figure 8.4 Filtering with higher helicity. (a) A phase sample with a phase retardation of 0.1π is filtered with $\ell = 1$ (b) and $\ell = 2$ (c). For $\ell = 1$ edges are isotropically enhanced. In the case of $\ell = 2$, the highest intensity can be observed in curved edges.

sector of φ where $\ell\varphi$ "counts" from 0 to 2π. For $\ell = 2$ this is not true; thus, filtering with a helicity of 2 does not highlight the region around point B. However, edges as around point C, where Φ again changes *within* a sector where 2φ changes from 0 to 2π, can give a nonzero signal. Filtering with even higher order helicities corresponds to selecting higher order moments. In digital spiral imaging [38] it is suggested to utilize all of these spectral moments for getting more information out of the images.

8.3.1 Isotropic Edge Enhancement

To realize a spiral phase filter it is, in principle, sufficient to display a circular spiral phase ramp with phase shift $\sim \exp(i\varphi)$ on a phase-modulating SLM and center it with respect to the zero Fourier components of the image wave, which is at the position of the focus of the illumination beam.

Practically speaking, in order to separate the correctly processed image wave from undesired reflections from the SLM cover glass, it is again advantageous to use the SLM in an off-axis configuration, which is achieved by numerically superposing the spiral phase mask with a blazed grating. This changes the direction of the processed image wave and separates it from the pure reflections. Such an off-axis spiral phase plate has a typical fork-like structure (see inset in Figure 8.5), which originates from the primary singularity of the on-axis vortex plate [16, 39]. Note that this central singularity is also maintained in the forked spiral phase structure. Light that is incident at this position is scattered at the singularity out of the imaging direction or even transferred into an evanescent wave [21]. Since for a perfectly centered spiral phase filter the zero Fourier component of the imaging wave coincides with the phase singularity, it is effectively removed from the imaging wave, giving a dark image background. This effect is also employed when using the singularity within vortex plates as efficient beam blockers in astronomic imaging, as, for example, in optical vortex coronagraphs [40].

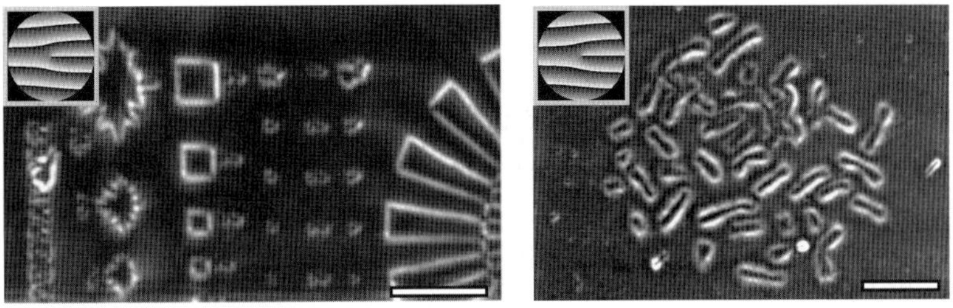

(a) (b)

Figure 8.5 Isotropic edge enhancement from spiral phase contrast: (a) A Richardson phase test pattern imaged with white light illumination from a xenon lamp. (b) Set of chromosomes imaged with laser light at 532 nm and a rotating diffuser. The scale bar corresponds to 10 μm for both images.

Two examples of isotropic edge contrast enhancement by Fourier filtering with an SLM-based off-axis spiral phase filter (indicated in the inset) but with different types of illumination are shown in Figure 8.5. The image of a Richardson test slide consisting of a phase pattern was taken with white light illumination from a xenon lamp with the condenser aperture almost closed (condenser lens 20× magnification with effective $NA_{cond} = 0.01$) in order to get sufficient spatial coherence from a point-like source. The objective lens had 63× magnification with $NA_{obj} = 0.95$. The set of chromosomes, on the other hand, was imaged with laser light ($\lambda = 532$ nm) with a diffuser plate rotating at kilo hertz rate in order to get rid of the speckle.

8.3.2
Pseudorelief Images

In the previous examples, the edge amplification of the spiral phase filter was demonstrated to be isotropic, which is to be expected, since the spiral phase plate – although it is not rotationally invariant – does not single out any specific direction or azimuthal phase angle. However, this symmetry can be broken *intentionally*, if one replaces the singularity in the center of the filter by a small circular disk of homogeneous phase retardation, which has to be similar in size to the central focused spot of the zero Fourier component. In the off-axis version, the central part is chosen to be a blazed grating (without the "pitchfork" singularity).

In this arrangement, the focused spot of the zero-order Fourier component of the image wave is not scattered out of the imaging beam any more, but develops into a plane wave in the image plane, which can now interfere with the remaining image wave in the camera plane. Changing the phase value of the central disk rotates the interference pattern. Alternatively, one may give the phases in the periphery a phase shift, which can easily be done by simply rotating the phase filter by some angle around its center.

Changing the relative phase between the periphery and the center of the spiral filter, either by changing the central disk or by rotating the phase plate, influences

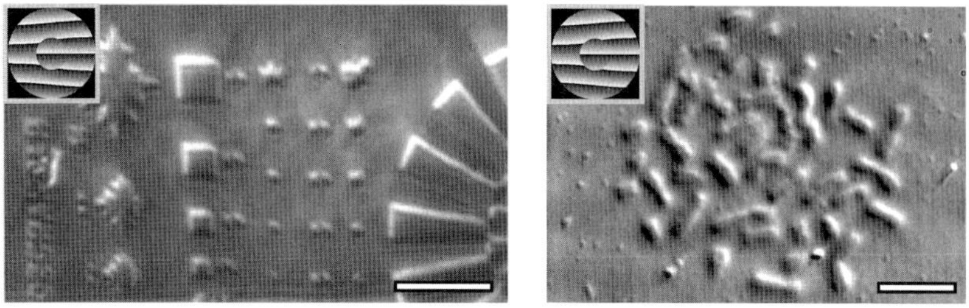

Figure 8.6 Spiral phase contrast filtering with relief effect. If the singularity in the center is replaced by a simple blazed grating, the plane-wave zero-order light can interfere with the filtered object field resulting in a shadow-like image (same samples and same illumination and imaging parameters as in Figure 8.5). Scale bar = 10 μm for both images.

the characteristic shadow effects [13]. The apparent direction of illumination gives the pictures a pseudoplasticity similar to the images recorded with DIC microscopy [12, 41]. Examples for this effect are shown in Figure 8.6, with all illumination and imaging parameters being the same as in Figure 8.5.

8.3.3 Spiral Fringe Metrology with SPC

In the previous section, we have shown how spiral phase filtering can be adapted to give the observed objects a three-dimensional appearance. The pseudorelief resembles images produced by DIC microscopy, but their physical origin is, nevertheless, different. This becomes clear when optically "thick" samples (i.e., samples showing an optical thickness in the order of one wavelength or more) such as the oil droplet in Figure 8.7 are imaged. What appears as shadow effect for thin objects evolves into a single, continuously spiraled interference fringe for thick objects. Figure 8.7 was taken with coherent light from a laser diode at 660 nm.

For optically thin phase objects SPC provides contrast enhancement; for optically thick objects, it may be employed to locally quantify the optical thickness of the sample, a goal that has already been addressed in microscopy in many ways, and is generally known under the name of quantitative phase microscopy (see for instance [42, 43] and references therein).

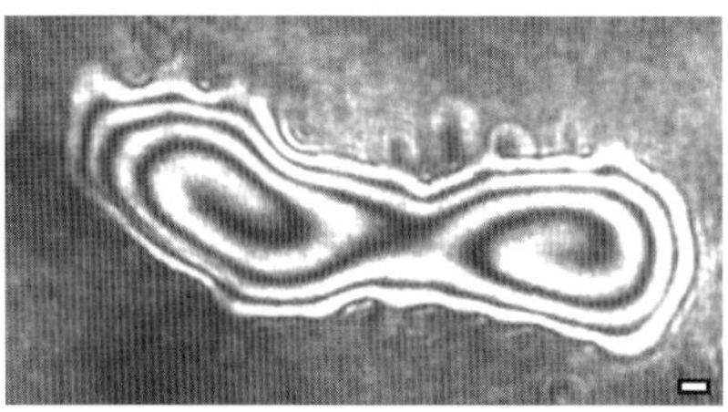

Figure 8.7 Spiral fringe: Oilsmear on a glass substrate. Scale bar = 10 μm for all images.

In the case of pure phase objects, the tangent along the spiral fringe is orthogonal to the local phase gradient. Following the spiral fringe once around corresponds to a optical thickness difference of $\lambda/\Delta n$ with Δn being the difference in refractive index between sample and solvent; thus, one can reconstruct the optical thickness from a single image [37]. For a known refractive index, the topography of the specimen can be measured, or vice versa, the refractive index can be determined with excellent accuracy [44].

8.4
Implementation and Performance

SPC can be emulated on an SLM screen, but it is also possible to upgrade a microscope for spiral phase imaging with a spiral phase plate [45]. In our case, this was a helical phase pattern imprinted on a photopolymer film coated on a $9 \times 9\,\text{mm}^2$ quartz glass plate. The optical thickness of the phase plate was chosen to correspond to a 2π step for a light wavelength of 650 nm. However, experimental practice showed that even with white light illumination quite good images could be taken, especially after introducing an appropriate line filter with a bandwidth of 12.5 nm into the optical path. The spiral phase plate has to be placed in an accessible Fourier plane of the optical path of the microscope and no other modifications are necessary. For illumination in transmission mode, it is possible to use the standard bright-field condenser, but the illumination aperture has to be almost closed in order to generate the necessary spatial coherence for the filtering. Spatial coherence in this case implies that (in the empty microscope) the illumination light focuses at a sufficiently small spot in the Fourier plane containing the spiral phase plate. The size of this spot as compared to the spatial extension of the Fourier image of the sample in the same plane determines the minimal size of the sample structures that appear with an intensified edge contrast in the image.

With regard to the resolution attainable in SPC, the following issues are important. It is fairly straightforward to find a generalization of the Rayleigh (or Sparrow) criterion for a spiral phase kernel [15, 45], with the NA of the microscope objective and of the illumination source being the essential parameters. The low effective NA (about 0.1) of the illumination required for spiral filtering thus means that the gain in contrast enhancement has to be traded off by a slight degradation in optical resolution. However, it is possible to play some tricks with a second (part of the) SLM panel in the illumination path, which significantly reduces this problem, as we have recently shown using an SLM [11].

When using SLMs for SPC, for example, parallel-aligned (PAL) LCD-panels, which allow good phase-only modulation, one has the option to change the parameters of the Fourier filter at a 60 Hz rate. Ferro-electric SLMs allow for faster switching rates, in the kilo Hertz regime. One can toggle between completely different microscopic methods such as bright field, dark field, DIC, and SPC. There exist different systems, with the pixels being optically or electronically addressed.

For the diffraction efficiency the fill factor of the panel is an important figure of merit, which, for reflective electronically addressed SLMs, may reach 90%. The practically achievable diffraction efficiency is decreased by a variety of factors. First, because of the pixelation of the SLM panel, the diffraction efficiency also depends on the largest spatial frequency of the displayed phase pattern, which is limited by the size of the pixels. And second, there is a nonnegligible (<40) loss of light by absorption (which affects the general brightness of the image, but *not* the image modulation contrast). In total, the *relative* efficiency, that is, the ratio of the shaped beam to the total light leaving the device, was typically very good, on the order of 80%.

Using the SLM in an "off-axis configuration," where a blazed grating is added to the Fourier filter, which spatially separates the first-order diffracted light that is then selectively used for imaging, the effect of the pixelation is less severe, since the first diffraction order by itself, when used as the imaging wave, does not contain the information on grating structure (such as the pixelation) that might disturb the picture, which would only show up in the entire diffraction pattern of the grating with all diffraction orders. The off-axis configuration slightly reduces the light efficiency, but we have shown that this is not a real problem and can be effectively countered by a somewhat brighter illumination source.

8.5 Conclusions

SPC microscopy has become a reliable and powerful tool, especially when implemented by means of a spatial light modulator for the add-on benefit of huge flexibility. Phase objects acquire good PC in transmission or reflection. The method requires a certain degree of spatial coherence, but unwanted effects such as speckles can largely be avoided by using a rotating diffuser plate. Depending on slight changes in the hologram on the SLM or the spiral phase plate relating to the central area of the filter, strong isotropic edge enhancement or relief-like shadow effects are possible. For larger microscopy samples spiral fringes appear, which immediately reveal whether the sample has a local depression or elevation, depending on the sense of rotation of the spirals. Quantitative reconstruction of optical thickness profiles is possible. It may be anticipated that the simplicity of the approach lends itself toward application in other areas of imaging.

References

1. Rost, F. (1996) *Fluorescence Microscopy*, vol. I, Cambridge University Press.
2. Bradbury, S. and Evennett, P. (1996) *Contrast Techniques in Light Microscopy*, BIOS Scientific Publishers, Oxford.
3. Zernike, F. (1934) Diffraction theory of the knife-edge test and its improved form, the phase-contrast method. *Mon. Not. R. Astron. Soc.*, **94**, 377–384.

4. Zernike, F. (1942) Phase contrast, a new method for the microscopic observation of transparent objects. *Physica*, **9**, 686–698.
5. Normanski, G. (1955) Interferometry with schlieren microscopy. *J. Phys. Radium*, **16**, 9–11.
6. Arnison, M., Larkin, K., Sheppard, C., Smith, N., and Cogswell, C. (2004) Linear phase imaging using differential interference contrast microscopy. *J. Opt. Soc. Am.*, **214**, (Part 1), 7–12.
7. Piekos, W. (1999) Diffracted-light contrast enhancement: a re-examination of oblique illumination. *Microsc. Res. Tech.*, **46**(4-5), 334–337.
8. Dodt, H., Frick, A., Kampe, K., and Zieglgansberger, W. (1998) NMDA and AMPA receptors on neocortical neurons are differentially distributed. *Eur. J. Neurosci.*, **10**(11), 3351–3357.
9. Dodd, J.G. (1977) Interferometry with schlieren microscopy. *Appl. Opt.*, **16**(2), 470–472.
10. Goodman, J.W. (2005) *Introduction to Fourier Optics*, Roberts and Company Publishers.
11. Maurer, C., Jesacher, A., Bernet, S., and Ritsch-Marte, M. (2008) Phase contrast microscopy with full numerical aperture illumination. *Opt. Express*, **16**(24), 19821–19829.
12. Fürhapter, S., Jesacher, A., Bernet, S., and Ritsch-Marte, M. (2005) Spiral phase contrast imaging in microscopy. *Opt. Express*, **13**(3), 689–694.
13. Jesacher, A., Fürhapter, S., Bernet, S., and Ritsch-Marte, M. (2005) Shadow effects in spiral phase contrast microscopy. *Phys. Rev. Lett.*, **94**(23), 233902.
14. Fürhapter, S., Jesacher, A., Maurer, C., Bernet, S., and Ritsch-Marte, M. (2007) *Spiral Phase Microscopy*, vol. 146, Pergamon Press. Imaging and Electron Physics, edition 2007.
15. Maurer, C., Jesacher, A., Bernet, S., and Ritsch-Marte, M. (2009) What spatial light modulators can do for optical microscopy. *Laser Photon Rev.*
16. Heckenberg, N.R., McDuff, R., Smith, C.P., and White, A.G. (1992) Generation of optical phase singularities by computer-generated holograms. *Opt. Letters*, **17**, 221–223.
17. Allen, L., Spreeuw, B.M.W.R.J.C., and Woerdman, J.P. (1992) Orbital angular-momentum of light and the transformation of Laguerre-Gaussian laser modes. *Phys. Rev. A: Gen. Phys.*, **45**(11), 8185–8189.
18. Allen, L., Padgett, M.J., and Babiker, M. (1999) The orbital angular momentum of light. *Prog. Optics*, **39**, 291–372.
19. Soskin, M.S., Gorshkov, V.N., Vasnetsov, M.V., Malos, J.T., and Heckenberg, N.R. (1997) Topological charge and angular momentum of light beams carrying optical vortices. *Phys. Rev. A: Gen. Phys.*, **56**(5), 4064–4075.
20. Swartzlander, G.A. and Law, C.T. (1992) Optical vortex solitons observed in kerr nonlinear media. *Phys. Rev. Lett.*, **69**(17), 2503–2506.
21. Roux, F.S. (2003) Optical vortex density limitation. *Opt. Commun.*, **223**(1-3), 31–37.
22. Che, Z., Segev, M., Wilson, D., Muller, R., and Maker, P. (1997) Self-trapping of an optical vortex by use of the bulk photovoltaic effect. *Phys. Rev. Lett.*, **78**(15), 2948–2951.
23. Gahagan, K.T. and Swartzlander, G.A. (1996) Optical vortex trapping of particles. *Opt. Lett.*, **21**(11), 827–829.
24. Paterson, L., MacDonald, M.P., Arlt, J., Sibbett, W., Bryant, P.E., and Dholakia, K. (2001) Controlled rotation of optically trapped microscopic particles. *Science*, **292**(5518), 912–914.
25. Parkin, S., Knoner, G., Singer, W., Nieminen, T.A., Heckenberg, N.R., and Rubinsztein-Dunlop, H. (2007) Optical torque on microscopic objects. *Laser Manipul. Cells Tissues*, **82**, 525–561.
26. Davis, J., McNamara, D., Cottrell, D., and Campos, J. (2000) Image processing with the radial Hilbert transform: theory and experiments. *Opt. Lett.*, **25**(2), 99–101.
27. Larkin, K.G., Bone, D.J., and Oldfield, M.A. (2001) Natural demodulation of two-dimensional fringe patterns. I. General background of the spiral phase quadrature transform. *J. Opt. Soc. Am.*, **18**, 1862–1870.

28. Arnison, M.R., Cogswell, C.J., Smith, N.I., Fekete, P.W., and Larkin, K.G. (2000) Using the Hilbert transform for 3D visualization of differential interference contrast microscope images. *J. Microsc.*, **199**, 79–84.
29. Khonina, S.N., Kotlyar, V.V., Shinkaryev, M.V., Soifer, V.A., and Uspleniev, G.V. (1992) The phase rotor filter. *J. Microsc.*, **39**(5), 1147–1154.
30. Jaroszewicz, Z. and Kolodziejczyk, A. (1993) Zone plates performing generalized Hankel-transforms and their metrological application. *Opt. Commun.*, **102**(5-6), 391–396.
31. Oemrawsingh, S.S.R., van Houwelingen, J.A.W., Eliel, E.R., Woerdman, J.P., Verstegen, E.J.K., Kloosterboer, J.G., and 't Hooft, G.W. (2004) Production and characterization of spiral phase plates for optical wavelengths. *Appl. Opt.*, **43** (3), 688–694.
32. Kotlyar, V.V., Kovalev, A.A., Khonina, S.N., Skidanov, R.V., Soifer, V.A., Elfstrom, H., Tossavainen, N., and Turunen, J. (2006) Diffraction of conic and Gaussian beams by a spiral phase plate. *Appl. Opt.*, **45**(12), 2656–2665.
33. Bernet, S., Jesacher, A., Fürhapter, S., Maurer, C., and Ritsch-Marte, M. (2006) Quantitative imaging of complex samples by spiral phase contrast microscopy. *Opt. Express*, **14**, 3792–3805.
34. Sakdinawat, A. and Liu, Y. (2007) Soft-x-ray microscopy using spiral zone plates. *Opt. Lett.*, **32**(18), 2635–2637.
35. Bokor, N. and Iketaki, Y. (2009) Laguerre-Gaussian radial Hilbert transform for edge-enhancement Fourier transform x-ray microscopy. *Opt. Express*, **17**(7), 5533–5539.
36. Bader, P., Zangerl, M., Bernet, S., and Ritsch-Marte, M. (2008) Imaging of laser induced ultrasound with a spiral phase contrast method. *J. Opt. A: Pure Appl. Opt.*, **10**(8), 085308.
37. Jesacher, A., Fürhapter, S., Bernet, S., and Ritsch-Marte, M. (2006) Spiral interferogram analysis. *J. Opt. Soc. Am.*, **23**, 1400–1409.
38. Torner, L., Torres, J., and Carrasco, S. (2005) Digital spiral imaging. *Opt. Express*, **13**(3), 873–881.
39. Mair, A., Vaziri, A., Weihs, G., and Zeilinger, A. (2001) Entanglement of the orbital angular momentum states of photons. *Nature*, **412**(6844), 313–316.
40. Swartzlander, G. (2001) Peering into darkness with a vortex spatial filter. *Opt. Lett.*, **26**(8), 497–499.
41. Fürhapter, S., Jesacher, A., Bernet, S., and Ritsch-Marte, M. (2005) Spiral interferometry. *Opt. Lett.*, **30**, 1953–1955.
42. Barty, A., Nugent, K., Paganin, D., and Roberts, A. (1998) Quantitative optical phase microscopy. *Opt. Lett.*, **23**(11), 817–819.
43. Lobera, J. and Coupland, J.M. (2008) Contrast enhancing techniques in digital holographic microscopy. *Meas. Sci. Technol.*, **19**(2), 025501.
44. Maurer, C., Bernet, S., and Ritsch-Marte, M. (2009) Refining common path interferometry with a spiral-phase Fourier filter. *J. Opt. A: Pure Appl. Opt.*, **11**(8), 094023-1–094023-7.
45. Maurer, C., Jesacher, A., Fürhapter, S., Bernet, S., and Ritsch-Marte, M. (2008) Upgrading a microscope with a spiral phase plate. *J. Microsc.*, **230**(1), 134–142.

9
Applications of Electromagnetic OAM in Astrophysics and Space Physics Studies

Bo Thidé, Nicholas M. Elias II, Fabrizio Tamburini, Siavoush M. Mohammadi, and José T. Mendonça

9.1
Introduction

To this day, most of the information we have about the Universe is obtained from the electromagnetic radiation from space that reaches our telescopes and other sensor systems on earth, and on board spacecraft orbiting the earth. The frequency spectrum of the signals thus received ranges from the highest (gamma, UV) through visible, infrared, and microwaves to radio at the lowest frequency, allowed by the propagation properties of the (plasma) medium in the signal path from the source to the observer. When an observer collects electromagnetic radiation from celestial sources with an earthbound instrument, the terrestrial atmosphere absorbs large portions of the upper range of the spectrum, while the ionospheric plasma sets a limit of $\sim$5–10 MHz (60–30 m vacuum wavelength), below which signals cannot penetrate down to the observer's instrument.

For the first time in history, humans now have the capability to leave the confines of planet Earth to deploy instruments in space, outside the atmosphere and the ionosphere. Observatories in space provide ample access to segments of the electromagnetic spectrum that until recently were totally unexplored. An interesting concept that has been discussed for quite some time is an observatory on the Moon [1]. Such a facility would open new frequency windows, facilitating observations of the Universe at both the submillimeter and the submegahertz ends of the electromagnetic spectrum. An observatory on the far side of the Moon will be shielded from much of the low-frequency interference from the Earth and its plasma envelope [2]. Like all space-borne, and many earthbound facilities, building a telescope on the lunar surface is an extremely costly endeavor and it is therefore of utmost importance that every ounce of information be squeezed out of the electromagnetic radiation collected.

Till date, virtually all radiation from nature has been analyzed only with respect to its intensity, spectral content, direction of arrival, and polarization. While polarization is a manifestation of the physical fact that electromagnetic radiation does not only carry linear momentum but also angular momentum, polarization measurements alone do not provide an exhaustive characterization of the total

Twisted Photons: Applications of Light with Orbital Angular Momentum.
Edited by Juan P. Torres and Lluis Torner

ISBN: 978-3-527-40907-5

angular momentum state of the electromagnetic radiation. Already, in the early 1900s, it was shown theoretically that collimated beams can carry angular momentum [3]. In the 1930s and 1940s, elegant Einstein-de Haas-type experiments for photons carried out by Beth [4] in optics and by Carrara [5] in radio showed that angular momentum can be transferred from electromagnetic beams to mechanical bodies. In the years that followed, very high angular momentum states of nuclei were discovered via their photon decays. Still, it was not until the 1990s that laser beams and microwave radio beams carrying both spin angular momentum (polarization) and orbital angular momentum (OAM) could be readily generated, controlled, and detected [6, 7]. Subsequently, it was demonstrated experimentally that individual photons can be endowed with OAM [8–10] and be entangled in these states [11, 12], showing that electromagnetic radiation can be characterized, analyzed, and utilized more fully than what was commonly known at the time.

Recently, the possibility of using the OAM degrees of freedom of light and radio in astrophysics and space physics has come to the fore [13–21]. It is therefore reasonable to assume that electromagnetic OAM is radiated by some astrophysical sources or is imparted upon radiation through interaction with plasma and fields in space, at least under certain conditions. The characterization of the OAM of light or radio beams intercepted by telescopes on Earth or in space can provide new and crucial information about the physical processes involved. Following this assumption, many authors have proposed studies with new instruments to detect and manipulate the electromagnetic OAM, henceforth referred to as POAM (photon orbital angular momentum), and also performed modifications to the telescope. This represents a single phase of a more complex evolution that is occurring now in astronomy.

As described by Harwit [13], POAM permits new types of measurements and paves the way for the utilization of topological and more generic degrees of freedom in astronomy and space sciences. This chapter describes and discusses some of these new opportunities.

9.2 Ubiquitous Astronomical POAM

POAM is a fundamental characteristic of individual photon wavefunctions (PWFs) as well as ensembles of PWFs. The latter are proportional to macroscopic electric fields. Even sources that possess no intrinsic POAM can exhibit "pointing POAM" if they are not located at the field-of-view center or "instrumental POAM" in the presence of high-order aberrations.

Existing astronomical instrumentation takes advantage of POAM for routine and specialized observations. We recommend reading Sections 2, 3, 4, 10, and 11 in [19] to understand the basic concepts and mathematics. Here, we summarize the behaviors of specific types of instruments described in Sections 5, 6, 7, 8, and 9 in [19].

As expected, propagation through free space induces no additional torque to electric fields. The same statement is true for propagation through an unaberrated telescope. Although images produced by an unaberrated telescope are undistorted due to torque, spatial resolution is lost because of the finite size of the entrance aperture.

Aberrated telescopes, on the other hand, induce torque and distort images. The perturbed component of the electric fields can be expressed in terms of an azimuthal Fourier series,

$$\Delta E(r, \psi; t) = \sum_{m=-\infty}^{\infty} \Delta E_m(r; t) e^{im\psi}$$

$$\leftrightarrows$$

$$\Delta E_m(r; t) = \frac{1}{2\pi} \int_0^{2\pi} \mathrm{d}\psi \, e^{-im\psi} \Delta E(r, \psi; t) \tag{9.1}$$

The complex phasors are called *vortices*, which are quantized spiral wave fronts and POAM basis functions. Aberrations are expressed in a similar manner, except that $\sin m\psi$ and $\cos m\psi$ functions are used instead of phasors. Since phasors are expressed in terms of sinusoids, aberrations are equivalent to the application of torque.

Consider a focal-plane coronagraph, which consists of a telescope followed by a reimaging system. The focal plane of the telescope contains a mask that blocks starlight to increase the contrast of faint companion stars or planets to the point where they can be detected in a reasonable amount of time. The telescope could have an exit pupil for adaptive optics, and the reimaging system could employ a high spatial frequency filter in its pupil for additional starlight suppression.

Assuming that the star is perfectly centered in the field of view, its electric fields are dominated by the $m = 0$ POAM state, that is, plane waves. Since the star is not a true point source, there are small contributions from the low-order POAM states, for example, $m = \pm 1, \pm 2$ due to pointing POAM (also called *structure POAM* in this case). The POAM spectrum of faint companions, however, is dominated by high-order POAM components. All telescopes are highly insensitive to high-order POAM states near the field-of-view center; so, if the low-order POAM states of the star are modulated to higher POAM states, the starlight is highly attenuated in the final image. The high-order POAM states of the faint companions are significantly less attenuated.

Now consider a long-baseline interferometer. Two small entrance apertures pass light to a beam combiner and finally to a detector. By changing the relative positions of the entrance apertures, such instruments can synthesize the entrance aperture of a much larger telescope, producing high-resolution images from measured visibilities.

For this analysis, we assume that (i) the line between the two apertures is perpendicular to the line of sight; and (ii) the midpoint between the two small entrance apertures is employed as the reference point for describing POAM in their plane. The apertures capture pieces of a vortex of radius $B/2$, where B

is their separation. Further, the phase difference of the vortex patches is zero for even POAM states and 180° for odd POAM states. Since only the even POAM states reach the detector, an interferometer thus acts as a crude POAM filter.

The fringes measured by the interferometer can be projected back onto the sky, indicating which source regions contribute to the measured visibilities. If the delay line tracks the fringe envelope peak of a small circular source at the field-of-view center, the phase differences mentioned above do not change and the peak of the sky fringe pattern is colocated with the source. On the other hand, if the delay line deviates 180° from the fringe envelope peak, the phase difference of the vortex patches is 180° for the even POAM states and 0° for the odd POAM states. Only the odd POAM states reach the detector, which means that the minimum (zero) of the sky fringe pattern is colocated with the source.

If we allow the two small entrance apertures to rotate, light from faint companions of the small central source is modulated as they move through the sky fringe pattern. The exact form of the modulation determines the location and brightness of the companions. Like coronagraphs, a "nulling interferometer" can be used to find faint companions.

For the last example, consider the same coronagraph as above, except that a rotating wedge replaces the focal-plane mask. The wedge is an opaque disk with a sector removed. The rotation modulates the POAM states, which can be measured with a single detector. What do we learn by modulating POAM states?

Assume that we are observing a binary star system and that their separation is smaller than the "λ/D" resolution of the entrance aperture. The telescope tracks them at the center of light, which means that each star has its own pointing/structure POAM. A standard telescope would see just a single star, since the angular information of the stars has been obscured by the individual Airy patterns. The coronagraph with rotating wedge modulates the POAM states, and the modulation pattern depends on the wedge angle as well as the angular separation and relative brightness of the stars. In principle, it should be possible to determine the separation and relative brightness, limited only by systematic and random errors and not by the telescope diameter (sub-Rayleigh imaging). Therefore, the oft-quoted best possible resolution "λ/D" is a myth.

9.3
Applications of POAM in Astronomy

We reiterate that POAM is a new degree of freedom, a ubiquitous key property encoded inside light and radio beams from celestial sources. Many astronomical applications can benefit from the POAM states of radiation. Among the properties already discussed, this section focuses on two more concrete subjects, that have already found a practical astronomical application and have been experimentally proved in the laboratory and/or at the telescope. One is sub-Rayleigh resolution, and the other is the practical handling of optical vortices (OVs) at the telescope obtained with stellar light, in view of stellar coronagraphy.

New detectors and photon counting techniques are revolutionizing observation. Astronomy at the quantum limit aims at characterizing the light in a deeper way than that usually obtained by analyzing the spectrum, intensity, and temporal variation of celestial body multiphoton properties, as described by Glauber correlation functions, in rapid phenomena in astrophysics, and also measures the POAM of light. For a review, see [22].

Most celestial sources are essentially variable faint photon sources located at infinity: their photon detection rate is of the order kilohertz to megahertz when analyzed with extremely large telescopes (ELTs). This makes the characterization of Glauber correlations and POAM detection of certain astrophysical fast phenomena, such as black hole accretion, quite difficult. POAM of light has been shown to be detectable at the single photon level as well [8–10], without being an intrinsic property of the photon itself. From a classical point of view, Heitler [23] gave a formulation of the POAM associated with a source by using multipolar expansions, but this approach implied the existence of a longitudinal component of the EM field in the EM wave, a phenomenon deeply rooted in the photon rest mass problem. POAM then seemed to be associated with only a spherical or Laguerre–Gaussian (LG) expansion of EM waves and to vanish at infinity. Only recently [24] have these assertions been corrected and extended through a classical approach of the EM source showing that an astronomical source can present a nonzero value of POAM [25].

In addition to the eventuality of detecting the POAM of light from astrophysical sources and the full characterization of the quantum properties of the photon stream, POAM can improve several fundamental techniques typical of classical astronomy. We now focus on two aspects that have been verified both theoretically and experimentally. The first is the dramatic improvement of the resolving power of a diffraction-limited telescope that permits breaking up of the Rayleigh criterion up to 1 order of magnitude when separating two nearby sources; the second is the actual production and control of the POAM of light from starlight for other applications such as stellar coronagraphy with ground-based telescopes, in view of an application to ELTs.

9.3.1 Sub-Rayleigh Resolution

The classical Rayleigh criterion that limits the resolving power of an astronomical diffraction-limited telescope can be overcome by separating two independent equally luminous monochromatic and white light sources at the diffraction limit. This was experimentally tested and numerically verified, using OVs instead of the usual Airy diffraction patterns [14].

The classical Rayleigh separability criterion states that the diffraction images of two point-like sources (Airy disks) are resolved when the maximum intensity of one source overlaps the first intensity minimum of the second equally bright source. The historical definition [26] refers to a symmetrical double-peaked profile with a central dip that is 5% lower than the intensity maxima, corresponding to a signal-to-noise ratio of $S/N = 60$. For a telescope having a diameter D, at a wavelength λ, the separation is obtained at $\theta_R = 1.22\lambda/D$ [27]. Techniques to overcome this limit have been discussed and developed in certain special cases [28, 29].

In this experiment, the diffraction pattern of one of the two sources crosses a phase-modifying device (PMD), an $l = 1$ fork hologram, on its center, generating the LG transform of its Airy disk. The second source, crossing the fork hologram in positions different from the optical center, acquires different POAM values and generates unsymmetric and distorted LG patterns. By adopting a signal-to-noise ratio of $S/N = 60$, one can formulate a separability criterion that uses the geometrical properties of the OVs. Instead of overlapping the maximum intensity of one source with the first minimum of the second one, this criterion is based on the asymmetric intensity distribution of the superposed LG patterns generated by the superposition of the symmetric OV of the on-axis source with the distorted vortex of the other off-axis source.

Figure 9.1 shows the intensity ratios of the main peaks produced by separated sources. The separation of the off-axis monochromatic source was tested in a range $0 \leq \delta \leq 700$ μm, with a step of 35 μm. The experimental data show a good agreement with the theoretical curve obtained from numerical simulations of the LG transform of an Airy disk. Here, the intensity ratios reach a minimum value of 0.48 when the separation is $\sim 0.42\delta_R$. The upper inset of the figure shows the plot of the positions of the main peak of the simulated LG modes in units of the separation δ_R versus the estimated POAM values. If one analyzes the relative intensities of the asymmetric peaks produced by the off-axis object, one may achieve an efficient sub-Rayleigh separability limit, that obviously depends on the S/N ratio of the data. The figure also reports two points obtained in white light that suggest a different slope at small separations with respect to the monochromatic behavior.

The historical definition of the Rayleigh criterion can be mimicked by assuming that two identical sources are just resolved when the intensities of the asymmetric peaks differ by at least 5%. In the monochromatic case, and with coherent laser light, a separability that is 50 times better than the Rayleigh limit can be reached. The results obtained in white light, instead, suggest the reaching of a separability about 10 times better then the Rayleigh limit. This lower resolution is mainly due to the nonperfect spatial filtering and lower degree of coherence of the sources. In the lower inset of Figure 9.1 we show a successful application of the separability criterion in white light, where we simulated an OV of a double star with an angular separation $\sim$10 times below the Rayleigh limit, as seen with a diffraction-limited telescope having the same focal ratio as our 122 cm Galileo telescope in Asiago. However, nothing in nature comes for free. What is gained in spatial resolution is unavoidably lost in intensity, being the maximum of an Airy diffraction pattern higher than that of its LG transform. Analogously, Fraunhofer diffraction patterns show that off-axis displacements of the input beam produce, also in this case, asymmetric diffraction patterns and the intensity profile along the direction of

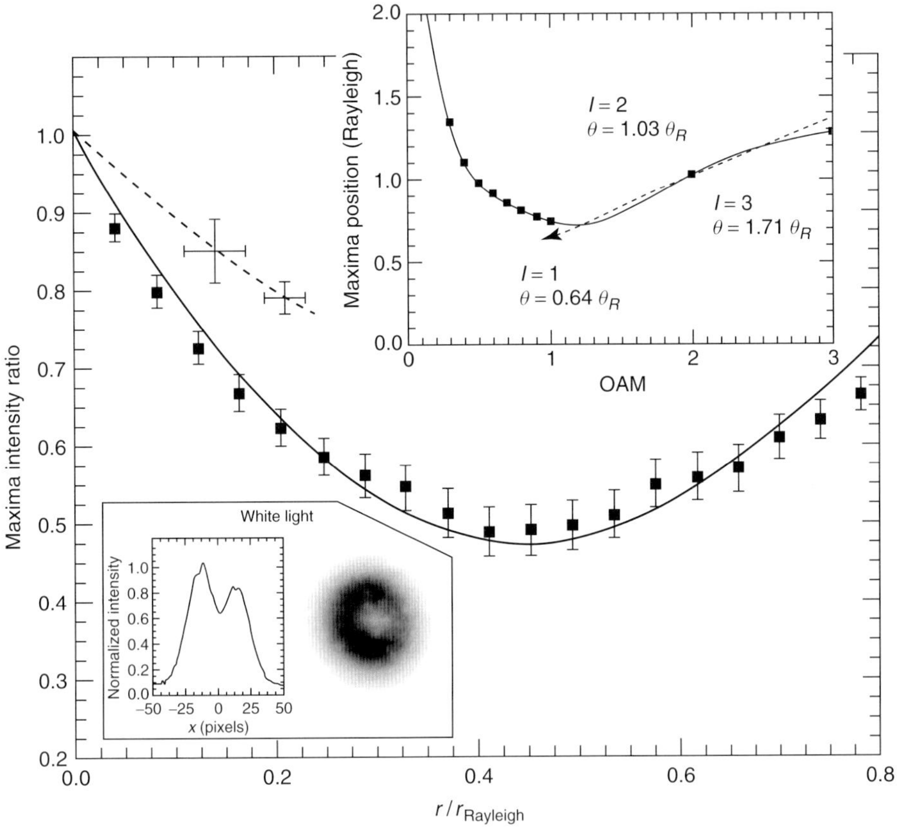

Figure 9.1 Ratio between the intensities of the peaks of the superposed LG modes versus the off-axis shift of the spot in units of the Rayleigh radius. Solid line: theoretical expectations for monochromatic light; the superposed dots and error bars are the experimental data. Dashed line: linear interpolation of the experimental data obtained in white light. A 5% difference (see text) between the intensities of the peaks implies, in both cases, a separability of at least 1 order of magnitude better than the Rayleigh limit. Upper inset: Positions of the maxima of the LG modes relative to the OV (in units of the Rayleigh radius) versus POAM. Triangles: angular separation values between two equally charged OVs. Lower inset: Simulation of two equally luminous stars in white light having an angular separation that is 10 times below the Rayleigh radius, as seen with a diffraction-limited telescope.

maximum asymmetry shows two different peaks [30]. The intensity distribution is in this case better described by Kummer beams, a class of beams with an OV nested inside, obtained by using the modified Bessel function of the first kind. For any integer value l of the topological charge, the intensity ratio R between the two maxima decreases exponentially with the off-axis displacement of the incident beam. In this case, it was shown that higher values of the topological charge can provide better resolutions in the relative positioning of two sources and this method could offer interesting applications in high-precision positioning systems.

This application would open new perspectives for the increase of the resolving power of diffraction-limited telescopes. One immediate application is the separation of close binary stellar systems with space telescopes and relative astrometry, by placing a PMD at the focal plane of a telescope.

9.3.2 Optical Vortices with Starlight

In the application of these techniques to ground-based telescopes, one must face the practical problems of how to actually obtain and handle OVs from the light of celestial objects. The main difficulties are the detrimental effects caused by the atmospheric turbulence, that is, the seeing. The turbulence both tilts the position of a star away from its actual position on the focal plane of the telescope and deforms the wave front, with the result of degrading the coherence of the light. One immediately understands that high-precision astrometry and other techniques such as stellar coronagraphy with OVs, which require extremely precise telescope pointing and tracking to put a star exactly in correspondence with the optical singularity, will be unavoidably affected. The tip-tilt will randomly move the star in a region around the optical singularity and the wave front distortion will introduce spurious additional POAM states that will affect a high-precision analysis of OVs for astrometry. In the visible band, the technique of adaptive optics still does not offer a valid stable solution similar to that operating in the infrared, and for the moment, only space telescopes could benefit from OV techniques.

Experimental studies of the detrimental effects of the atmospheric turbulence in the generation, detection, and manipulation of OAM of light from stars were undertaken [18]. With an $l = 1$ blazed fork hologram at the focal plane of the Asiago 122 cm telescope, OVs from the stellar system Rasalgethi (α Herculis) and from the single star Arcturus (α Bootis) were obtained. An analysis was made of the structure of the OVs obtained from nonmonochromatic starlight under very poor seeing conditions using a fast CCD camera to obtain speckle patterns and carry out the lucky imaging technique, one of the speckle imaging techniques used in modern astronomy, as an alternative to adaptive optics. The choice of a fork hologram as a phase-modulating device was made because it has the advantage of generating OVs with the same ℓ at all wavelengths for on-axis polychromatic sources (Figure 9.2). Monochromatic on-axis beams produce OVs with OAM indices $\ell = ml$, where m is the diffraction order of the grating, while adding (or subtracting) a quantity of the OAM value of the impinging light. For off-axis sources, ℓ decreases as the star moves away from the center of the hologram. H is a grating with a number l of dislocations on its center and our $l = 1$ fork hologram, H, is blazed at the first diffraction order with 20 lines/mm and has an active area of 2.6×2.6 mm^2. During the experiment, H was placed at the F/16 Cassegrain focus of the T122 Asiago telescope.

The spatial coherence of the incident beam is generally assumed to hold only for stellar sources when observed from space. But when ground-based telescopes are used, fairly dramatic effects are introduced by atmospheric turbulence, and spatial

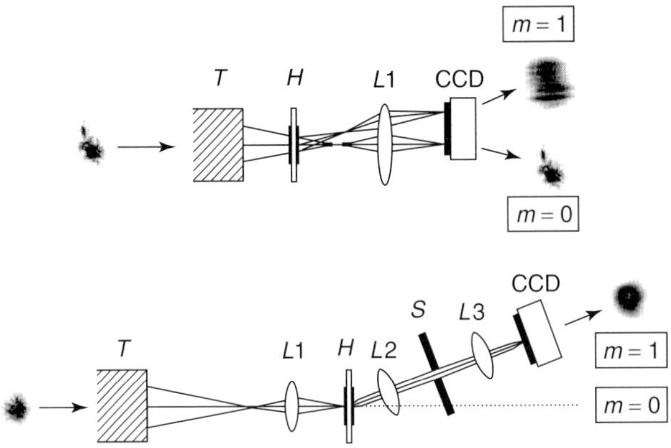

Figure 9.2 Optical setups, without (a) and with (b) spatial filter. T, telescope; $L1, L2, L3$, lenses; H, $l = 1$ fork hologram; S, spatial filter (slit). Stellar speckle patterns are sketched on the left of the optical setups, while the output images at the zeroth and first diffraction order are on the right.

coherence is unavoidably lost. However, for exposures shorter than the turbulence timescale (~10–100 ms in the optical/near-infrared), we can obtain a group of bright speckles that represent the interference image produced by the coherent wave fronts generated by the random distribution of the atmospheric irregularities. Adaptive optics in the visible band do not offer the required corrections to the wave front so as to obtain a diffraction-limited image. A single nearly diffraction-limited stellar image can occasionally be produced when most of the stellar light falls in a single bright speckle. Fried [31] coined the term *lucky exposures* to describe high-quality short exposures occurring in such a fortuitous way. This is the basis of *lucky imaging* [32]. To obtain symmetric OVs from stellar sources, we impose the additional restriction of selecting only those exposures where the star to be made faint is centered with the hologram dislocation.

Simultaneous observations were made of both the speckle patterns and the OVs generated by the multiple system Rasalgethi (α Her) and then by the single star Arcturus (α Boo). In the latter case, we applied a chromatic filter to limit the dispersion of the diffraction grating and recover the donut pattern of the OV. The α Her is a visual binary composed of two unresolved binary systems presently separated by $4''.7$: α Her A, formed by an M5 Ib-II semiregular variable ($m_V = 2.7$–4.0) and a fainter companion separated by $0''.19$ [33]; and α Her B, containing a G0 II-III giant ($m_V = 5.4$) and a fainter secondary separated by $0''.0035$ [34]. α Boo, instead, is a single star having visual magnitude $m_V = 0.04$ and spectral type K1.5 III.

The sequence of frames in Figure 9.3 clearly shows the detrimental effects caused by the atmospheric turbulence, which randomly distorts the wave front. On the left side of each frame, the OV generated by the first diffraction order of H is shown, while the corresponding speckle pattern is shown on the right. The OVs produced by nonmonochromatic light beams crossing the fork hologram show intensity

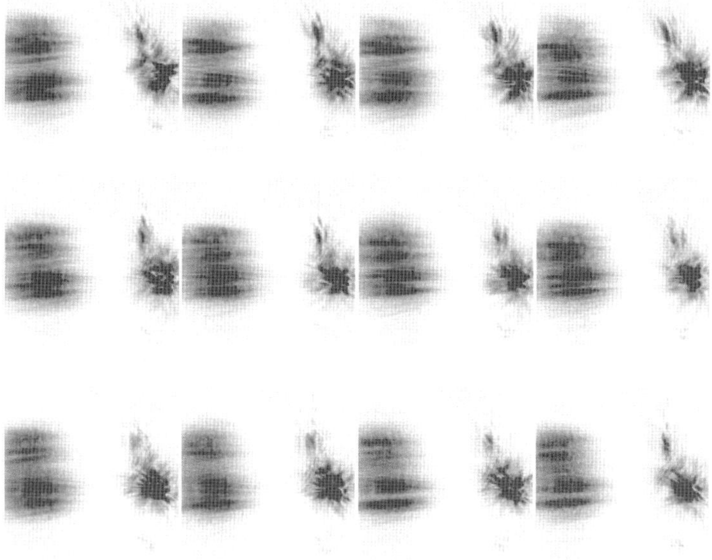

Figure 9.3 Snapshot of the speckles of Rasalgethi and their OVs. The distortion of the OV due to the atmospheric seeing of the double system is clearly visible. Sometimes in the satellite vortex, B apparently dissipates the central dark strip (see text). This effect can be ascribed to the composite structure of the double system's speckles that may fuse together because of the bad seeing conditions.

patterns that appear as rings stretched along the dispersion direction with a central dark strip. The spectral dispersion also causes a partial filling of the central dark zone. Thus, if we want to use nonmonochromatic OVs produced by fork holograms for OV coronagraphy, we must limit the spectral range and/or restore the donut shape. In our case, we used a variable spatial filter, S, made by a slit placed on the Fourier plane of the collimating lens $L2$ to limit the dispersion of the light at the first diffraction order. This adjustable slit works as a tunable bandpass filter with flat spectral response and has been used only for single on-axis stars.

The single star α Boo was set at the center of the hologram and the spatial filter S was introduced to produce a nearly monochromatic circularly symmetric OV. We adopted a slit width of 0.1 mm, corresponding to a 300 Å bandpass filter width in the visible spectrum, that ensured enough S/N ratio for the 70 ms exposures. Figure 9.4 shows the OV obtained by summing the selected (2%) lucky frames. The central region of the OV is not totally dark because of the loss of the starlight coherence due to the extremely poor seeing conditions and the presence of residual chromatism. We however notice that the contrast between the dark center and the bright ring improves down to 52% with respect to the previous unfiltered symmetric OVs. The lack of light detection in the other orders of diffraction, except the first, indicate that the intrinsic OAM already present in the starlight was not detectable with our instrument.

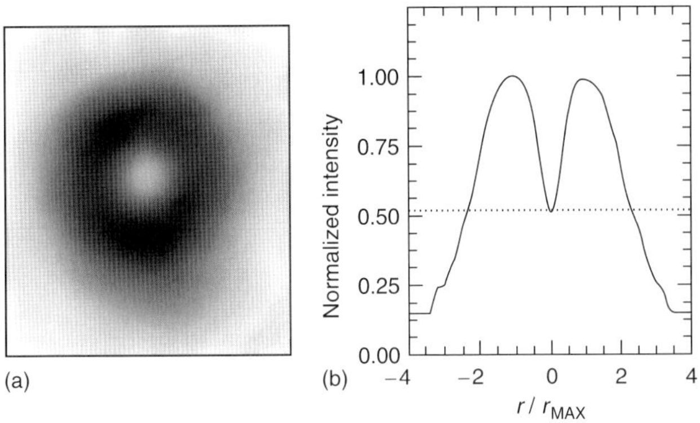

Figure 9.4 (a) The optical vortex (OV) obtained by summing the selected 2% good frames (see text). (b) Normalized intensity profile of the OV across the direction perpendicular to the dispersion. The abscissa is the width of the vortex normalized to the position of the maxima of the ideal OV.

A test was performed with the same hologram to identify the difficulties that an experimenter will experience when performing OV coronagraphy [29]. To this end, the double system STF1258 UMa, whose components are separated by $9''.9$ and have visual magnitudes $m_{V,A} = 7.72$, $m_{V,B} = 7.87$ was selected. STF1258 was imaged through a 100 Å bandpass red filter centered at 6532 Å with and without a Lyot stop. The exposure times were 0.01 s.

Figure 9.5 shows the results of $\ell = 1$ OV coronagraphy test. Figure 9.5a shows the intensity profiles of the two stars without and with the Lyot mask, obtained by averaging over a 40-pixels strip. Figure 9.5b shows the corresponding snapshots. The on-axis component B of the binary system appears to have been made faint by a factor $\sim$1.7, close to the factor 2 derived from numerical simulations. The partial obscuration of the on-axis star is due to the fact that we were using an $\ell = 1$ mask. Total obscuration can be mainly achieved with even ℓ-valued spiral phase plates (SPPs).

What is immediately evident is that OV-based techniques crucially depend on the positioning of the source with respect to the optical singularity of the PMD, which requires high precision in pointing and tracking of the telescope and in the adaptive optics system.

9.4
Applications of POAM in Space Physics

In space physics, usually defined as the astrophysics of the $\sim$1 AU region of space around Earth, which is accessible to our space probes, radio and radar methods are of central importance. Their usage ranges from the passive reception of radio emissions from the Sun and the planets to active sounding out of the ionosphere, the Sun, and planetary objects (radar astronomy). On the borderline between

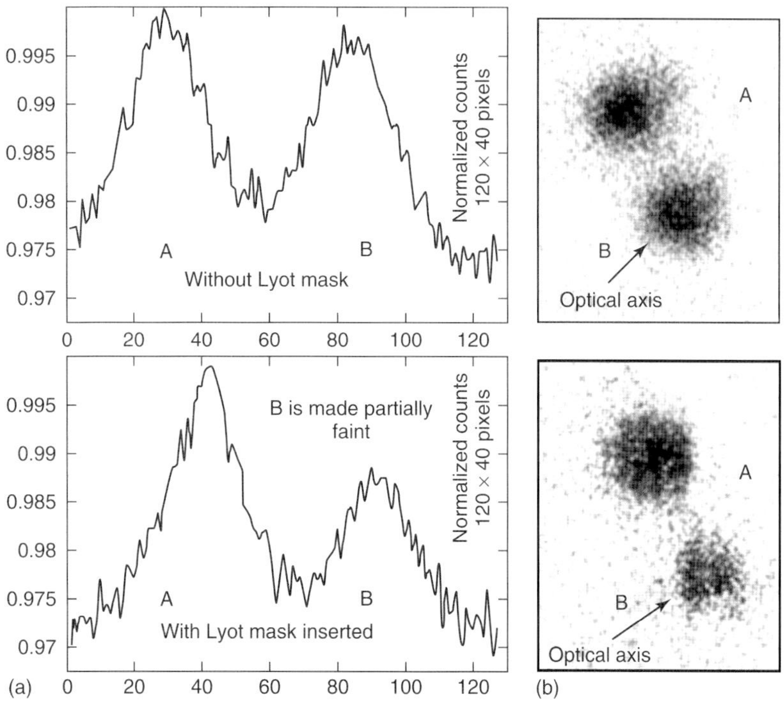

Figure 9.5 $\ell = 1$ optical vortex coronagraphy test of STF1258 UMa: (a) the averaged intensity profiles of the double star before (up) and after (down) the insertion of the Lyot stop. (b) The corresponding snapshots (see text).

radio-based space physics and astrophysics, we find astroparticle physics, which studies radio pulses generated when ultra-high energy particles enter the denser part of the atmosphere [35] or the Moon [36]. Hence, there is a certain overlap between radio-based space physics and radio astronomy. They cross-fertilize each other and they experience similar problems and limitations.

One area of common interest is to investigate the effect on cosmic radio signals from near and far as they propagate to the receiving telescope or other radio sensor instruments. For example, irregularities of the phase screen type distort both the amplitude and phase. Turbulence and irregularities can occur in the ionospheric plasma. They can occur naturally not only as a result of a natural external perturbation, but also when the ionosphere is perturbed in a controlled, repeatable manner, allowing systematic stimulus – response type experiments to be made, using the near-Earth space as a giant radio laboratory [37].

The possibility of studying space plasma vorticity remotely by measuring the OAM of radio beams interacting with the vortical plasma was pointed out in [17]. Extrapolating inferences from the studies of optical effects associated with POAM [38] to the plasma physics domain, one can conclude that there should be a coupling between plasma vorticity and radio beam OAM. The study of these effects is believed to pave the way for improved self-calibration and radio interference mitigation techniques in radio astronomy, and for finding new methods of generating electromagnetic beams that carry OAM. It is known that when branch points are present in the phase of an EM field the usefulness of least mean-square error wave front reconstructing systems is limited [39].

The nonlinear interaction of EM beams carrying OAM and a medium can excite a number of parametric processes [42–48]. For instance, the exchange of angular momentum between electromagnetic and electrostatic waves in a plasma, due to the stimulated Raman and Brillouin backscattering processes, was predicted [40]. In the process, OAM-carrying phonons and plasmons would be generated (Figure 9.6). Such OAM states could be probed with a radar or radar-like beam of photons carrying various combinations of SAM and OAM to produce fundamental plasma wave interaction studies of a new kind [17].

The interaction between EM waves carrying angular momentum and charges can lead to the generation of azimuthal currents [49], which in turn can give rise to magnetic fields in plasmas [50, 51]. This is the well-known inverse Faraday effect (IFE) predicted by Pitaevskii [52] and Pershan [53] and first observed in a plasma by Deschamps [54].

The plasmas in the Earth's surroundings have frequencies ranging from a few kilohertz to tens of megahertz. The frequencies of the radio beams that interact with these plasmas are therefore of the order of 1 GHz or lower. In 2007, it was demonstrated that radio beams of the phase-mode class, emitted from certain circular arrays and fed in a certain manner, carry POAM [16]. The relatively low radio frequencies allow the radio sensors to be fully digital. Facilities of this kind therefore enable systematic, fundamental POAM experiments to be performed

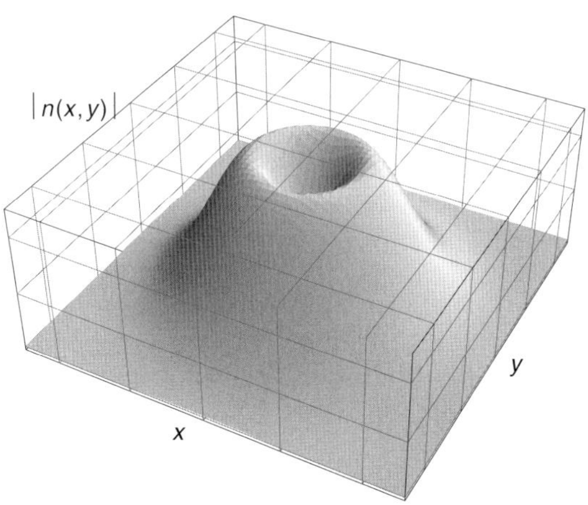

Figure 9.6 A phonon (ion-acoustic wave) in a plasma can carry OAM (but not spin angular momentum (SAM)). Source: From [40]. (Please find a color version of this figure on the color plates.)

Table 9.1 Scaling of the normalized POAM, obtained in computer simulations, as a function of OAM state number l for a right-hand circular polarized beam ($s = -1$) formed by a ring array of 10 crossed dipoles. Array radius $D = \lambda$, antennas 0.1λ over perfect ground, polar angle $\theta = 0$. Source: From [16].

l	s	$j=l+s$	Normalized POAM
0	−1	−1	−1.019
1	−1	0	−0.022
2	−1	1	0.971
3	−1	2	1.81

digitally, under full software control. This is a major step forward from analog techniques. The development of nanotechnology antennas and faster digitizers is likely to push the upper frequency limit for which digital OAM experiments can be performed, eventually reaching the optical regime.

Table 9.1 shows the results of numerical experiments involving the generation of OAM radio beams with the help of conventional phase-mode circular arrays [16]. The predicted normalized POAM is compared with the values measured in the experiment. In Figure 9.7, the angular momentum radiation pattern from such an antenna [41] is plotted. Note: This plot is different from an ordinary antenna pattern plot which describes the angular distribution of linear momentum (Poynting vector)!

Finally, we point out that an electron–neutrino beam, propagating in a background plasma with vorticity, can be decomposed into orbital momentum (OAM) states, similar to the OAM photon states [55]. This result may be relevant to the understanding of neutrino sources in astrophysics.

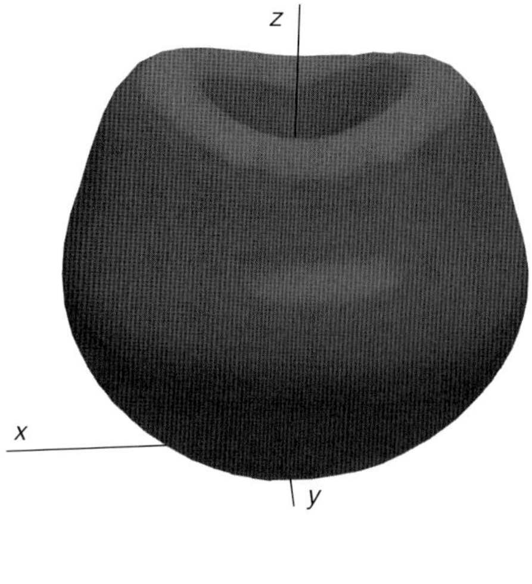

Figure 9.7 Plot of $|\mathbf{J}^{em}|$, that is, the intensity distribution of POAM, generated by a circular antenna array in the xy plane. Source: From [41]. (Please find a color version of this figure on the color plates.)

9.A
Appendix: Theoretical Foundations

9.A.1
Classical Field Picture

A consistent, straightforward analysis, based on Maxwell's equations

$$\nabla \cdot \mathbf{E} = \frac{\rho}{\varepsilon_0} \tag{9.A.1a}$$

$$\nabla \times \mathbf{E} = -\frac{\partial \mathbf{B}}{\partial t} \tag{9.A.1b}$$

$$\nabla \cdot \mathbf{B} = 0 \tag{9.A.1c}$$

$$\nabla \times \mathbf{B} = \mu_0 \mathbf{j} + \varepsilon_0 \mu_0 \frac{\partial \mathbf{E}}{\partial t} \tag{9.A.1d}$$

will yield a large number of conservation laws [25, 56]. The two most well-known ones are the conservation of energy (Poynting's theorem)

$$\frac{\mathrm{d}U^{\mathrm{mech}}}{\mathrm{d}t} + \frac{\mathrm{d}U^{\mathrm{em}}}{\mathrm{d}t} + \oint_{S'} \mathrm{d}^2x' \hat{\mathbf{n}}' \cdot \mathbf{S} = 0 \tag{9.A.2}$$

and the conservation of linear momentum (integrated Poynting flux)

$$\frac{\mathrm{d}\mathbf{p}^{\mathrm{mech}}}{\mathrm{d}t} + \frac{\mathrm{d}\mathbf{p}^{\mathrm{em}}}{\mathrm{d}t} + \oint_{S'} \mathrm{d}^2x' \hat{\mathbf{n}}' \cdot \mathbf{T} = 0 \tag{9.A.3}$$

Here U^{mech} is the mechanical energy,

U^{em} is the electromagnetic field energy,

$\mathbf{S} = \varepsilon_0 c^2 \mathbf{E} \times \mathbf{B}$ is the Poynting vector,

$\mathbf{p}^{\mathrm{mech}}$ is the mechanical linear momentum,

$\mathbf{p}^{\mathrm{em}} = \frac{1}{c^2} \int_{V'} \mathrm{d}^3x' \mathbf{S}$ is the EM field linear momentum and (9.A.4)

$\mathbf{T} = \frac{\varepsilon_0}{2} \mathbf{1} \left(\mathbf{E} \cdot \mathbf{E} + c^2 \mathbf{B} \cdot \mathbf{B}\right) - \varepsilon_0 \left(\mathbf{EE} + c^2 \mathbf{BB}\right)$
is the linear momentum flux tensor (9.A.5)

Likewise, one can derive the conservation law for angular momentum

$$\frac{\mathrm{d}\mathbf{J}^{\mathrm{mech}}}{\mathrm{d}t} + \frac{\mathrm{d}\mathbf{J}^{\mathrm{em}}}{\mathrm{d}t} + \oint_{S'} \mathrm{d}^2x' \hat{\mathbf{n}}' \cdot \mathbf{M} = 0 \tag{9.A.6}$$

where

$\mathbf{J}^{\mathrm{mech}}$ is the mechanical angular momentum,

$$\mathbf{J}^{\text{em}} = \frac{1}{c^2}\int_{V'} (\mathbf{x}' - \mathbf{x}_0) \times \mathbf{S} \text{ is the EM field angular momentum around } \mathbf{x}_0 \text{ and} \tag{9.A.7}$$

$$\mathbf{M} = (\mathbf{x} - \mathbf{x}_0) \times \mathbf{T} \text{ is the angular momentum flux tensor around } \mathbf{x}_0. \tag{9.A.8}$$

The conservation law (9.A.6) is the angular analog of the linear momentum conservation law (9.A.3), and describes how an electromagnetic source radiates angular momentum. It is easy to see that the angular momentum has the same $1/r^2$ falloff as the linear momentum and therefore, is also transmitted over very large distances.

Expressed in the vector potential $\mathbf{A}$, the field angular momentum can often be written as

$$\mathbf{J}^{\text{em}} = \varepsilon_0 \int_{V'} \mathrm{d}^3x' (\mathbf{E} \times \mathbf{A}) + \varepsilon_0 \int_{V'} \mathrm{d}^3x' E_i[(\mathbf{x}' - \mathbf{x}_0) \times \nabla]A_i \tag{9.A.9}$$

For a single Fourier component, the origin chosen at $\mathbf{x}_0$, and in the real-valued representation, one finds that this can be rewritten as

$$\mathbf{J}^{\text{em}} = -\mathrm{i}\frac{\varepsilon_0}{2\omega}\int_{V'} \mathrm{d}^3x' (\mathbf{E}^* \times \mathbf{E}) + \frac{\varepsilon_0}{2\hbar\omega}\int_{V'} \mathrm{d}^3x' E_i \hat{L} E_i \tag{9.A.10}$$

where

$$\hat{L} = -\mathrm{i}\hbar\mathbf{x}' \times \nabla \tag{9.A.11}$$

is the quantum OAM operator. Not surprisingly, in beam geometry, the first term on the right-hand side of Eq. (9.A.10) can be identified as the SAM and the second term as the OAM.

A more satisfactory procedure to explore the electromagnetic field in a search for conservation laws and related useful properties, is to find all symmetries of a Lagrangian for the EM fields and use mathematical tools such as group theory and Noether's theorem [57] to derive all conserved quantities. A Lie group analysis of the first-order continuous symmetries will show that the classical electromagnetic field has 23 constants of motion. Three of them are the components of the electromagnetic angular momentum.

9.A.2
Photon Picture

At the single photon level, the detectability of OAM emitted by a faint astronomical source can be faced with a fully relativistic quantum mechanical approach, by adopting the PWF formalism [58–64], which is completely equivalent to QED [65], without requiring particular boundary conditions for the field quantization describing OAM states [66]. In addition, the PWF ↔ QED correspondence for free photons sheds light on the intrinsicality of the photon OAM [11, 24, 67, 68]. The problem of writing down a wavefunction for the photon, that is, describing the EM field with a first-quantization formalism, has its origin at the birth of quantum mechanics, leaving unsettled the issue of the localization of the photon that makes the issue of definition of a PWF controversial among many authors. But this is not crucial for the present discussion, and the reader can find more information in [8, 16, 59]. The application of the PWF formalism to OAM detection and manipulation and to Glauber correlations [60, 63, 64, 69] in astronomy provides a useful and complete description of those scenarios where the photons are either propagating in free space or in linear media, in situations when photons maintain a constant number during their evolution; that is, when there are no creation or annihilation processes occurring in the field and the OAM value is maintained.

Historically, the first attempts at constructing a PWF can be found in the unpublished works by Majorana [70]. The quantum states of the electromagnetic field were described within a first-quantization formalism, taking the form of a Dirac-like equation obtained from the Riemann–Silberstein (RS) formulation of Maxwell's equations. This approach is justified by the fact that the Maxwell equations present an intrinsic mathematical structure similar to that of a quantum wave function in relativistic theory [71–73]. Conversely, the same procedure that is followed to write the Dirac equation can be used to derive the Maxwell equations. Even if the Dirac equation was formulated to describe the relativistic electron, a particle with nonzero rest mass, $\hbar/2$ spin, and elementary charge e, Majorana was the first to extend the Dirac equation to particles with arbitrary spin and mass, in a more general infinite-spin component formalism [74]. A subclass is the well-known group of Weyl–Majorana equations that describe massless neutral spinors.

Adopting the Majorana formulation and the RS vector can, without loss of generality, be defined as

$$\mathbf{F} = \frac{\mathbf{E}}{c} \pm \mathrm{i}\mathbf{B} \tag{9.A.12}$$

and the Maxwell equations in the vacuum become

$$\nabla \cdot \mathbf{F} = 0, \quad \mathrm{i}\nabla \times \mathbf{F} = \pm\frac{1}{c}\frac{\partial \mathbf{F}}{\partial t} \tag{9.A.13}$$

By applying the quantization rule, $\mathbf{p} \leftrightarrow \hat{\mathbf{p}} \equiv -\mathrm{i}\hbar\nabla$, $\mathbf{F}$ will be structured in the form of a wavefunction for the photon

$$\mp\frac{\mathrm{i}\hbar}{c}\frac{\partial}{\partial t}\mathbf{F} + \mathrm{i}\hat{\mathbf{p}} \times \mathbf{F} = \mathbf{0} \tag{9.A.14}$$

and the other group of Maxwell equations, $\nabla \cdot \mathbf{F} = \mathbf{0}$, describe the transversality of the fields with respect to the propagation direction, $\hat{\mathbf{p}} \cdot \mathbf{F} = 0$. By introducing the complex-valued 3×3 matrices

$$\hat{s}_x = \begin{pmatrix} 0 & 0 & 0 \\ 0 & 0 & -\mathrm{i} \\ 0 & \mathrm{i} & 0 \end{pmatrix}, \quad \hat{s}_y = \begin{pmatrix} 0 & 0 & \mathrm{i} \\ 0 & 0 & 0 \\ -\mathrm{i} & 0 & 0 \end{pmatrix}, \quad \hat{s}_z = \begin{pmatrix} 0 & -\mathrm{i} & 0 \\ \mathrm{i} & 0 & 0 \\ 0 & 0 & 0 \end{pmatrix} \tag{9.A.15}$$

one obtains the Dirac-like equation

$$\mathrm{i}\hbar\frac{\partial}{\partial t}\mathbf{F} = \hat{H}\mathbf{F} \tag{9.A.16}$$

where $\hat{H} = \pm c\hat{\mathbf{s}} \cdot \hat{\mathbf{p}}$ and $\hat{\mathbf{s}} = (\hat{s}_x, \hat{s}_y, \hat{s}_z)$ [58, 60, 70]. This Hamiltonian has eigenvalues $\pm c\,p, 0$. The eigenvalue 0 is forbidden by the transversality condition [75].

A covariant formulation for **F** is obtained via the existing isomorphism between the algebras of the group $SL(2C)$ of unimodular 2×2 matrices in the complex field and the proper orthochronous Lorentz group $SO(1,3)$, the group of 4×4 (pseudo)orthogonal real matrices that leave the Minkowski metric $\eta^{\mu\nu} = \text{diag}(1, -1, -1, -1)$ invariant. Since the two algebras are isomorphic, the two groups then satisfy a local isomorphism. More precisely, when extended to a global isomorphism, the correspondence is reduced to a $2 \to 1$ homomorphism. In fact, consider the matrix

$$\overline{x} = \begin{pmatrix} x_0 + x_3 & x_1 - \mathrm{i}x_2 \\ x_1 + \mathrm{i}x_2 & x_0 - x_3 \end{pmatrix} \tag{9.A.17}$$

built with the space–time coordinates of an event, $\{x\}_i$, $(i = 0, 1, 2, 3)$. Transform $\overline{x}$ by an $SL(2C)$ transformation $\overline{x'} = A^{-1}\overline{x}A$, where $A \in SL(2C)$, so that $\det \overline{x'} = \det \overline{x}$. Then the homomorphism is $2 \to 1$ because A corresponds to a Lorentz transformation Λ on the four coordinates but both $\pm A$ correspond to the same Λ. The group $SL(2C)$ has two nonequivalent, fundamental representations $\left(\frac{1}{2}, 0\right)$ and $\left(0, \frac{1}{2}\right)$ that can be also considered as two spinorial representations of $SO(1,3)$, for example, the chiral and antichiral Weyl spinors.

Tensorial higher-dimensional products of the fundamental representations of the Lorentz group are divided into two classes: the tensorial representations, from the product of even times the fundamental representations $(\frac{1}{2}, 0)$ and/or $(0, \frac{1}{2})$, and the spinorial representations, from the product of odd ones and the RS vector. The wavefunction of the photon can be cast in a six-dimensional representation $\left(\frac{1}{2}\frac{1}{2}, 0\right) \oplus \left(0, \frac{1}{2}\frac{1}{2}\right)$ by using the definition of the Faraday electromagnetic tensor $F^{\mu\nu}$. We recall that the double antisymmetric tensor components are $F^{0i} = -E_i/c$ and $F^{ij} = -\varepsilon^{ijk}B_k$, where ε^{ijk} is the totally antisymmetric Ricci (Levi–Civita) tensor. Its self-dual and antiself-dual parts $\left(F^{+}_{\mu\nu}, F^{-}_{[\mu\nu]}\right)$ written in covariant spinorial notation become

$$F_{(\alpha\beta)} = (\sigma^{[\mu}\overline{\sigma}^{\nu]}F^{+}_{[\mu\nu]})\alpha\beta \tag{9.A.18}$$

$$\overline{F}_{(\dot{\alpha}\dot{\beta})} = (\overline{\sigma}^{[\mu\sigma\,\nu]}F^{-}_{[\mu\nu]})_{\dot{\alpha}\dot{\beta}} \tag{9.A.19}$$

where $F_{(\alpha\beta)} \in \left(\frac{1}{2}, \frac{1}{2}, 0\right)$ and $\overline{F}_{(\dot{\alpha}\dot{\beta})} \in \left(0, \frac{1}{2}\frac{1}{2}\right)$, but

$$F_{(\alpha\beta)} \propto \left((\sigma^{[0}\overline{\sigma}^{i]})F^{+}_{0i}\right)_{\alpha\beta} \equiv \left(\overline{\sigma}^{i}F^{+}_{i}\right)_{\alpha\beta} \tag{9.A.20}$$

$$\overline{F}_{(\dot{\alpha}\dot{\beta})} \propto \left((\overline{\sigma}^{[0}\sigma^{i]})F^{-}_{0i}\right)_{\dot{\alpha}\dot{\beta}} \equiv \left(\sigma^{i}F^{-}_{i}\right)_{\dot{\alpha}\dot{\beta}} \tag{9.A.21}$$

The 2×2 matrices $(\sigma^{\mu})_{\gamma}{}^{\dot{\beta}}$ and $(\overline{\sigma}^{\mu})_{\dot{\alpha}}{}^{\gamma}$ are defined by the Pauli matrices $\overline{\sigma}^{0} = \sigma^{0} = 1$, $\overline{\sigma}^{i} = -\sigma^{i}$ and σ^{i}. From these definitions we obtain the RS tensor expressed in positive and negative helicity wavefunctions of the photon $F_{(\alpha\beta)}$ and $\overline{F}_{(\dot{\alpha}\dot{\beta})}$ in covariant notation

$$F^{\pm}_{i} = \frac{E_i}{c} \pm \mathrm{i}B_i \tag{9.A.22}$$

For the PWF the Dirac-like equation can be derived as

$$(\overline{\sigma}^{\mu}\partial_{\mu})_{\dot{\alpha}}{}^{\beta}F_{(\beta\alpha)} = (\overline{\sigma}^{\mu}\partial_{\mu}\sigma^{\nu}\overline{\sigma}^{\lambda})_{\dot{\alpha}\alpha}F^{+}_{\nu\lambda} = 0 \tag{9.A.23}$$

and its complex conjugate. By saturating Eq. (9.A.23) with $(\sigma^{\tau})^{\alpha\dot{\alpha}}$ one gets, for $\tau = 0$, the transversality condition and, for $\tau = i$, the second Maxwell set of Eq. (9.A.13) that is the Dirac equation (9.A.16). The same results are obtained by saturating the complex conjugate with $(\overline{\sigma}^{\tau})^{\dot{\alpha}\alpha}$.

The RS vector is used to describe the EM field vorticity [76–79]. RS vortices are defined by the condition

$$\mathbf{F}(\mathbf{r}, t) \cdot \mathbf{F}(\mathbf{r}, t) = 0 \tag{9.A.24}$$

since the loci of points satisfying this condition are lines in space where the phase of the field is singular, surrounded by zones where the phase gradient vector is circulating. LG beams represent a particular class of RS vortices in which the field has spatial symmetry and the RS vortex lines remain stationary. Exact solutions of electromagnetic waves carrying angular momentum have been described in the momentum representation and then cast in terms of PWF in the RS formalism [79]. This representation is simply equivalent to the description obtained in QED by quantizing the field [65], for example, in paraxial approximation [66]. At the single photon level, the meaning of an OAM-carrying PWF is that it represents the probability amplitude of finding a photon in a certain eigenstate of momentum, helicity, and OAM.

The angular momentum J of a particle is given by the sum of the OAM l and the intrinsic (spin) angular momentum s, and the wavefunction of a particle with spin S is a $2S$-rank symmetric spinor with $2S + 1$ components. The EM field is a vectorial field, therefore, the photon is assigned a spin 1 that can be represented by a rank 2 spinor. OAM is related to the spatial dependence of the wavefunction. For this reason, a clear unequivocal separation of l and S requires the independence of the intrinsic (spin) and spatial-extrinsic (OAM) properties of the wavefunction. In the case of the PWF this is not possible because of the transversality condition of the field, which imposes a dependence on the momentum, forbidding the separation between S and L.

Another important point is that one cannot apply the concept of spin in terms of angular momentum of a particle in its rest frame. The photon is moving at the speed of light. The only proper property of the photon is then the sum $j = l + s$, being the two quantities that are somehow entangled. SAM and OAM of the photon become meaningless if defined separately. They are auxiliary concepts that describe the PWF generating the spin-to-OAM conversion when traversing inhomogeneous media. Spin and polarization are related to the helicity states of photons and a change of polarization might induce spin-to-OAM conversion. We notice that, by definition, the intrinsic properties of a quantum particle are those characteristics that do not depend on the choice of reference frame and coordinates. Those quantities are simply rest mass, electric charge, and spin. S and L cannot evolve separately in the case of the photon. Otherwise, we would find a paradox in the mathematical structure of the PWF. In fact, if the OAM were an intrinsic property of the photon, it should then be a fortiori related to the intrinsic component represented by the spin S alone as the spin calculated with QED at the single photon level would imply an $S - L$ separability. In that case, the Dirac-like equation for the RS field would admit an infinite spectrum of intrinsic angular momentum states [74] for the PWF, also at the single photon level without coordinate dependence.

The RS formalism is complex because F is complex by construction. The analytic continuation of the RS formalism gives, through the conservation equations, a handle on symmetries "hidden" inside the Maxwell equations, by any combination of the translation of the RS components in space and time and this also includes the OAM of light. Any conservation law has the form of a time translation plus a spatial translation

$$\frac{1}{c}\partial_t a(i) + \nabla \cdot a(i+1) = 0 \tag{9.A.25}$$

where i is the tensor rank, $a(i)$ the mixed space–time component of a general function of the RS and $a(i+1)$ the purely spatial component. For $i = 0$ we have the conservation law of a scalar $a(0)$ and the correspondent current vector term $a(1)$. Higher tensorial ranks involve the conservation of vectors and tensors, that express the coherence functions of the field, such as for the Kujawski tensor [80] for single- and multiphoton correlations. By applying the prescriptions of quantum mechanics, the operator ∂_t corresponds to energy and the generator of spatial translation (∇) to momentum. OAM, instead, is related to the referring of the field with respect to an event in space–time with a 4-vector (r_0, r)

$$\frac{1}{c}\partial_t (r \times a(i)) + \nabla \times a(i+1) = 0 \tag{9.A.26}$$

and with the equation of motion for r, we can recover the well-known SAM-to-OAM conversion process.

Of course, measurable physical quantities are supposed to be expressed in terms of sequences of digits which, in theoretical models for these observables, are represented by real numbers. For this reason, in the analytical continuation of Maxwell equations one takes the real part of the solution, apparently losing one part of the information. Maxwell equations can be cast in terms of a Dirac-like equation, and this first-quantization formulation is fully equivalent to the subclass of solutions of quantum electrodynamics [65]. Here, by applying the quantum correspondence principle, fields become operators that are complex-valued functions in Hilbert spaces. The symmetries found with the analytical continuation of the RS formalism then describe the property of the EM field in terms of operators and, as dictated by quantum mechanics, observables are real numbers obtained via the product of the field and its complex conjugate (within certain boundary conditions). One important point we want to discuss is the mathematical correspondence between the complex formalism of the analytical continuation that the Dirac equation of the EM field can be written in the real Majorana formalism by imposing a nonsingular unitary transformation $M \rightarrow UMU^{-1}$ that swaps the matrices $\alpha_2 = \begin{pmatrix} 0 & \sigma^2 \\ \sigma^2 & 0 \end{pmatrix}$, related to the momentum with those related to the mass term $\beta = \begin{pmatrix} I & 0 \\ 0 & -I \end{pmatrix}$ so that both disappear by imposing $m = 0$. Finally, by changing the sign of the other two α matrices, one obtains the real-valued Dirac–Majorana equation for the EM field for both helicity states of the photon

$$(\partial_t + \alpha \cdot \nabla + \beta m) F = 0 \rightarrow \left(\partial_t + \hat{\alpha} \cdot \nabla\right) F = 0 \tag{9.A.27}$$

where $\hat{\alpha_1} = -\alpha_1$, $\hat{\alpha_2} = \beta$ and $\hat{\alpha_3} = -\alpha_3$ [81]. This would indicate that in any case, because of the zero rest mass of the photon, the group of solutions and symmetries obtained with the full complex approach is isomorphic to those obtained with a real formulation of the RS vector times a unitary transformation. We argue that the symmetries derived from the complex RS formalism actually represent properties of the field that are directly expressed in terms of quantum mechanical operators by the PWF formalism.

References

1. Kassim, N.E. and Weiler, K.W. (eds) (1990) *Low Frequency Astronomy from Space*, Springer-Verlag, New York.
2. Jester, S. and Falcke, H. (2009) Science with a lunar low-frequency array: from the dark ages of the Universe to nearby exoplanets. *New Astron. Rev.*, **53**, 1–26.
3. Abraham, M. (1914) Der Drehimpuls des Lichtes. *Phys. Z.*, **XV**, 914–918.
4. Beth, R.A. (1936) Mechanical detection and measurement of the angular momentum of light. *Phys. Rev.*, **50**(2), 115–125.
5. Carrara, N. (1949) Torque and angular momentum of centimetre electromagnetic waves. *Nature*, **164** (4177), 882–884.
6. Allen, L., Beijersbergen, M.W., Spreeuw, R.J.C. and Woerdman, J.P. (1992) Optical angular momentum of light and the transformation of Laguerre-Gauss laser modes. *Phys. Rev. A*, **45**, 8185–8189.
7. Kristensen, M. and Woerdman, J.P. (1994) Is photon angular momentum conserved in a dielectric medium? *Phys. Rev. Lett.*, **74** (14), 2171.
8. Arnaut, H.H. and Barbosa, G.A. (2000) Orbital and intrinsic angular momentum of single photons and entangled pairs of photons generated by parametric down-conversion. *Phys. Rev. Lett.*, **85** (2), 286–289.
9. Leach, J., Padgett, M.J., Barnett, S.M., Franke-Arnold, S. and Courtial, J. (2002) Measuring the orbital angular momentum of a single photon. *Phys. Rev. Lett.*, **88** (25), 257901(4).
10. Leach, J., Courtial, J., Skeldon, K., Barnett, S.M., Franke-Arnold, S. and Padgett, M.J. (2004) Interferometric methods to measure orbital and spin, or the total angular momentum of a single photon. *Phys. Rev. Lett.*, **92** (1), 013601(4).
11. Mair, A., Vaziri, A., Weihs, G. and Zeilinger, A. (2001) Entanglement of the orbital angular momentum states of photons. *Nature*, **412**, 313–316.
12. Vaziri, A., Weihs, G. and Zeilinger, A. (2002) Superpositions of the orbital angular momentum for applications in quantum experiments. *J. Opt. B: Quant. Semiclass. Opt.*, **4**, S47–S51.
13. Harwit, M. (2003) Photon orbital angular momentum in astrophysics. *Astrophys. J.*, **597** (2), 1266–1270.
14. Tamburini, F., Anzolin, G., Bianchini, A. and Barbieri, C. (2006) Overcoming the Rayleigh criterion limit with optical vortices. *Phys. Rev. Lett.*, **97** (25), 163903(4).
15. Dravins, D. (2007) Photonic astronomy and quantum optics, in *High Time Resolution Astrophysics*, Astrophysics and Space Science Library, vol. 351 (eds D. Phelan, O. Ryan and A. Shearer), Springer, Netherlands, pp. 95–132.
16. Thidé, B., Then, H., Sjöholm, J., Palmer, K., Bergman, J.E.S., Carozzi, T.D.,

Istomin, Y.N., Ibragimov, N.H. and Khamitova, R. (2007) Utililization of photon orbital angular momentum in the low-frequency radio domain. *Phys. Rev. Lett.*, **99** (8), 087701.

17. Thidé, B. (2007) Nonlinear physics of the ionosphere and LOIS/LOFAR. *Plasma Phys. Controlled Fusion*, **49** (8), B103–B107.
18. Anzolin, G., Tamburini, F., Bianchini, A., Umbriaco, G. and Barbieri, C. (2008) Optical vortices with starlight. *Astron. Astrophys.*, **488**, 1159–1165.
19. Elias, N.M. II (2008) Photon orbital angular momentum in astronomy. *Astron. Astrophys.*, **492**, 883–922.
20. Berkhout, G.C.G. and Beijersbergen, M.W. (2008) Method for probing the orbital angular momentum of optical vortices in electromagnetic waves from astronomical objects. *Phys. Rev. Lett.*, **101**, 100801.
21. Berkhout, G.C.G. and Beijersbergen, M.W. (2009) Using a multipoint interferometer to measure the orbital angular momentum of light in astrophysics. *J. Opt. A: Pure Appl. Opt.*, **11**, 094021 (7pp).
22. Dravins, D. *et al.* (2005) QuantEYE: quantum optics instrumentation for astronomy, OWL instrument concept study. ESO document OWL-CSR-ESO-00000-0162.
23. Heitler, W. (1936) *The Quantum Theory of Radiation*, Clarendon Press, Oxford.
24. Molina-Terriza, G., Torres, J.P. and Torner, L. (2002) Management of the angular momentum of light: Preparation of photons in multidimensional vector states of angular momentum. *Phys. Rev. Lett.*, **88** (1), 013601.
25. Thidé, B. (2010) *Electromagnetic Field Theory*, Dover Publications, New York, in press. ISBN: 978-0-486-4773-2. On-line internet version: http://www.plasma.uu.se/CED/Book.
26. Danjon, A. and Couder, A. (1935) *Lunettes et Teléscopes*, Éditions de la Revue d'Optique Théorique et Instrumentale, Paris.
27. Hecht, E. (2002) *Optics*, 4th edn, Addison-Wesley, San Francisco, CA.
28. Müller, H., Chiow, S.-W., Long, Q., Vo, C. and Chu, S. (2005) Active sub-Rayleigh alignment of parallel or antiparallel laser beams. *Opt. Lett.*, **30**, 3323–3325.
29. Swartzlander, G.A. Jr. (2001) Peering into darkness with a vortex filter. *Opt. Lett.*, **26** (8), 497–499.
30. Anzolin, G., Tamburini, F., Bianchini, A. and Barbieri, C. (2009) Method to measure off-axis displacements based on the analysis of the intensity distribution of a vortex beam. *Phys. Rev. A*, **79** (3), 033845.
31. Fried, D.L. (1978) Probability of getting a lucky short-exposure image through turbulence. *J. Opt. Soc. Am. (1917–1983)*, **68**, 1651–1658.
32. Law, N.M., Mackay, C.D. and Baldwin, J.E. (2006) Lucky imaging: high angular resolution imaging in the visible from the ground. *Astron. Astrophys.*, **446**, 739–745.
33. McAlister, H.A., Hartkopf, W.I., Sowell, J.R., Dombrowski, E.G. and Franz, O.G. (1989) ICCD speckle observations of binary stars. IV – measurements during 1986–1988 from the Kitt Peak 4 m telescope. *Astron. J.*, **97**, 510–531.
34. Halbwachs, J.L. (1981) List of estimated angular separations of spectroscopic binaries. *Astron. Astrophys. Suppl. Ser.*, **44**, 47.
35. Falcke, H. *et al.* (2005) Detection and imaging of radio flashes from cosmic ray air showers. *Nature*, **435**, 313–316.
36. Stål, O., Bergman, J., Thidé, B., Daldorff, L.K.S. and Ingelman, G. (2007) Prospects for lunar satellite detection of radio pulses from ultrahigh energy neutrinos interacting with the Moon. *Phys. Rev. Lett.*, **98** (7), 071103.
37. Thidé, B., Sergeev, E.N., Grach, S.M., Leyser, T.B. and Carozzi, T.D. (2005) Competition between Langmuir and upper-hybrid turbulence in a high-frequency-pumped ionosphere. *Phys. Rev. Lett.*, **95** (25), 255002(4).
38. Torner, L., Torres, J.P. and Carrasco, S. (2005) Digital spiral imaging. *Opt. Express*, **13** (3), 873–881.
39. Fried, D.L. (1998) Branch point problems in adaptive optics. *J. Opt Soc Am. A*, **15** (10), 2759–2768.
40. Mendonça, J.T., Thidé, B. and Then, H. (2009) Stimulated Raman and Brillouin

backscattering of collimated beams carrying orbital angular momentum. *Phys. Rev. Lett.*, **102** (18), 18505(4).

41. Mohammadi, S.M., Daldorff, L.K.S., Bergman, J.E.S., Karlsson, R.L., Thidé, B., Forozesh, K. and Carozzi, T.D. (2009) Orbital angular momentum in radio – a system study. *IEEE Trans. Ant. Prop.*, **58** (2), 565–572.
42. Petrov, D.V., Molina-Terriza, G. and Torner, L. (1999) Vortex evolution in parametric wave mixing. *Opt. Commun.*, **162** (6), 357–366.
43. Tabosa, J.W.R. and Petrov, D.V. (1999) Optical pumping of orbital angular momentum of light in cold cesium atoms. *Phys. Rev. Lett.*, **83**, 4967–4971.
44. Arlt, J., Dholakia, L., Allen, L. and Padgett, M.J. (1999) Parametric down-conversion for light beams possessing orbital angular momentum. *Phys. Rev. A*, **59** (5), 3950–3952.
45. Caetano, D.P., Almeida, M.P. and Souto Ribero, P.H. (2002) Conservation of orbital angular momentum in stimulated down-conversion. *Phys. Rev. A*, **66** (4), 041801(R).
46. Molina-Terriza, G., Torres, J.P. and Torner, L. (2003) Orbital angular momentum of photons in noncollinear parametric downconversion. *Opt. Commun.*, **228**, 155–169.
47. Ren, X.-F., Guo, G.-P., Yu, B., Li, J. and Gua, G.-C. (2004) The orbital angular momentum of down-converted photons. *J. Opt. B: Quant. Semiclass. Opt.*, **6**, 243–247.
48. Huguenin, J.A.O., Martinelli, M., Caetano, D.P., Coutinho dos Santos, B., Almeida, M.P., Souto Ribeiro, P.H., Nussenzveig, P. and Khoury, A.Z. (2006) Orbital angular momentum exchange in parametric down conversion. *J. Mod. Opt.*, **53** (5-6), 647–658.
49. Sadovskii, A.I. (1899) *Uchen. Zapis. Imp. Yurev. Univ.*, **52** (1), 1.
50. Shvets, G., Fisch, N.J. and Rax, J.-M. (2002) Magnetic field generation through angular momentum exchange between circularly polarised radiation and charged particles. *Phys. Rev. E*, **65**, 046403.
51. Istomin, Y.N. (2002) The ponderomotive action of a powerful electromagnetic wave on ionospheric plasma. *Phys. Lett. A*, **299**, 248–257.
52. Pitaevskii, L.P. (1961) Electric forces in an transparent dispersive medium. *Sov. Phys. JETP, Engl. Transl.*, **12** (5), 1008–1013.
53. Pershan, P.S. (1963) Nonlinear properties of solids: Energy considerations. *Phys. Rev.*, **130** (3), 919–929.
54. Deschamps, J., Fitaire, M. and Lagoutte, M. (1970) Inverse Faraday effect in a plasma. *Phys. Rev. Lett.*, **25**, 1330–1332.
55. Mendonça, J.T. and Thidé, B. (2008) Neutrino orbital angular momentum in a plasma vortex. *Europhys. Lett.*, **84**, 41001.
56. Schwinger, J., DeRaad, L.L. Jr., Milton, K.A. and Tsai, W.-Y. (1998) *Classical Electrodynamics*, Perseus Books, Reading, MA.
57. Noether, E. (1918) Invariante Variationsprobleme. *Nachr. Ges. Wiss. Göttingen*, **1** (3), 235–257. English transl.: (1971) Invariant variation problems. Trans. Theor. Stat. Phys., **1**, 186–207.
58. Kobe, D.H. (1999) A relativistic Schrödinger equation for a photon and its second quantization. *Found. Phys.*, **29**, 1203.
59. Bialynicki-Birula, I. (1994) On the wave function of the photon. *Acta Phys. Pol., A*, **86**, 97–116.
60. Bialynicki-Birula, I. (1996) Photon wave function, in *Progress in Optics*, vol. XXXVI (ed. E. Wolf), Elsevier, Amsterdam, Holland, pp. 245–294.
61. Esposito, S. (1999) Covariant Majorana formulation of electrodynamics. *Found. Phys.*, **28**, 231–244.
62. Gersten, A. (2001) Maxwell equations as the one-photon quantum equation. *Found. Phys.*, **31**, 1211–1231.
63. Raymer, M.G. and Smith, B.J. (2005) The Maxwell wave function of the photon, in *Society of Photo-optical Instrumentation Engineers (SPIE) Conference Series*, vol. 5866 (eds C. Roychoudhuri and K. Creath). Bellingham, WA, U.S.A., pp. 293–297. (Invited).
64. Smith, B.J. and Raymer, M.G. (2006) Two-photon wave mechanics. *Phys. Rev. A*, **74** (6), 062104.

65. Tamburini, F. and Vicino, D. (2008) Photon wave function: a covariant formulation and equivalence with QED. *Phys. Rev. A*, **78** (5), 052116.

66. Calvo, G.F., Picón, A. and Bagan, E. (2006) Quantum field theory of photons with orbital angular momentum. *Phys. Rev. A*, **73**, 013805–1 (9 pages).

67. Arlt, J., Dholakia, K., Allen, L. and Padgett, M.J. (1998) The production of multiringed Laguerre-Gaussian modes by computer-generated holograms. *J. Mod. Phys.*, **45**, 1231–1237.

68. O'Neil, A.T., MacVicar, I., Allen, L. and Padgett, M.J. (2002) Intrinsic and extrinsic nature of the orbital angular momentum of a light beam. *Phys. Rev. Lett.*, **88** (5), 053601.

69. Smith, B.J. and Raymer, M.G. (2007) Photon wave functions, wave-packet quantization of light, and coherence theory. *New J. Phys.*, **9**, 414.

70. Mignani, R., Recami, E. and Baldo, M. (1974) About a Dirac-like equation for the photon according to Ettore Majorana. *Lett. Nuovo Cimento*, **11** (12), 568–572.

71. Riemann, B. (1901) *Die Partiellen Differential-gleichungen der Mathematische Physik*, vol. 2 (ed. H. Weber, Vieweg, Braunschweig.

72. Silberstein, L. (1907) Elektromagnetische Grundgleichungen in bivektorieller Behandlung. *Ann. Phys. (Leipzig)*, **327** (3), 579–586.

73. Bergman, J.E.S., Mohammadi, S.M., Carozzi, T.D., Daldorff, L.K.S., Thidé, B., Karlsson, R.L. and Eriksson, M. (2008) Conservation laws in generalized Riemann-Silberstein electrodynamics. **arXiv**, 0803.2383.

74. Majorana, E. (1932) Teoria relativistica di particelle con momento intrinseco arbitrario. *Nuovo Cimento*, **9**, 335–344.

75. Berestetskii, V.B., Lifshitz, E.M. and Pitaevskii, V.B. (1971) in *Relativistic Quantum Theory, volume 1 of Course of theoretical physics – Pergamon International Library of Science, Technology, Engineering & Social Studies* (eds L.D. Landau and E.M. Lifshitz), Pergamon Press, Oxford.

76. Berry, M.V. (2004) The electric and magnetic polarization singularities of paraxial waves. *J. Opt. A: Pure Appl. Opt.*, **6**, 475–481.

77. Berry, M.V. (2004) Riemann-Silberstein vortices for paraxial waves. *J. Opt. A: Pure Appl. Opt.*, **6**, S175–S177.

78. Radozycki, T. (2004) Quantum effects in the evolution of vortices in the electromagnetic field. *Phys. Rev. E*, **69** (6), 066616.

79. Bialynicki-Birula, I. and Bialynicki-Birula, Z. (2006) Beams of electromagnetic radiation carrying angular momentum: the Riemann-Silberstein vector and the classical-quantum correspondence. *Opt. Commun.*, **264**, 342–351.

80. Kujawski, A. (1966) Relativistic formulation of correlation theory of electromagnetic fields. *Nuovo Cimento*, **44B**, 326–337.

81. Itzykson, C. and Zuber, J.-B. (1980) *Quantum field theory*, International series in pure and applied physics, McGraw-Hill, New York.

82. Pike, E.R. and Sarkar, S. (1995) The Oxford University Press, New York.

10
Optical Vortex Cat States and their Utility for Creating Macroscopic Superpositions of Persistent Flows

Ewan M. Wright

10.1
Introduction

The aim of this chapter is to introduce the idea of optical vortex cat states and demonstrate their utility for creating macroscopic superpositions of persistent flows for a Bose–Einstein condensate (BEC) in a toroidal or ring trap. Weakly interacting atomic BECs are considered, and a persistent flow is defined as one in which all N atoms are in the same rotational or flow state characterized by a winding number q, and net orbital angular momentum (OAM) $Nq\hbar$ around the axis of the ring BEC. Thus, even for a modest condensate of say 10^3 atoms, persistent flows with different winding numbers are macroscopically distinguishable. The relevance of macroscopic superposition states lies in the fact that they are important for testing the range of validity of the quantum theory, as stressed by Leggett [1], and macroscopic superpositions of persistent flows may allow quantum-limited angular momentum measurements, leading to the potential for ultra-precise gyroscopes. Furthermore, macroscopic superpositions of persistent flows would be cold atom analogs of those observed in superconducting quantum interference devices [2–4].

If the stated aim was instead to create a BEC in which *each atom* was in a superposition of flow states with differing winding numbers, then a number of schemes already exist for producing such states, including stirring using rotating light-shift potentials based on a combination of laser fields [5, 6], "phase engineering" involving a Gaussian laser beam whose center is rotated, which couples the external motion to the internal state via Rabi oscillations [7, 8], and most significantly for this work, vortex coupling in which two-photon stimulated Raman transitions are driven using Laguerre–Gaussian (LG) fields to transfer OAM from the LG beam photons to the trapped atoms [9–11]. The LG fields are examples of optical vortices, which carry OAM associated with the spiral transverse phase profile of their electric field envelopes [12]. Experiments from NIST have already demonstrated quantized rotation of trapped atoms using LG fields [13], and persistent flow of a BEC in a toroidal trap [14]. Simula *et al.* [15] have shown using three-dimensional numerical simulations based on the Gross–Pitaevskii equation that the experiment on quantized rotation of trapped atoms may be cast in terms

Twisted Photons: Applications of Light with Orbital Angular Momentum.
Edited by Juan P. Torres and Lluis Torner

ISBN: 978-3-527-40907-5

of coherent Bragg scattering of the atoms from a chiral light-shift potential. In vortex coupling, the fields driving the two-photon stimulated Raman transitions are typically LG fields of different winding numbers, and together they produce a rotating light-shift potential, which can be made to stir the BEC in one or other direction and excite a persistent flow. The direction of the stirring, and hence of the excited persistent flow, is determined by the sign of the frequency detuning factor between the fields and the LG field winding numbers. Furthermore, Dowling and coworkers [10, 11] have shown how arbitrary superpositions of flow states can be generated and detected.

A number of theoretical works have appeared involving the creation of macroscopic superpositions of BECs [16–23]. Of particular interest, here, is the recent work on macroscopic superposition states in ring superlattices in which an array of BECs trapped in optical potentials are coupled via tunneling and formed into a ring, thereby creating a discrete analog of a ring BEC [18–23]. These authors have proposed a detailed scheme for creating macroscopic superpositions of different persistent flows on the ring lattice, and also discussed at length the barriers to creating such states [21]. Very recently, Dagnino *et al.* [24] have studied vortex nucleation in rotating BECs as a case study of symmetry breaking in quantum systems. They argue that the strongly correlated many-body state at nucleation is generally a superposition of the presence and absence of vortices, akin to a cat state. The proposal to be advanced here has some similarities with this work, but the correlated many-body state does not rely on nucleation for its production.

The central idea of the proposal advanced here is to create macroscopic superpositions of persistent flows on ring BECs, as opposed to a ring superlattice, using two-photon stimulated Raman transitions driven by one field that is a classical optical vortex, and a second quantized field that is a cat state of optical vortices. More specifically, we consider a cat state involving a superposition of two coherent state optical vortices of opposite winding number $\ell = \pm 1$, with each quasi-classical coherent state having a large mean photon number. The physics of how a macroscopic superposition of persistent flows may be created is as follows: if each of the two quasi-classical coherent state field components of the cat state was used individually in conjunction with the classical vortex to realize a vortex coupler as described above, one could create two macroscopically distinguishable persistent flows from an initially nonrotating BEC, since the two coherent states produce distinct light-shift potentials and the resultant stirring and flow. However, by using the optical vortex cat state, the initially nonrotating BEC will be exposed to "quantum stirring" in that the cat state will expose the BEC to a quantum superposition of both distinct light-shift stirring potentials, but with the caveat that if the direction of the persistent flow was measured, one or the other of the two macroscopically distinguishable persistent flows would be realized, but not a combination. Thus, the approach taken here is based on the realization that if one can create an optical vortex cat state then one can use quantum stirring to create a macroscopic superposition of persistent flows. The fact that the cat state nature of the driving quantized optical field can be transferred to the atoms via vortex coupling is not physically surprising in the view of previous studies of quantum atom optics involving BECs in quantized light fields [16, 25]. More specifically, Kanamoto *et al.* [26] have previously considered vortex coupling using quantized fields but concentrated on the distinction between the number states and the coherent states.

These ideas are substantiated in the remainder of this chapter. In Section 10.2, a proposal for generating optical vortex cat states based on electromagnetically induced transparency (EIT)-enhanced Kerr nonlinearities due to atoms loaded into hollow core photonic-crystal fibers (PCFs) [27, 28] are described. The stupendous progress in the areas of EIT-enhanced nonlinearities [29–32] and few photon nonlinear optics in hollow core fibers loaded with atoms [33–36] holds the promise that the proposal presented here may come to fruition in the future. Also, some limitations to the proposal and some other possibilities for generating optical vortex cat states are discussed. In Section 10.3, the idea of quantum stirring is developed and it is shown that, given one can create an optical vortex cat state, one can generate macroscopic superpositions of persistent flows. A summary and some closing thoughts are given in Section 10.4.

10.2 Optical Vortex Cat States

10.2.1 Linear Fiber Propagation

In this section, a scheme based on nonlinear fiber optics for generating a quantum mechanical superposition of macroscopically distinguishable optical vortices of winding numbers $\ell = \pm 1$ is described. To proceed, some basic properties of the modes of optical fibers that shall be used in the development below are first reviewed. It is well known that optical fibers can support a variety of transverse modes, and for weak guiding conditions in which the refractive index difference between the core and cladding medium is small, these modes are termed linearly polarized (LP) modes since their electric fields are LP in the plane perpendicular to the fiber axis [37, 38]. For concreteness the z axis is chosen as the fiber axis, and the linear polarization state of the propagating field is designated using the real unit vector $\mathbf{e}_{LP}$. These modes, furthermore, form a complete set of expansion functions for the fields propagating in the fiber. For a cylindrically symmetric fiber the lowest-order mode is designated LP_{01}, and in cylindrical coordinates (r, θ, z), this mode $U_{01}(r, z) = u_0(r)\exp(i\beta_0 z)$ typically has a Gaussian-like transverse profile $u_0(r)$, with β_0 being the propagation constant of the mode. The next highest mode is labeled LP_{11} and is doubly degenerate

$$U_{11}(r,\theta,z) = \begin{cases} U_1(r,\theta,z) = u_1(r)\cos(\theta)\exp(i\beta_1 z) \\ U_2(r,\theta,z) = u_1(r)\sin(\theta)\exp(i\beta_1 z) \end{cases} \tag{10.1}$$

Here $\beta_1 = n_1\omega/c$ is the propagation constant of the LP_{11} mode for a field of center frequency ω, with n_1 being the corresponding mode refractive index and $v = c/n_1$ the phase velocity, $u_1(r)$ is the transverse mode profile, which we take as real without loss of generality, and the factors $\sin(\theta)$ and $\cos(\theta)$ give each of the modes a distinct azimuthal lobe structure, one with a dipole-like pattern aligned along the x axis the other along the y axis. Figure 10.1a illustrates each of these azimuthal lobes within the core of the fiber, and due to their azimuthal structure, these modes have small spatial overlap. We note that we can also form optical vortex modes of opposite winding numbers $\ell = \pm 1$ from linear combinations of the LP_{11} modes

$$\begin{aligned} U_{\ell=\pm 1}(r,\theta,z) &= \left(U_1(r,\theta,z) \pm i U_2(r,\theta,z)\right) \\ &= u_1(r)\exp(i\beta_1 z)\exp(\pm i\theta) \end{aligned} \tag{10.2}$$

The transverse mode profile $u_1(r)$ has a null at the origin by virtue of the phase singularity present there, and has a single radial peak at $r = r_1$. The intensity profile of the peak is therefore donut shaped. Our goal here is to devise a scheme to generate a cat state of macroscopically distinguishable states having the optical vortex mode profiles $U_{\ell=\pm 1}(r,\theta,z)$ using nonlinear propagation in the fiber.

10.2.2
Quantum Fiber Propagation

For the propagation of an optical field of center frequency ω in a weakly guiding fiber that supports LP modes, we may decompose the quantized electric field operator into positive and negative frequency components as $\hat{\mathbf{E}}(\mathbf{r},t) = [\hat{\mathbf{E}}^{(+)}(\mathbf{r},t) + \hat{\mathbf{E}}^{(-)}(\mathbf{r},t)]$, where the operator for the positive frequency component of the vector electric

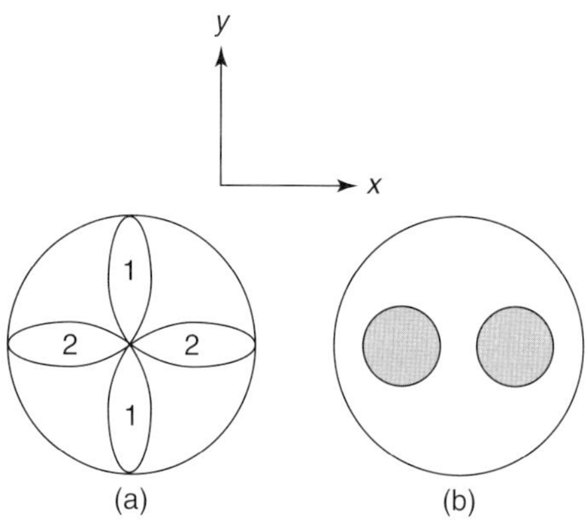

Figure 10.1 This figure shows the fiber core in the transverse xy plane looking back along the z axis. (a) illustrates the different azimuthal mode patterns for the modes labeled $j = 1, 2$, the lobes associated with the respective modes being marked. (b) shows an example of a nonuniform distribution of the nonlinearity, with the nonlinearity concentrated in the shaded regions. With this geometry mode, $j = 1$ will experience little nonlinearity in comparison to mode $j = 2$ due to the difference in their spatial overlaps with the nonlinear regions.

field is [39]

$$\hat{\mathbf{E}}^{(+)}(\mathbf{r}, t) = \sum_j i\sqrt{\frac{\hbar\omega}{2\epsilon V}}\mathbf{e}_{\mathrm{LP}}U_j(r, \theta, z)\hat{a}_j e^{-i\omega t} = (\hat{\mathbf{E}}^{(-)}(\mathbf{r}, t))^{\dagger} \tag{10.3}$$

with $V = A_{\mathrm{eff}}l$ the quantization volume, A_{eff} being the effective area of the fiber and $l = v\tau$ the pulse length for a pulse of duration τ, $\epsilon = \epsilon_0 n_1^2$, and $\hat{a}_j$ and $\hat{a}_j^{\dagger}$ are the annihilation and creation operators for the corresponding LP modes. Here we use a two-mode model and restrict the sum over modes to those modes labeled $j = 1, 2$ in Eq. (10.1), giving for the electric field operator

$$\hat{\mathbf{E}}^{(+)}(\mathbf{r}, t) = i\sqrt{\frac{\hbar\omega}{2\epsilon V}}\mathbf{e}_{\mathrm{LP}}u_1(r)\left[\cos(\theta)\hat{a}_1 + \sin(\theta)\hat{a}_2\right]e^{i(\beta_1 z - \omega t)} \tag{10.4}$$

This two-mode approximation is possible under the assumption that only those two modes will be externally excited, with all other LP modes remaining in their quantum mechanical ground state. Furthermore, we consider here, quasi-continuous wave fields, as opposed to ultrashort picosecond or femtosecond duration pulses, and ignore the linear dispersive properties of the fiber for simplicity in notation.

The initial quantum field state at the input to the optical fiber is taken as a product of coherent states [39]

$$|\Psi_{\mathrm{IN}}\rangle = |\alpha_1\rangle_1|\alpha_2\rangle_2 \tag{10.5}$$

where the subscripts $j = 1, 2$ refer to the two LP modes, and the coherent state is given in terms of number states as

$$|\alpha_j\rangle = e^{-|\alpha_j|^2}\sum_{n=0}^{\infty}\frac{\alpha_j^n}{\sqrt{n!}}|n\rangle_j \tag{10.6}$$

More specifically we consider the specific case $\alpha_1 = \alpha, \alpha_2 = i\alpha$, $|\Psi_{\mathrm{IN}}\rangle = |\alpha\rangle_1|i\alpha\rangle_2$, and without loss of generality, we take α to be real. Then using the property $\hat{a}_j|\alpha_j\rangle = \alpha_j|\alpha_j\rangle$ for coherent states, we find for the expectation value for the electric field operator at the fiber input $z = 0$

$$\begin{aligned}\langle\Psi_{\mathrm{IN}}|\hat{\mathbf{E}}^{(+)}(\mathbf{r}, t)|\Psi_{\mathrm{IN}}\rangle &= i\sqrt{\frac{\hbar\omega}{2\epsilon V}}\alpha\mathbf{e}_{\mathrm{LP}}u_1(r)e^{i\theta}e^{-i\omega t}\\ &= i\sqrt{\frac{\hbar\omega}{2\epsilon V}}\alpha\mathbf{e}_{\mathrm{LP}}U_{\ell=+1}(r, \theta, z = 0)e^{-i\omega t}\end{aligned} \tag{10.7}$$

For a quantized field with a large number of photons, $|\alpha|^2 \gg 1$ as assumed here, this initial condition corresponds to a quasi-classical field having an optical vortex profile $U_{\ell=+1}(r, \theta, z = 0)$ of winding number $\ell = 1$.[1)] Such classical optical vortices are routinely generated [12] using a range of techniques including spiral phase plates, holographic methods, and mode converters; so there is no issue with generating this initial condition for the quantum field at the input to the fiber.

1) We could alternatively have chosen $\ell = -1$ for this purpose by setting $\alpha_1 = \alpha$, $\alpha_2 = -i\alpha$, the key is that the input field is a classical optical vortex of a well-defined winding number.

10.2.3 Optical Vortex Cat State via Self-Phase Modulation

Next we consider nonlinear propagation in the optical fiber, and to proceed, we first examine the case of a single-mode field. In particular, the idea is to take advantage of the fact that optical fibers display a nonlinear optical Kerr effect, whereby the refractive index experienced by a propagating field has a shift proportional to the field intensity $\Delta n = n_2 I$, $n_2 > 0$ being the nonlinear Kerr coefficient [38, 40]. As a result an intense classical optical field propagating in the fiber will experience a phase shift that depends on the field intensity as well as the fiber length, an effect known as self-phase modulation (SPM). In the quantum field description, this means that the SPM experienced by a field described by a number state $|n\rangle$ will depend on the photon number n. In related work, Yurke and Stoler [41] considered propagation of a single-mode coherent state through an amplitude dispersive medium and Milburn and Holmes [42, 43] considered the quantum dynamics of an anharmonic oscillator, both of these models being intimately related to nonlinear propagation in an optical fiber with SPM. Using the results of these papers, we find that upon propagating a distance L under the action of SPM in the fiber, a single-mode field with initial state $|\Psi_{\text{IN}}\rangle$ will be transformed according to $|\Psi_{\text{OUT}}\rangle = \exp(-i\gamma(\hat{a}^\dagger\hat{a})^2)|\Psi_{\text{IN}}\rangle$, where the nonlinear parameter $\gamma = \hbar\omega^2 n_2 L/2vV$ [44] (see also Section 2 of [45]). In particular, for an initial single-mode coherent state $|\Psi_{\text{IN}}\rangle = |\alpha\rangle$, we obtain the exact result

$$|\Psi_{\text{OUT}}\rangle = e^{-|\alpha|^2}\sum_{n=0}^{\infty}\frac{\alpha^n}{\sqrt{n!}}e^{-i\gamma n^2}|n\rangle \tag{10.8}$$

The remarkable result exposed by Yurke and Stoler [41] is that for $\gamma = \pi/2$ the output state is a quantum superposition of the coherent states $|\pm\alpha\rangle$

$$|\Psi_{\text{OUT}}\rangle = \frac{1}{\sqrt{2}}[e^{-i\pi/4}|\alpha\rangle + e^{i\pi/4}|-\alpha\rangle] \tag{10.9}$$

which represents a cat state since the two coherent state components are macroscopically distinguishable for $|\alpha|^2 >> 1$.

We now apply these ideas to our two-mode fiber problem. In particular, what we like to arrange is that the mode $j = 1$ experiences a negligible Kerr effect but mode $j = 2$ experiences SPM over the fiber length with $\gamma = \pi/2$. This is possible in principle since the two degenerate LP modes have different spatial structures (Figure 10.1a), so that if the distribution of nonlinearity within the fiber core can be tailored so that the nonlinear regions overlap one mode but not the other, then the two modes will experience very different Kerr effects. For example, if only the shaded regions in the fiber core in Figure 10.1b are nonlinear, then only the mode $j = 2$ will experience significant nonlinearity.[2] Such a tailoring of the spatial

2) In general, the nonuniform fiber profile can also cause the two transverse modes $j = 1, 2$ to have different propagation constants, but we shall neglect that complication here for simplicity in presentation.

distribution of the nonlinearity is possible in principle using hollow core PCFs loaded with atoms, discussed in Section 10.2.4. In general we introduce a function $F(x,y)$, which is unity in regions that are nonlinear and zero otherwise, in terms of which the nonlinear parameter for each LP mode involves the spatial overlap of each mode with the nonlinearity

$$\gamma_j = \left(\frac{\hbar\omega^2 n_2 L}{2vV}\right)\frac{\int dxdy|u_j(x,y)|^4 F(x,y)}{\int dxdy|u_j(x,y)|^2}, \qquad j=1,2 \tag{10.10}$$

Thus, by judicious choice of $F(x,y)$, we can engineer the situation in which γ_1 is very small compared to γ_2 (see Figure 10.1). Then using the incident state at the fiber input $|\Psi_{\text{IN}}\rangle = |\alpha\rangle_1|i\alpha\rangle_2$ representing a quasi-classical field with an optical vortex profile of winding number $\ell = 1$, the quantum state at the output of the fiber $z = L$ is given by

$$|\Psi_{\text{OUT}}\rangle = e^{-i(\gamma_1(\hat{a}_1^\dagger\hat{a}_1)^2+\gamma_2(\hat{a}_2^\dagger\hat{a}_2)^2)}|\Psi_{\text{IN}}\rangle \tag{10.11}$$

and choosing $\gamma_1 = 0, \gamma_2 = \pi/2$, we obtain

$$|\Psi_{\text{OUT}}\rangle = \frac{1}{\sqrt{2}}[e^{-i\pi/4}|\alpha\rangle_1|i\alpha\rangle_2 + e^{i\pi/4}|\alpha\rangle_1|-i\alpha\rangle_2] = \sum_{\ell=\pm1}\frac{e^{-i\ell\pi/4}}{\sqrt{2}}|\alpha\rangle_1|i\ell\alpha\rangle_2 \tag{10.12}$$

The electric field expectation values with respect to each of the two quasi-classical coherent state components $|\alpha\rangle_1|i\ell\alpha\rangle_2$ comprising the output quantum field state $|\Psi_{\text{OUT}}\rangle$ are

$$\begin{aligned}
{}_2\langle i\alpha|_1\langle\alpha|\hat{\mathbf{E}}^{(+)}(\mathbf{r},t)|\alpha\rangle_1|i\alpha\rangle_2 &= i\sqrt{\frac{\hbar\omega}{2\epsilon V}}\alpha\mathbf{e}_{\text{LP}}U_{\ell=+1}(r,\theta,z=L)e^{-i\omega t}\\
{}_2\langle -i\alpha|_1\langle\alpha|\hat{\mathbf{E}}^{(+)}(\mathbf{r},t)|\alpha\rangle_1|-i\alpha\rangle_2 &= i\sqrt{\frac{\hbar\omega}{2\epsilon V}}\alpha\mathbf{e}_{\text{LP}}U_{\ell=-1}(r,\theta,z=L)e^{-i\omega t}
\end{aligned} \tag{10.13}$$

The first term in the output quantum field state therefore, corresponds to a quasi-classical coherent state for a field with a vortex profile $U_{\ell=+1}(r,\theta,z=L)$ with winding number $\ell = 1$, and the second term corresponds to a quasi-classical coherent state for a field with a vortex profile $U_{\ell=-1}(r,\theta,z=L)$ with winding number $\ell = -1$. The output quantum field state $|\Psi_{\text{OUT}}\rangle$, therefore, corresponds to a cat state of macroscopically distinguishable field states corresponding to optical vortices with opposite winding numbers $\ell = \pm1$. We have therefore demonstrated that it is possible in principle to generate an optical vortex cat state using nonlinear fiber optics. We remark that due to the fact that coherent states with differing amplitudes are not strictly orthogonal, the two optical vortex components of the above cat state are not strictly orthogonal. However, for $|\alpha|^2 >> 1$ the overlap between the two coherent state components will be very small and the two components may be treated as orthogonal for all practical purposes.

10.2.4 Photonic-Crystal Fibers

There are currently significant barriers to creating optical vortex cat states in optical fibers as described above. Key amongst these, in order of severity, are tailoring the spatial distribution of the nonlinearity in the fiber core, producing a large enough Kerr effect to realize the condition $\gamma_2 = \pi/2$, and the detrimental effects of dissipation on the cat state [43, 46], and each of these are addressed in order below. Briefly, proposed here is generating optical vortex cat states based on EIT-enhanced Kerr nonlinearities due to atoms loaded into hollow core PCFs [27, 28].

The first barrier to realizing the proposed scheme is tailoring the spatial profile of the Kerr nonlinearity, but PCFs [27, 28] offer a viable experimental solution to this. These are not conventional step-index fibers, but are rather formed using arrays of tiny air gaps formed through the fiber, which define the linear guiding structure (Figure 1 of [28]). These fibers have LP modes of the type required here (Figure 12 of [28]). A key aspect of PCFs is that the air gaps can be filled with nonlinear media, for example, liquids or vapors, and this has been used with great success, for example, in supercontinuum generation in PCF [47]. Then by selectively loading air gaps in chosen spatial regions throughout the PCF, we can realize the spatial distribution of the nonlinear medium alluded to earlier, so that mode $j = 1$ experiences a negligible Kerr effect and mode $j = 2$ a sizeable Kerr effect. For example, one can envisage a PCF with small air gaps used to define the linear guiding properties of the fiber along with two adjacent hollow cores positioned as in Figure 10.1b that will be loaded with nonlinear media. PCFs, therefore, offer a viable solution to the problem of tailoring the distribution of nonlinearity in the fiber core.

The second barrier to generating cat states is that the nonlinear Kerr coefficient n_2 in conventional silica fibers is orders of magnitude too small, long fiber fibers are needed leading to losses, and dissipation kills the cat [46]. A means of greatly boosting the Kerr effect is required, and this can be done using EIT-enhanced nonlinearities utilizing atomic vapors, for example, rubidium loaded into the hollow cores of the PCF. This sounds quite esoteric, but there has recently been stupendous progress in the areas of EIT-enhanced nonlinearities [29–32] and experiments on few photon nonlinear optics in hollow core fibers loaded with atoms [33–36]. Recalling the expression for the nonlinear coefficient for mode $j = 2$

$$\gamma_2 = \left(\frac{\hbar\omega^2 n_2 L}{2vV}\right)\frac{\int dxdy|u_2(x,y)|^4 F(x,y)}{\int dxdy|u_2(x,y)|^2} \tag{10.14}$$

we see that in addition to the enhancement of n_2, the nonlinear coefficient can be boosted by reducing the light velocity v, suggesting the use of slow light techniques [30, 32], and also designing the PCF to have a small mode volume $V = A_{\text{eff}} v\tau$, meaning small mode area A_{eff} and short pulse duration τ.

The final barrier, to be discussed here, to generating an optical vortex cat state using the proposed scheme is that any dissipation due to the fiber or nonlinear medium has a very detrimental effect on the ability to create or sustain a cat state

[43, 46], and this is a severe problem. On the other hand, EIT-enhanced Kerr effects raise the possibility of creating the cat states using shorter PCF lengths, thereby potentially reducing the detrimental effects of dissipation. Furthermore, there are proposals for realizing SPM in atomic vapors with no associated linear medium absorption utilizing two-level EIT and dark states [48], though the linear absorption of the fiber glass would remain. However, since the atomic vapor would supply an optical nonlinearity orders of magnitude larger than the fiber glass the fiber lengths needed would be greatly reduced with respect to conventional fibers. Whether or not dissipation remains a limiting factor to the generation of optical vortex cat states using nonlinear fiber optics will depend crucially on advances in PCF materials and fabrication to reduce fiber losses. In addition, as the state of the art evolves over the coming years in the area of few photon nonlinear optics in fibers, and with concomitant large enhancements in the attainable Kerr effects, it is hopeful that the conditions can be met for generating an optical vortex cat state $\gamma_2 = \pi/2$ with sufficiently small dissipation.

In spite of there being significant barriers to currently realizing optical vortex cat states using nonlinear PCFs, it is a worthy goal due to its possible applications including the generation of macroscopic persistent flows, which are discussed later. To highlight their utility, another potential application in the area of quantum cryptography is described. In the proposed scheme, to generate an optical vortex cat state the input quantum state to the PCF is a quasi-classical coherent state with winding number $\ell = +1$ or alternatively $\ell = -1$. One can view the two options for the initial winding number $\ell = \pm 1$ as encoding one bit of information that can be measured reliably [49], and each bit can be sent as a single coherent state pulse of duration τ, the winding number being varied between pulses to produce an input data stream. After propagation through the PCF, each initial coherent state pulse representing a single bit is transformed into the optical vortex cat state composed of two quasi-classical coherent states with opposite winding number, and this cat state is used to transmit each bit of information through a communication channel, for example, a linear fiber or free space [50]. Since the quantum field propagating in the communication channel is a cat state of both winding numbers, an eavesdropper trying to measure the winding number by intercepting the pulse in the communication channel will find ± 1 with equal probability, so that they will be unable to infer the initial winding number for any given pulse by this strategy: The initial bit of information is effectively hidden from the eavesdropper by virtue of the "quantum shroud" provided by the cat state. It may be argued that the eavesdropper just needs to measure the full quantum state of the field more carefully for a given pulse, but this seems an impossible task given that they only have a single pulse to work with, and no-cloning theorems would seem to rule out the possibility of proliferating copies of the quantum field state to overcome this. The question is then how can the initial winding number of a given pulse be retrieved reliably at the receiver following the communication channel? The key is to use a PCF loaded with atoms just before the receiver, which is identical to the input PCF but with the important difference that the sign for γ_2 is reversed compared to the input. Since the sign of the SPM in the second PCF is of opposite sign to that in the input PCF, the second PCF unwinds the intricate phases underlying the cat state and retrieves the initial quasi-classical state, with well-defined winding number, from the cat state [41]. So far we tacitly assumed $n_2 > 0$ and $\gamma_2 > 0$, but the sign of these can in principle be reversed by changing the operating conditions of the EIT enhancement, and the sign of $\gamma_2 = \pm 1$ for the input PCF and $\gamma_2 = \mp 1$ for the second PCF could be varied between pulses to provide a cipher for encoding and decoding the data stream. Thus, after the second PCF the field state will again be a quasi-classical coherent state with the same winding number as the input, so the bit encoded in the winding number of any single pulse can again be reliably measured at the receiver [49]. The eavesdropper may use the strategy of inserting their own PCF along the communication channel to intercept the data stream, but unless they know the cipher based on the values of γ_2 used at the input, they will not be able to reconstruct the initial data stream.

10.2.5
Other Schemes

Before moving onto the generation of macroscopic persistent flows, some other potential schemes for generating optical vortex cat states are discussed. First, it is important to acknowledge that there is currently a great deal of research directed toward creating tailor-made quantum superpositions of optical vortices with OAM for use in quantum information processing and communication, but these typically involve small numbers of photons to minimize the detrimental effects of decoherence on these entangled states [51–56]. In contrast, cat states with large average photon numbers are required for the purpose of creating macroscopic superpositions of persistent flows for BECs with a large number of atoms. In a recent paper, Glancy and Macedo de Vasconcelos [45] have reviewed a number of methods to produce cat states of coherent states, including the Kerr effect, degenerate optical parametric oscillators, backaction evasion measurement, and photon subtraction. Each of these methods could in principle be used to create optical vortex cat states if the two coherent state components of the cat state can be produced with suitable entangled spatial beam properties [57, 58]. A virtue of the method proposed here is that it automatically produces a cat state of optical vortices.

Finally, recent work by De Martini and coworkers [59] offers the possibility for creating a novel optical vortex cat state. In particular, they describe the creation of a microscopic–macroscopic field state, where a macrostate with $>10^4$ photons is entangled with a microstate containing one photon using a quantum-injected quantum parametric amplifier. Furthermore, they have shown that these microscopic–macroscopic states have high resilience against the detrimental effects of decoherence [60, 61]. If the microstate and macrostate could also have entangled spatial properties, such as opposite winding numbers, such a state would offer a very intriguing possibility for creating macroscopic superpositions of persistent flows.

10.3 Macroscopic Superposition of Persistent Flows

In this section, we describe how the optical vortex cat state can be utilized to generate a macroscopic superposition of persistent flows. First, we develop the light-shift potential arising from a vortex coupler driven by optical vortices, followed by discussions of the ring BEC geometry and our model for a BEC on a ring. Finally, we show that our model leads to the generation of macroscopic persistent flows.

10.3.1 Optical Light-Shift Potential

The proposal here is to use vortex coupling in which two-photon stimulated Raman transitions are driven by optical vortices to transfer OAM from the photons to the trapped atoms [9–11, 15], but with the caveat that one driving field is a classical vortex of winding number $-p$, $|p| > 1$, whereas the other field is an optical vortex cat state involving winding numbers $\ell = \pm 1$. Vortex couplers can be configured such that they either involve a change in the internal atomic states [9–11] or not [13, 15]. We consider the case that the internal atomic states are not changed by the stimulated Raman transitions in which case the system of cold atoms may be treated as a scalar BEC subjected to a light-shift potential [15]. In particular, we consider the case that the applied optical vortices are concentric and copropagating along the z axis, and that the output from the PCF is imaged so that the transverse radial peak of the intensity mode profile of the cat state overlaps the ring BEC. Likewise, we assume that the mode profile of the classical optical vortex is arranged so that the ring BEC coincides with a radial peak. For example, the classical optical vortex may take the form of an LG beam of winding number $-p$ and radial index zero, which has a single radial peak [12]. The combination of optical vortices then interacts with the atoms in the BEC that are trapped circumferentially on a ring lying in the xy plane with a fixed value of z. The operator for the positive frequency component of the vector electric field now has a c-number contribution due to the classical field in addition to the quantum field in Eq. (10.4)

$$\hat{\mathbf{E}}^{(+)}(\mathbf{r},t) = i\sqrt{\frac{\hbar\omega}{2\epsilon V}}\mathbf{e}_{\mathrm{LP}}u_1(r)\left[\cos(\theta)\hat{a}_1 + \sin(\theta)\hat{a}_2\right]e^{i(\beta_1 z-\omega t)} + i\mathbf{e}_p u_p(r)e^{-ip\theta}E_p e^{i(\beta_p z-\omega_p t)} \tag{10.15}$$

with $\mathbf{e}_p$ the polarization state of the applied classical field of amplitude E_p and mode profile $u_p(r)$, both taken as real without loss of generality; ω_p is the field frequency; and β_p the wavevector directed along the z axis. For off-resonance conditions between the light fields and the atom, the light-shift potential operator is generally given by [62]

$$\hat{V}(\mathbf{r},t) = -\frac{1}{2}\hat{\mathbf{E}}^{(-)}(\mathbf{r},t)\cdot\boldsymbol{\alpha}_R\cdot\hat{\mathbf{E}}^{(+)}(\mathbf{r},t) \tag{10.16}$$

where $\boldsymbol{\alpha}_R$ is the real part of the atomic polarizability tensor operator. By judicious choice of the atomic transitions and field polarization states, we assume that the only matrix element that survives out of this tensor product is $(\mathbf{e}_{\mathrm{LP}} \cdot \boldsymbol{\alpha}_R \cdot \mathbf{e}_p) = \alpha_R$.[3] In this limit, we obtain the light-shift potential operator as

$$\hat{V}(\mathbf{r}, t) = -\frac{1}{2}\alpha_R E_p \sqrt{\frac{\hbar\omega}{2\epsilon V}} u_1(r) u_p(r) e^{ip\theta} \left[\cos(\theta)\hat{a}_1 + \sin(\theta)\hat{a}_2\right] e^{-i\Delta\omega t} + h.c. \tag{10.17}$$

where $\Delta\omega = \omega - \omega_p$ is the detuning factor, h.c. means Hermitian conjugate, and the position of the ring is chosen such that $\exp(i(\beta_1 - \beta_p)z) = 1$ without loss of generality.

10.3.2
Ring Trap and Quantum Stirring

Our basic model for the ring BEC is that the atoms are trapped in a toroidal trap with very tight confinement along the z axis and a ring profile in the xy plane. In particular, we assume that the atoms are trapped, by other external laser and/or magnetic fields, on a ring of radius R, and that the trapping is sufficiently tight, such that the atomic gas is rendered quasi-one-dimensional around the ring [14, 63–65]. Then, using X to denote the coordinate along the ring, we may write $X = L\theta/2\pi$, with $L = 2\pi R$ being the ring circumference. In this ring geometry the light-shift potential operator evaluated along the ring circumference becomes

$$\begin{aligned} \hat{V}(X, t) = &-\frac{1}{2}\alpha_R E_p \sqrt{\frac{\hbar\omega}{2\epsilon V}} u_1(R) u_p(R) e^{i2\pi pX/L} \\ &\times \left[\cos(2\pi X/L)\hat{a}_1 + \sin(2\pi X/L)\hat{a}_2\right] e^{-i\Delta\omega t} + h.c. \end{aligned} \tag{10.18}$$

Then for the two coherent state components $|\alpha\rangle_1 |i\ell\alpha\rangle_2$ of the optical vortex cat state in Eq. (10.12), we obtain with $\ell = \pm 1$

$$\begin{aligned} {}_2\langle i\ell\alpha|_1\langle\alpha|\hat{V}(X, t)|\alpha\rangle_1|i\ell\alpha\rangle_2 &= -\alpha_R \alpha E_p \sqrt{\frac{\hbar\omega}{2\epsilon V}} u_1(R) u_p(R) \\ &\quad \times \cos\left(\frac{2\pi(p+\ell)X}{L} - \Delta\omega t\right) \\ &= V_{\mathrm{LS}} \cos\left(\frac{2\pi(p+\ell)X}{L} - \Delta\omega t\right) = V_\ell(X, t) \end{aligned} \tag{10.19}$$

This yields two macroscopically distinguishable quasi-classical light-shift potentials $V_\ell(X, t)$ of depth

$$V_{\mathrm{LS}} = -\alpha_R \alpha E_p \sqrt{\frac{\hbar\omega}{2\epsilon V}} u_1(R) u_p(R) \tag{10.20}$$

3) Selecting only one term from the atomic polarizability tensor is not absolutely necessary but greatly reduces the notational complexity of the ensuing development.

which rotate with different angular velocities $\Omega_\ell = \Delta\omega/|p+\ell|$ for $\ell = \pm 1$, and therefore produce different stirring. Thus, by using the optical vortex cat state a BEC trapped on the ring will experience "quantum stirring" in that the cat state will expose the BEC to a quantum superposition of both distinct light-shift stirring potentials, but with the caveat that if the direction of the persistent flow is measured one or other of the two macroscopically distinguishable persistent flows will be realized, not a combination.

10.3.3
Matter Waves on a Ring

We assume that the cold atoms are prepared on a ring in a N-body quantum state $|N(t=0)\rangle$, and that at $t = 0$ the light-shift potential above is initiated. The quantum state of the combined atom-light system at $t = 0$ is taken to be $\Psi(t=0)\rangle = |N(t=0)\rangle|\Psi_{\mathrm{OUT}}\rangle$. Assuming further that the light field is largely unaffected by the presence of the atoms, then the state of the system for $t > 0$ can be written as

$$|\Psi(t)\rangle = \frac{1}{\sqrt{2}}\left[e^{-i\pi/4}|N_{\ell=+1}(t)\rangle|\alpha\rangle_1|i\alpha\rangle_2 + e^{i\pi/4}|N_{\ell=-1}(t)\rangle|\alpha\rangle_1|-i\alpha\rangle_2\right] \quad (10.21)$$

where $|N_{\ell=\pm1}(t)\rangle$ are the N-body atom states corresponding to the two components of the optical vortex cat state, and $|N_{\ell=\pm1}(t=0)\rangle = |N(t=0)\rangle$, that is, they have the same initial condition.

Here, we consider the simplest example of noninteracting atoms on a ring, in which case the Hamiltonian operator in a second quantized form for the system of atoms may be written as [66, 67]

$$\hat{H}(t) = \int_0^L dX\left[\frac{\hbar^2}{2m}\left(\frac{\partial\hat{\psi}^\dagger}{\partial X}\right)\left(\frac{\partial\hat{\psi}}{\partial X}\right) + \hat{V}(X,t)\hat{\psi}^\dagger\hat{\psi}\right] \quad (10.22)$$

where $\psi(X,t)$ and $\psi^\dagger(X,t)$ are the atomic annihilation and creation field operators with boson commutation relations, and m is the atomic mass. Evaluating the Hamiltonian operator with respect to the state vector (Eq. (10.21)) yields

$$H(t) = \langle\Psi(t)|\hat{H}(t)|\Psi(t)\rangle = \sum_{\ell=\pm1}\int_0^L dX\langle N_\ell(t)|\left[\frac{\hbar^2}{2m}\left(\frac{\partial\hat{\psi}^\dagger}{\partial X}\right)\left(\frac{\partial\hat{\psi}}{\partial X}\right) + V_\ell(X,t)\hat{\psi}^\dagger\hat{\psi}\right]|N_\ell(t)\rangle \quad (10.23)$$

where we used the fact that the quasi-classical coherent states comprising the optical vortex cat state are orthogonal for all intents and purposes here. Since the total Hamiltonian $H(t)$ is the sum of the Hamiltonians, one for each ℓ, this means that we can solve for the N-atom dynamics for each optical vortex component separately. To proceed, we express the atomic state vectors as [66]

$$|N_\ell(t)\rangle = \frac{1}{\sqrt{N!}}\int_0^L dX_1\ldots\int_0^L dX_N\psi_\ell(X_1,\ldots,X_N,t)\hat{\psi}^\dagger(X_1,t)\ldots\hat{\psi}^\dagger(X_N,t)|0\rangle \quad (10.24)$$

and the equation of motion $i\hbar\frac{\partial\psi_\ell}{\partial t} = \frac{\delta H}{\delta\psi_\ell^*}$ yields the pair of Schrödinger equations

$$i\hbar\frac{\partial\psi_\ell}{\partial t} = \sum_{j=1}^{N}\left[-\frac{\hbar^2}{2m}\frac{\partial^2}{\partial X_j^2} + V_\ell(X_j,t)\right]\psi_\ell(X_1,\ldots,X_N,t), \quad \ell = \pm 1 \tag{10.25}$$

along with the initial conditions

$$\psi_{\ell=\pm1}(X_1,\ldots,X_N,t=0) = \psi_0(X_1,\ldots,X_N) \tag{10.26}$$

The key observation at this point is that the N-atom dynamics is governed by a different light-shift potential $V_\ell(X,t)$ for each winding number component present in the optical vortex cat state.

10.3.4 Macroscopic Superposition of Persistent Flows

For our model of noninteracting bosons on a ring, we write the N-atom wavefunctions ψ_ℓ in the Hartree form appropriate to a BEC in which all atoms occupy the same state [66, 67]

$$\psi_\ell(X_1,\ldots,X_N,t) = \prod_{j=1}^{N}\phi_\ell(X_j,t) \tag{10.27}$$

leading to the equation of motion for the single-particle orbitals $\phi_\ell(X,t)$

$$i\hbar\frac{\partial\phi_\ell}{\partial t} = \left[-\frac{\hbar^2}{2m}\frac{\partial^2}{\partial X^2} + V_{LS}\cos\left(\frac{2\pi(p+\ell)X}{L} - \Delta\omega t\right)\right]\phi_\ell(X,t), \quad \ell = \pm 1 \tag{10.28}$$

We consider the specific case that the initial BEC is in the nonrotating ground state $\phi_\ell(X,t=0) = \phi_0 = 1/\sqrt{L}$ for $t < 0$, and use the following ansatz for the single-particle orbital for $t > 0$

$$\phi_\ell(X,t) = \phi_0\left[a_0^\ell(t) + a_+^\ell(t)e^{i(2\pi(p+\ell)X/L-\Delta\omega t)} + a_-^\ell(t)e^{-i(2\pi(p+\ell)X/L-\Delta\omega t)}\right] \tag{10.29}$$

This ansatz is based on the assumption that only persistent flows with winding numbers $\pm(p+\ell)$ will be appreciably excited by the light-shift potentials, coupling to other winding numbers being phase-mismatched for $|p| \gg 1$, and we note that LG beams can now be prepared with very large values of winding number [68]. For the coherent state component of the cat state with winding number ℓ, the amplitude $a_0^\ell(t)$ appearing in Eq. (10.29) is that for the BEC to stay in its nonrotating state with winding number $q = 0$, $a_+^\ell(t)$ the amplitude for the BEC to be transferred to the flow state with winding number $q = (p+\ell)$, and $a_-^\ell(t)$ the amplitude for the BEC to be transferred to the flow state with winding number $q = -(p+\ell)$. Substituting the ansatz (Eq. (10.29)) in the Schrödinger equation (Eq. (10.28)), we obtain the coupled-amplitude equations

$$
\begin{aligned}
i\hbar\frac{da_0^\ell}{dt} &= \frac{V_{\mathrm{LS}}}{2}(a_+^\ell + a_-^\ell) \\
i\hbar\frac{da_+^\ell}{dt} &= \underbrace{\left[\frac{\hbar^2}{2m}\frac{4\pi^2(p+\ell)^2}{L^2} - \hbar\Delta\omega\right]}\, a_+^\ell + \frac{V_{\mathrm{LS}}}{2}a_0^\ell \\
i\hbar\frac{da_-^\ell}{dt} &= \left[\frac{\hbar^2}{2m}\frac{4\pi^2(p+\ell)^2}{L^2} + \hbar\Delta\omega\right] a_-^\ell + \frac{V_{\mathrm{LS}}}{2}a_0^\ell
\end{aligned}
\tag{10.30}
$$

These equations govern the quantum dynamics of the vortex coupler with the applied optical vortex cat state, and determine how the initial nonrotating BEC can be transformed into a quantum superposition of persistent flows.

To proceed we choose the detuning factor $\Delta\omega = \Delta\omega_r$ such that the underbraced term in the coupled-amplitude equations (Eq. (10.30)) vanishes for the $\ell = 1$ optical vortex component of the cat state

$$
\hbar\Delta\omega_r = \frac{\hbar^2}{2m}\frac{4\pi^2(p+1)^2}{L^2}
\tag{10.31}
$$

and for $\ell = 1$ the coupled-amplitude equations (10.30) become

$$
\begin{aligned}
i\hbar\frac{da_0^1}{dt} &= \frac{V_{\mathrm{LS}}}{2}(a_+^1 + a_-^1) \\
i\hbar\frac{da_+^1}{dt} &= \frac{V_{\mathrm{LS}}}{2}a_0^1 \\
i\hbar\frac{da_-^1}{dt} &= 2\hbar\Delta\omega_r a_-^1 + \frac{V_{\mathrm{LS}}}{2}a_0^1
\end{aligned}
\tag{10.32}
$$

Then under the assumption that $\hbar|\Delta\omega_r| \gg |V_{\mathrm{LS}}|$, the amplitude $a_-^1(t) \approx 0$ will remain small as it is highly phase-mismatched, and resonant coupling occurs only between $a_0^1(t)$ and $a_+^1(t)$ giving the reduced equations

$$
i\hbar\frac{da_0^1}{dt} = \frac{V_{\mathrm{LS}}}{2}a_+^1, \quad i\hbar\frac{da_+^1}{dt} = \frac{V_{\mathrm{LS}}}{2}a_0^1
\tag{10.33}
$$

These equations are easily solved giving

$$
|a_0^1(t)|^2 = \cos^2\left(\frac{V_{\mathrm{LS}}t}{2\hbar}\right), \quad |a_+^1(t)|^2 = \sin^2\left(\frac{V_{\mathrm{LS}}t}{2\hbar}\right)
\tag{10.34}
$$

Thus, if the optical vortex cat state persists for the pulse duration $t = \tau$ such that $|V_{\mathrm{LS}}\tau/2\hbar| = \pi/2$, then the component of the optical vortex cat state with $\ell = 1$ will result in a persistent flow with winding number $q = (p+1)$.[4)]

4) We remark that if we had chosen the detuning factor $\hbar\Delta\omega_r = -\frac{\hbar^2}{2m}\frac{4\pi^2(p+1)^2}{L^2}$ to create resonant coupling between the amplitudes $a_0^1(t)$ and $a_-^1(t)$ in Eqs. (10.30) we would create a persistent flow with winding number $q = -(p+1)$ instead.

To complete the analysis, we next have to examine what happens to the BEC under the action of the component of the optical vortex cat state with winding number $\ell = -1$ with the same resonant condition as Eq. (10.31). In this case, the coupled-amplitude equations (10.30) become

$$\begin{aligned} i\hbar\frac{da_0^{-1}}{dt} &= \frac{V_{\rm LS}}{2}(a_+^{-1} + a_-^{-1}) \\ i\hbar\frac{da_+^{-1}}{dt} &= \left(\frac{\hbar^2}{2m}\frac{16\pi^2 p}{L^2}\right)a_+^{-1} + \frac{V_{\rm LS}}{2}a_0^{-1} \\ i\hbar\frac{da_-^{-1}}{dt} &= \left(\frac{\hbar^2}{2m}\frac{8\pi^2(p^2+1)}{L^2}\right)a_-^{-1} + \frac{V_{\rm LS}}{2}a_0^{-1} \end{aligned} \tag{10.35}$$

For this case, there is no resonant coupling between $a_0^{-1}(t)$ and $a_\pm^{-1}(t)$, and for large $|p|$, we can ensure that the phase-mismatches are large enough that any coupling is strongly inhibited and the atomic gas remains dominantly in its ground state with $|a_0^{-1}|^2 \approx 1, |a_\pm^{-1}|^2 \approx 0$.

10.3.5 Discussion

Bringing the above results together, we have a scheme such that for the $\ell = 1$ component of the optical vortex cat state the BEC can be completely transferred from its initial nonrotating ground state to a persistent flow of winding number $q = (p+1)$, whereas for the $\ell = -1$ component the BEC remains in its ground state with winding number $q = 0$. The quantum state of the ring BEC following the optical vortex fields will therefore be a macroscopic superposition of persistent flows with winding numbers $q = 0, (p+1)$, and a measurement of the flow of the trapped gas will find the BEC in one or other of the persistent flows with equal probability. We have thus theoretically demonstrated a scheme to generate macroscopic superpositions of persistent flows in a ring BEC. We remark that we could have easily have retained atom–atom interactions in this analysis, and in the same Hartree approximation the coupled-amplitude equations (10.30) would acquire nonlinear terms. Even in the presence of such nonlinear terms resonant coupling between persistent flow states can still occur [9], so the scheme proposed here would still be viable.

The above results have some similarity to the recent work of Dagnino *et al.* [24], where nucleation in a rotating BEC was shown to lead to a quantum superposition of the absence and presence of vortices. In their case, the many-body system becomes strongly correlated in the vicinity of the critical rotation frequency when nucleation occurs, whereas in the present proposal the strong correlations arise from the use of the optical vortex cat state in the vortex coupling. In both cases, the net result is a quantum state of the BEC, which is a quantum superposition of the absence and presence of BEC vortices.

Finally, it is also pointed out that the interesting microscopic–macroscopic quantum field states discovered by De Martini *et al.* [59] could also be used to generate macroscopic superpositions of persistent flows on ring BECs, in particular, if the microstate containing one photon and the macrostate with many photons had opposite winding numbers. Then in combination with the applied classical optical vortex, the microstate would produce negligible vortex coupling, leaving the BEC in its nonrotating ground state, whereas the macrostate could be designed to transfer the BEC to a prescribed persistent flow. Thus, if their spatial profiles can be suitably entangled, the microscopic–macroscopic quantum field states could produce macroscopic superpositions of persistent flows of the same nature as those discussed here using optical vortex cat states.

10.4 Summary and Conclusions

In summary, a scheme to generate optical vortex cats states based on nonlinear fiber optics using PCFs has been proposed. Although the proposal is clearly beyond the current state of the art, one can hope that future advances in PCF fabrication and EIT-enhanced Kerr nonlinearities will allow optical vortex cat states to be realized in the future. It is also demonstrated that if an optical vortex cat state can be realized, it can be used to generate macroscopic superpositions of a persistent flow in a ring BEC using vortex coupling based on two-photon stimulated Raman transitions. Such macroscopic superpositions of persistent flows are of interest as they are cold atom analogs of those observed in superconducting quantum interference devices [2–4]. Furthermore, the macroscopic superpositions of persistent flows described here are realizations of atomic NOON states, which have potential applications in precision measurements [69].

In this chapter, the focus has been on the underlying concepts and much work remains to be done to substantiate these ideas, for example, more detailed aspects of the PCF design, quantum nonlinear propagation in the PCF structure, and the fundamental limitations to the generation of optical vortex cat states arising from dissipation. The resilience to decoherence of the microscopic–macroscopic quantum field states discovered by De Martini *et al.* [59–61] means that they may actually be better candidates for generating cat states in the short term. Likewise, for the ring BEC, more detailed consideration of the trapping geometry and sources of decoherence, for example, trap fluctuations, which could kill the macroscopic superpositions, will have to be considered in detail.

This work is supported in part by the Joint Services Optical Program (JSOP).

References

1. Legget, A.J. (2002) Testing the limits of quantum mechanics: motivation, state of play, prospects. *J. Phys. Condens. Matter*, **14**, R415–R451.
2. Nakamura, Y., Pahkin, Yu.A., and Tsai, J.S. (1999) Coherent control of macroscopic quantum states in a single-Cooper-pair box. *Nature*, **398**, 786–788.
3. Friedman, J.R., Patel, V., Chen, W., Tolpygo, S.K., and Lukens, J.E. (2000) Quantum superposition of distinct

macroscopic states. *Nature*, **406**, 43–46.
4. van der Wal, C.H., ter Haar, A.C.J., Wilhelm, F.K., Schouten, R.N., Harmans, C.J.P.M., Orlando, T.P., Lloyd, S., and Mooij, J.E. (2000) Quantum superposition of macroscopic persistent-current states. *Science*, **290**, 773–777.
5. Marzlin, K.-P. and Zhang, W.P. (1998) Laser-induced rotation of a trapped Bose-Einstein condensate. *Phys. Rev. A*, **57**, 3801–3804.
6. Marzlin, K.-P. and Zhang, W.P. (1998) Quantized circular motion of a trapped Bose-Einstein condensate: coherent rotation and vortices. *Phys. Rev. A*, **57**, 4761–4769.
7. Matthews, M.R., Anderson, B.P., Haljan, P.C., Hall, D.S., Weiman, C.E., and Cornell, E.A. (1999) Vortices in a Bose-Einstein condensate. *Phys. Rev. Lett.*, **83**, 2498–2501.
8. Willimas, J.E. and Holland, M.J. (1999) Preparing topological states of a Bose-Einstein condensate. *Nature*, **401**, 568–572.
9. Marzlin, K.-P., Zhang, W.P., and Wright, E.M. (1997) Vortex coupler for atomic Bose-Einsten condensates. *Phys. Rev. Lett.*, **79**, 4728–4731.
10. Kapale, K.T. and Dowling, J.P. (2005) Vortex phase qubit: generating arbitrary, counter-rotating, coherent superpositions in Bose-Einstein condensates via optical angular momentum beams. *Phys. Rev. Lett.*, **77**, 173601.
11. Thanvanthri, S., Kapale, K.T., and Dowling, J.P. (2008) Arbitrary coherent superpositions of quantized vortices in Bose-Einstein condensates via orbital angular momentum of light. *Phys. Rev. A*, **77**, 053825.
12. For a collection of seminal papers on LG beams see. (2003) Allen, L., Barnett, S. M., and Padgett, M. J., (eds) *Optical Angular Momentum*, Institute of Physics Publishing, London.
13. Anderson, M.F., Ryu, C., Clade, P., Natarajan, V., Vaziri, A., Helmerson, K., and Phillips, W.D. (2006) Quantized rotation of atoms from photons with orbital angular momentum. *Phys. Rev. Lett.*, **97**, 170406.
14. Ryu, C., Anderson, M.F., Clade, P., Natarajan, V., Helmerson, K., and Phillips, W.D. (2007) Observation of persistent flow of a Bose-Einstein condensate in a toroidal trap. *Phys. Rev. Lett.*, **99**, 260401.
15. Simula, T.P., Nygaard, N., Hu, S.X., Collins, L.A., Schneider, B.I., and Molmer, K. (2008) Angular momentum exchange between coherent light and matter fields. *Phys. Rev. A*, **77**, 015401.
16. Cirac, J.I., Lewenstein, M., Molmer, K., and Zoller, P. (1998) Quantum superposition states of Bose-Einstein condensates. *Phys. Rev. A*, **57**, 1208–1218.
17. Sorensen, A., Duan, L.-M., Cirac, J.I., and Zoller, P. (2001) Many-body entanglement with Bose-Einstein condensates. *Nature*, **409**, 63–66.
18. Dunnignham, J. and Hallwood, D. (2006) Creation of macroscopic superpositions of flow states with Bose-Einstein condensates. *Phys. Rev. A*, **74**, 023601.
19. Hallwood, D.W., Burnett, K., and Dunningham, J. (2006) Macroscopic superpositions of superfluid flows. *New J. Phys.*, **8**, 180.
20. Dunnignham, J., Burnett, K., Roth, R., and Phillips, W.D. (2006) Creation of macroscopic superpositions from arrays of Bose-Einstein condensates. *New J. Phys.*, **8**, 182.
21. Hallwood, D.W., Burnett, K., and Dunningham, J. (2007) The barriers to producing multiparticle superposition states in rotating Bose-Einstein condensates. *J. Mod. Opt.*, **54**, 2129–2148.
22. Nunnenkamp, A., Rey, A.M., Burnett, K., Satija, I.I., and Clark, C.W. (2007) Entanglement and the Mott transition in a rotating bosonic ring lattice. *Phys. Rev. A*, **75**, 063616.
23. Nunnenkamp, A., Rey, A.M., and Burnett, K. (2008) Generation of macrosocpic states in ring superlattices. *Phys. Rev. A*, **77**, 023622.
24. Dagninio, D., Barberan, N., Lewenstein, M., and Dalibard, J. (2009) Vortex nucleation as a case study of symmetry breaking in quantum systems. *Nat. Phys.*, **5**, 431–437.
25. Moore, M.G., Zobay, O., and Meystre, P. (1999) Quantum optics of Bose-Einstein

condensates in quantized light fields. *Phys. Rev. A*, **60**, 1491–1506.

26. Kanamoto, R., Wright, E.M., and Meystre, P. (2007) Quantum dynamics of Raman-Coupled Bose-Einstein condensates using Laguerre-Gaussian beams. *Phys. Rev. A*, **75**, 063623.
27. Russell, P. (2003) Photonic-crystal fibers. *Science*, **299**, 358–362.
28. For a recent review with extensive references see Russell, P.St.J. (2006) Photonic-crystal fibers. *J. Lightwave Tech.*, **24**, 4729–4749.
29. Harris, S.E. and Hau, L.V. (1999) Nonlinear optics at low light levels. *Phys. Rev. Lett.*, **82**, 4611–4614.
30. Lukin, M.D. and Imamoglu, A. (2000) Nonlinear optics and entanglement of ultraslow single photons. *Phys. Rev. Lett.*, **84**, 1419–1422.
31. Lukin, M.D. and Imamoglu, A. (2001) Controlling photons using electromagnetically induced transparency. *Nature*, **413**, 273–276.
32. For a recent review with extensive references see Fleischhauer, M., Imamoglu, A., and Marangos, J.P. (2005) Electromagnetically induced transparency: optics in coherent media. *Rev. Mod. Phys.*, **77**, 633–673.
33. Ghosh, S., Bhagwat, A.R., Kyle Renshaw, C., Goh, S., Gaeta, A.L., and Kirby, B.J. (2006) Low-light-level optical interactions with Rubidium vapor in a photonic band-gap fiber. *Phys. Rev. Lett.*, **97**, 023603.
34. Spillane, S.M., Pati, G.S., Salit, K., Hall, M., Kumar, P., Beausoleil, R.G., and Shahriar, M.S. (2008) Observation of nonlinear optical interactions of ultralow levels of light in a tapered optical nanofiber embedded in a hot Rubidium vapor. *Phys. Rev. Lett.*, **100**, 233602.
35. Bajcsy, M., Hofferberth, S., Balic, V., Peyronel, T., Hafezi, M., Zibrov, A.S., Vuletic, V., and Lukin, M.D. (2009) Efficient all-optical switching using slow light within a hollow fiber. *Phys. Rev. Lett.*, **102**, 203902.
36. Hafezi, M., Chang, D.E., Gritsev, V., Demler, E., and Lukin, M.D. (2009) Photonic quantum transport in a nonlinear optical fiber. *arXiv:0907.5206 [quant-phys]*, 29 Jul.
37. See, for example, Snyder, A.W. and Love, J.D. (1983) *Optical Waveguide Theory*, Chapman & Hall, London.
38. For an overview of linear and nonlinear fiber optics see Agrawal, G.P. (2001) *Nonlinear Fiber Optics*, Academic Press, San Diego.
39. For an overview of field quantization and quantum states of the field see Chaps. 1 & 2 of Walls, D.F. and Milburn, G.J. (1994) *Quantum Optics*, Springer-Verlag, Berlin.
40. See, for example, Boyd, R.W. (2003) *Nonlinear Optics*, Academic Press, Boston.
41. Yurke, B. and Stoler, D. (1986) Generating quantum mechanical superpositions of macroscopically distinguishable states via amplitude dispersion. *Phys. Rev. Lett.*, **57**, 13–16.
42. Milburn, G.J. (1986) Quantum and classical Liouville dynamics of the anharmonic-oscillator. *Phys. Rev. A*, **33**, 674–685.
43. Milburn, G.J. and Holmes, C.A. (1986) Dissipative quantum and classical Liouville mechanics of the anharmonic oscillator. *Phys. Rev. Lett.*, **56**, 2237–2240.
44. Kitagawa, M. and Yamamoto, Y. (1986) Number-phase minimum-uncertainty state with reduced number uncertainty in a Kerr nonlinear interferometer. *Phys. Rev. A*, **34**, 3974–3988.
45. Glancy, S. and Macedo de Vasconcelos, H. (2008) Methods for producing optical coherent state superpositions. *J. Opt. Soc. Am. B*, **25**, 712–733.
46. Kennedy, T.A.B. and Drummond, P.D. (1988) Quantum-field superpositions via self-phase modulation of coherent wavepackets. *Phys. Rev. A*, **38**, 1319–1326.
47. Dudley, J.M., Genty, G., and Coen, S. (2006) Supercontinuum generation in photonic crystal fiber. *Rev. Mod. Phys.*, **78**, 1135–1184.
48. Wong, V., Boyd, R.W., Stroud, C.R., Bennick, R.S., Aronstein, D.L., and Park, Q.-Han. (2001) Absorptionless self-phase modulation via dark-state electromagnetically induced transparency. *Phys. Rev. A*, **65**, 013810.

49. Leach, J., Padgett, M.J., Barnett, S.M., Franke-Arnold, S., and Courtial, J. (2002) Measuring the orbital angular momentum of a single photon. *Phys. Rev. Lett.*, **88**, 257901.
50. Gibson G., Courtial J., Padgett M.J., Vasnetsov M., Pas'ko V., Barnett S.M., and Franke-Arnold S., Free-space information transfer using light beams carrying orbital angular momentum *Opt. Express* **12**, 5448–5456.
51. Arnaut, H.H. and Barbosa, G.A. (2000) Orbital and intrinsic angular momentum of single photons and entangled pairs of photons generated by parametric down-conversion. *Phys. Rev. Lett.*, **85**, 286–289.
52. Mair, A., Vaziri, A., Weihs, G., and Zellinger, A. (2001) Entanglement of orbital angular momentum states of photons. *Nature*, **412**, 313–316.
53. Molina-Terriza, G., Torres, J.P., and Torner, L. (2002) Management of the angular momentum of light: preparation of photons in multidimensional vector spaces of angular momentum. *Phys. Rev. Lett.*, **88**, 013601.
54. Molina-Terriza, G., Torres, J.P., and Torner, L. (2007) Twisted photons. *Nat. Phys.*, **3**, 305–310.
55. Courtial, J. and O'holleran, K. (2007) Experiments with twisted light. *Eur. Phys. J. Spec. Top.*, **145**, 35–47.
56. Lassen, M., Leuchs, G., and Anderson, U.L. (2009) Continuous variable entangling and squeezing of orbital angular momentum states. *Phys. Rev. Lett.*, **102**, 163602.
57. Wagner, K., Janousek, J., Delaubert, V., Zou, H., Harb, C., Treps, N., Morizur, J.F., Lam, P.K., and Bachor, H.A. (2008) Entangling the spatial properties of laser beams. *Science*, **321**, 541–543.
58. Janousek, J., Wagner, K., Morizur, J.-F., Treps, N., Lam, P.K., Harb, C.C., and Bachor, H.-A. (2009) Optical entanglement of co-propagating modes. *Nat. Photon.*, **3**, 399–402.
59. De Martini, F., Sciarrino, F., and Vitelli, C. (2008) Entanglement test on a microscopic-macroscopic system. *Phys. Rev. Lett.*, **100**, 253601.
60. De Martini, F., Sciarrino, F., and Spagnolo, N. (2009) Decoherence, environment-induced superselection, and classicality of a macroscopic quantum superposition generated by quantum cloning. *Phys. Rev. A*, **79**, 052305.
61. De Martini, F., Sciarrino, F., and Spagnolo, N. (2009) Anomalous lack of decoherence of the macroscopic quantum superpositions based on phase-covariant quantum cloning. *Phys. Rev. Lett.*, **103**, 100501.
62. See for example Sec. III of Jessen, P.S. and Deutsch, I.H. (1996) Optical Lattices. *Adv. At. Mol. Opt. Phys.*, **37**, 95–138.
63. Wright, E.M., Arlt, J., and Dholakia, K. (2000) Toroidal optical dipole traps for atomic Bose-Einstein condensates using Laguerre-Gaussian beams. *Phys. Rev. A*, **63**, 013608.
64. Gupta, S., Murch, K.W., Moore, K.L., Purdy, T.P., and Stamper-Kurn, D.M. (2005) Bose-Einstein condensation in a circular waveguide. *Phys. Rev. A*, **95**, 143201.
65. Arnold, A.S., Garvie, C.S., and Riis, E. (2006) Large magnetic storage ring for Bose-Einstein condensates. *Phys. Rev. A*, **73**, 041606.
66. For a review of many-body theory as applied to atomic BECs see, for example, Chap 8 of Meystre, P. (2001) *Atom Optics*, Springer-Verlag, New York.
67. Pethick, C.J. and Smith, H. (2008) *Bose-Einstein Condensation in Dilute Gases*, Cambridge University Press, Cambridge.
68. Clifford, M.A., Arlt, J., Courtial, J., and Dholakia, K. (1998) High-order Laguerre-Gaussian modes for studies of cold atoms. *Opt. Commun.*, **156**, 300–306.
69. Hallwood, D.W., Stokes, A., Cooper, J.J., and Dunningham, J. (2009) Measuring atomic NOON states and using them to make precision measurements. *New. J. Phys.*, **11**, 103040.

11
Experimental Control of the Orbital Angular Momentum of Single and Entangled Photons

Gabriel Molina-Terriza and Anton Zeilinger

In this book, the reader will find an opportunity for studying the different facets of the orbital angular momentum (OAM) of light, from defining the basic properties of the OAM of electromagnetic fields to the most novel applications in fields as diverse as biomedicine to telecommunication technologies. This chapter gives the interested reader an opportunity to understand the enormous possibilities of applying the concept of OAM at the quantum level; that is, when nonclassical features of the electromagnetic field become relevant. Most notable examples are the cases of single or entangled photons. This area of research, which lies at the intersection of the fields of singular optics [1] and quantum optics [2], started as a purely intellectual subject with questions regarding the very meaning of the angular momentum of photons or their conservation in matter-mediated interactions and has evolved to have a weight on its own with very important applications in the field of quantum information [3].

In this chapter, we do not give a complete introduction to the subject of OAM or the quantum optical aspects of it, as there are a few very complete reviews on the subject [1, 4, 5], and it will be redundant with other chapters of this book. Rather, we will try another different approach. Our objective is to make the reader capable of understanding the quantum optics and quantum information experiments regarding the OAM of photons, when he/she completes reading this chapter. We then start with a small introduction on the subject, review some of the techniques used to control the OAM of a single photon, and then explain how to create and control OAM correlations between different photons. We end the chapter with a few interesting applications in the field of quantum information and with a discussion on the future possibilities.

11.1
Introduction to the Photon OAM

From the point of view of quantum optics, it is possible to choose a series of physical quantities such as energy, momentum, and angular momentum to

Twisted Photons: Applications of Light with Orbital Angular Momentum.
Edited by Juan P. Torres and Lluis Torner

ISBN: 978-3-527-40907-5

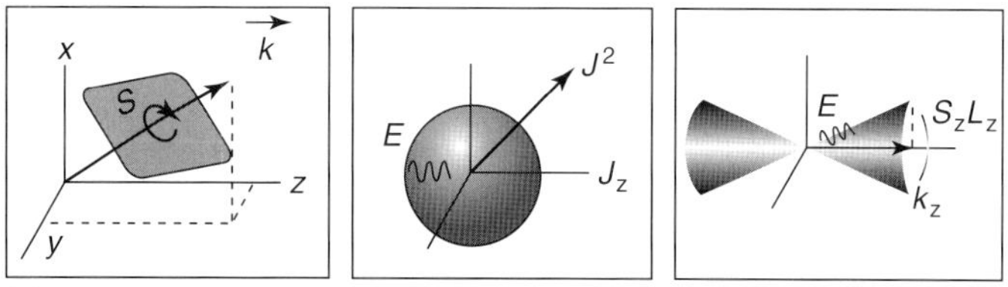

(a) (b) (c)

Figure 11.1 Different sets of modes that can be used for describing the electromagnetic field, and their respective set of parameters. (a) Plane wave, (b) multipolar mode, and (c) cylindrical mode. (Please find a color version of this figure on the color plates.)

describe the quantum excitations of the electromagnetic field in vacuum [6]. It is well known that not all the possible physical quantities can be used at the same time in this description, as some of them cannot be measured coincidentally without producing a mutual disturbance in the measurement. This would be the same kind of uncertainty principle as the quantum measurements of the position and momentum of an electron, which cannot be perfectly defined simultaneously. One of such possible set of measurements that has been extendedly used both in theoretical and experimental quantum optics is the set consisting of energy E, linear momentum direction $\hat{k}$, and the transversal polarization s (also called *spin angular momentum* (SAM)[1)]).

This set of quantities gives rise to a family of modes of the electromagnetic field. In this case, the modes are the well-known "plane waves," which are parameterized with the corresponding quantities mentioned above; that is, energy (temporal frequency of the mode or modulus of the wave-vector), linear momentum direction (propagation direction or wave-vector direction), and transversal polarization (helicity or polarization). In Figure 11.1a, we summarize the characteristics of a plane wave. In order to measure and control the plane wave set of modes, we have a series of well-studied experimental techniques such as spectral filters, lenses, polarizers, and waveplates.

Nevertheless, the plane waves mentioned above are not the only possible set of modes or physical properties that can be used to manage the quantum information that a photon carries. Two other possible sets are represented in Figure 11.1b,c. The first of those families of modes is parameterized by the energy (E), total angular momentum of the field (J^2), the z component of the angular momentum (J_z), and the parity of the field (P), which define the electromagnetic multipolar modes. It is clear from the way we have presented this set of modes that they are spherically symmetric electromagnetic modes. This set of electromagnetic modes is very important for processes such as light interactions with small

1) It is very important not to confound this SAM with the total angular momentum. Also it should be noted that there is a subtle difference between this polarization angular momentum and the intrinsic spin of the photon, which is a characteristic of the photon particle, as well as its mass and charge.

material particles, but they have not been extensively used in the field of quantum information with photons, because they are difficult to generate, control, and measure. Following the same logic, we could also define a set of electromagnetic, cylindrically symmetric modes, which are defined by using the energy, the z component of the linear momentum (k_z), the z component of the OAM (L_z), and the z component of the SAM (S_z). One of the most interesting properties of this set of modes is that they naturally lead to the set of paraxial modes with OAM. When the z component of the linear momentum is very close to the modulus of the linear momentum $k_z \simeq \|\mathbf{k}\| = E/hc$, the transversal components of the linear momentum are much smaller than the longitudinal one and the expressions for the electromagnetic modes become simplified with the paraxial approximation. Moreover, experiments in optics and quantum optics are typically performed fulfilling the paraxial approximation.[2)]

With this hierarchy of physical properties and set of modes, we have been able to reach the well-known paraxial modes with OAM, starting from the plane waves and passing through the multipolar modes. As in any basis decomposition, of course, we can relate the modes of one basis with the modes of the other. This scheme has explicitly been derived for the important case of the relation between cylindrically and spherically symmetric modes [7].

One important property of the cylindrically symmetric paraxial modes is that the OAM and SAM are independently determined, which in experimental terms means that we are able to control the spatial properties and the polarization of the field independently. Owing to the versatility of the techniques to spatially control the optical beams, the OAM of light has also found very important applications in the field of quantum information. An advantage with respect to other techniques to control the state of photons is that it allows to automatically codify the information in a large alphabet, which permits the access to high-dimensional Hilbert spaces. The possibilities that this fact opens are explored in the last sections of this chapter. Meanwhile, in the next section we show which kind of techniques can be used to control the spatial properties of photons and, thus, their OAM.

11.2 Control of the OAM State of a Single Photon

The experimental tools to control the OAM of photons can be roughly divided into those which allow us to measure that we just have a single excitation of the electromagnetic mode, that is, a single photon, and those which allow us to control the modes themselves. The first ones are self-explanatorily called *single photon detectors*, and among the most popular ones we encounter the avalanche photo diodes (APDs), when used in the Geiger mode. There are a few others which can also be used such as the photo multiplier tubes (PMTs), but the APDs are more

2) Notable examples where the paraxial condition is not valid are experiments in nanophotonics and, in general, interactions of optical fields with particles that are much smaller than the wavelength.

common in the kind of experiments we describe below owing to their low noise and high efficiencies in the red- and near-infrared range of wavelengths. An important drawback of the APDs is that they cannot usually distinguish between one single photon and multiple photon signals. This means that the APDs are used mainly when the intensity of the signal is very low, that is, the number of photons per second is less and typically below a few megahertz. Recently, new detectors have been developed which allow the exact determination of photon numbers. The most notable example are superconducting transition-edge detectors (STEDs) [8].

Before properly describing the experimental tools to specifically address the OAM paraxial modes, we have to give further details about the structure of such modes. Then, mathematically the paraxial monochromatic OAM modes can be written as [9]

$$\mathbf{E_m}(r, \phi; z) = f_m(r; z) \exp(im\phi)\, \mathbf{u} \tag{11.1}$$

where $\mathbf{E_m}$ represents the vector amplitude associated with the mode, $\mathbf{u}$ is a transverse polarization vector ($\hat{\mathbf{u}} \perp \hat{\mathbf{z}}$), and (r, ϕ, z) are the cylindrical coordinates. We are distinguishing between the transversal and the longitudinal coordinates because, within the paraxial approximation, when the field structure is defined at a given transversal plane (say $z = 0$), it defines the whole three-dimensional structure of the field. Then, from now on when the z coordinate is not expressly written, we mean that $z = 0$. The index m defines the quantity of OAM the mode contains and distinguishes different modes. It is easy to check that these modes are orthogonal, but there are many different radial modes $f_m(r; 0)$ which can be used. One could then further refine the set of modes by using a radial and orthogonal basis. One possibility is to use the well-known Laguerre–Gaussian (LG) basis, which besides the azimuthal index m is labeled with a radial index p, that is, $\mathrm{LG}_{m,p}$. In experimental quantum optics, one very important mode is the $\mathrm{LG}_{0,0}$ or Gaussian mode, which is expressed as

$$\mathbf{E_0}(r, \phi) = \frac{2}{\pi w^2} \exp\left(-\frac{r^2}{w^2}\right) \mathbf{u} \tag{11.2}$$

The Gaussian mode is a very good approximation of the lowest mode that can propagate in usual optical fibers. In particular, it is a good approximation to the only propagating mode of commercial single mode fibers (SMFs). Then, the output of an SMF is always a Gaussian mode and its intensity is proportional to the projection of the Gaussian mode onto the initial spatial state of the input field. In particular, if a single photon is present in the field and propagated through an SMF, the operation that the fiber does can be expressed as $|0\rangle\langle 0|$, where $|0\rangle$ represents the spatial quantum state of a single photon being in the Gaussian mode. We thus have a way to uniquely identify the Gaussian mode, that is, one of the set of LG modes.

The next step in our program of experimentally controlling the OAM state of a single photon is preparing transformations of the spatial state of photons. In that way we are able to prepare or measure any LG mode by transforming it into or from a Gauss mode. There are many experimental techniques available to that purpose, from active real-time spatial light modulators (SLMs) to passive refractive or diffractive devices such as prisms. The most basic and simple to explain device is the spiral phase plate [10, 11]. The spiral phase plate introduces an azimuthal-dependent retardation to the field. The main effect is multiplying the field by a term of the kind $\exp(iM\phi)$, where the value of M depends on the construction of the spiral phase plate. There are many ways of experimentally building a spiral phase plate and we leave the interested reader to find the most suitable one for their purposes in the specialized literature. The effect of a spiral phase plate onto a single photon spatial state $|\Psi_i\rangle$ can be expressed with an unitary operation, that is, $|\Psi_o\rangle = U_M|\Psi_i\rangle$. By means of applying phase plates to our initial Gaussian beam, we can construct our set of experimentally addressable single photon OAM states:

$$|n\rangle = U_n|0\rangle \tag{11.3}$$

In Figure 11.2 we plot the amplitude of the far field of the first three modes. In the figure, we show a comparison between some of the modes and their corresponding LG mode, that is, an LG mode with the same $m, p = 0$, and an appropriate beam width. It can be seen that both sets are very similar.

In the same way that we can experimentally prepare the so-called *computational basis* $|0\rangle, |1\rangle, |-1\rangle, |2\rangle, \ldots$, we can also rotate a given state into a proper superposition of OAM states. Experimentally we should only find the proper spatial transformation, by means of diffractive elements or otherwise, to implement the proper Hilbert space rotation. In Figure 11.2, it is shown how the superposition of the states look like. From these images, it can be appreciated that some of the transformations can be easily approximated by displacing the spiral phase plates [12, 13]. Other more complicated transformations require the use of specific shape changing elements. In general, any superposition state can be created with an appropriate control of the spatial shape of the mode [14, 15].

We have shown that one can control the relative amplitudes of a single photon state of the kind $|\Psi\rangle = \sum a_n|n\rangle$. In terms of quantum information, this is the complete control you can have over a pure state. One can produce single photon mixture states, that is, density matrices, by, for example, probabilistically preparing different spatial operations. Nevertheless, a much more interesting issue in terms of quantum control of photons is the preparation of OAM states of two or more photons. By properly preparing pure two-photon states, one can completely control the state of a single photon, including mixed states. Then, in the next section, we review some of the experimental techniques to produce multiphoton OAM states.

11.3 Control of the OAM State of Multiple Photons

It is easily proven that, for completely controlling a two-photon state, new experimental tools are needed when compared to the case of a single photon. One way of showing this is by fixing the dimension of the Hilbert space to N, that is, the

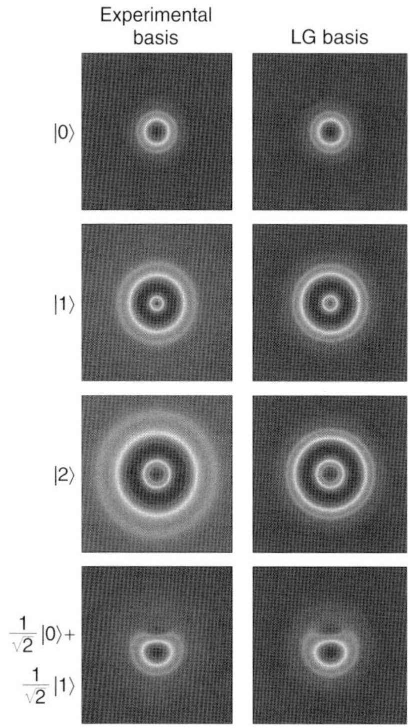

Figure 11.2 Orbital angular momentum modes: first column, experimentally achievable modes as explained; second column, Laguerre–Gaussian basis. The first three rows represent the amplitude distribution of the three lowest modes and last row shows the amplitude distribution of a superposition of two modes. Note that the three pure modes have an amplitude distribution that is rotationally invariant with respect to the propagation direction (perpendicular to the page). This is a trademark of the OAM eigenmodes. Also note the singularity in the center of the two higher-order pure modes. This singularity moves off-center in the superposition. (Please find a color version of this figure on the color plates.)

basis states are $|0\rangle, |1\rangle, \ldots, |N-1\rangle$. In this case, we will need $2(N-1)$ parameters to control an arbitrary one photon pure state: $(N-1)$ relative amplitudes and $(N-1)$ relative phases. If we have two photons and prepare them independently, we will be able to control $4(N-1)$ parameters. But, in this last case, the dimension of the Hilbert space is much larger, as our N-dimensional two-photon state will have the following elements in the basis: $|0\rangle_A|0\rangle_B, |0\rangle_A|1\rangle_B, \ldots, |N-1\rangle_A|N-1\rangle_B$, which are N^2 elements. Then, in order to prepare an arbitrary two-photon state, we would need $2(N^2-1)$ parameters. By comparing the two numbers, it is clear that we need new resources to arbitrarily control a two-photon state than independent control of each separate photon. If we turn to mixed states, we realize that the difference in parameter counting would be even larger. Actually, it can be proven that in order to produce arbitrary two-photon states, it is needed a nontrivial two-photon operator, that is, an element that acts on two photons at the same time.

Interestingly enough, this is also enough to produce arbitrary states of more than two photons.

Generally speaking, any multiparticle state that cannot be written as simply as the direct product of the independent particle states is called an *entangled state*. For pure states this statement takes the following mathematical form: if $|\Psi\rangle_{AB} \neq |\psi\rangle_A|\phi\rangle_B$, then we say that the particles A and B are in the entangled state $|\Psi\rangle_{AB}$. Then, the question for experimentalists is how to prepare and control multiphoton entangled states.

For the case of OAM-entangled two-photon states, a breakthrough in this subject was made by Mair and coworkers [12]. Measuring the OAM correlations between two photons generated from a spontaneous parametric down-conversion source (SPDC), they showed that photons could be entangled in their OAM degree of freedom. Since then, other schemes such as cold atomic ensembles are being explored [16], but up to now SPDC-based sources are still the most widely used and reliable sources of OAM-entangled photon generation.

The SPDC process is a second-order nonlinear effect where a high-frequency pump beam (typically in the violet – ultraviolet range of wavelengths) is converted into two low-frequency beams, which we call *signal* and *idler* (typically in the red- or near-infrared) by mediation of a crystal. In conventional systems pumped with continuous wave light, this effect is very small, which means that usually only a small fraction of the pump beam is converted into signal and idler. As an example, a typical experiment with a nonlinear crystal pumped with a continuous wave beam with a 400 nm wavelength, and using a few milliwatt of power, could produce around $10^5 - 10^6$ pairs of signal and idler photons around 800 nm. Quantum mechanically one could describe the process as the transformation of a pump photon into two signal and idler photons. For this process to be possible the so-called *phase matching* conditions have to be fulfilled, which can be basically described as energy and momentum conservation of the interacting photons. The present crystal growth technology has eased the task of finding suitable nonlinearities and geometries for the fulfillment of the phase matching conditions. In particular, by means of the periodically poling technique, one can generate photons in a crystal of potassium tytanil phosphate (KTP) in a collinear geometry (all the beams propagate in the same direction), in the absence of walk-off (the beams are not affected by birefringence as they propagate along one of the principal axes of the crystal) and the generated photons may have different polarizations (the so-called *Type II* condition), which allows for efficient separation of the signal and idler. As we would see shortly, these conditions offer the best chance of generating and controlling OAM states. Under these conditions, the state of the generated photons $|\Psi\rangle_{s,i}$ can be well approximated by the following expression: [17, 18]

$$|\Psi\rangle_{s,i} = N_{\text{SPDC}} \int d\mathbf{q}_s, d\mathbf{q}_i \exp\left(-A|\mathbf{q}_s - \mathbf{q}_i|^2 - B|\mathbf{q}_s + \mathbf{q}_i|^2\right) |\mathbf{q}_s\rangle_s |\mathbf{q}_i\rangle_i \qquad (11.4)$$

where the states $|\mathbf{q}_s\rangle$ ($|\mathbf{q}_i\rangle$) represent a signal (and similarly for the idler) photon in a plane wave mode with longitudinal wave-vector $k_z \simeq 2\pi n_s/\lambda_s$ and transversal momentum $\mathbf{q}_s = (q_s^x, q_s^y)$, A and B are two possibly complex constants that will allow us to tune the momentum correlations among the photons. The normalization factor is $N_{\text{SPDC}} = \frac{|AB|^{1/2}}{\pi}$.

We can control this two-photon state with parameters A and B, which, in the simplest case, depend on the experimental conditions in the following way:

$$A = \frac{w_p^2}{4} \quad B = \frac{\alpha L}{4k_p^0} \tag{11.5}$$

where L is the length of the nonlinear crystal, α is a fitting constant to approximate the phase matching *sinc* function by a Gaussian function (typically $\alpha = 0.455$), w_p is the pump beam width considered at the center of the crystal, and $k_p^0 = \omega_p n_p / c$, with ω_p and n_p being the corresponding angular frequency and refractive index of the pump beam, respectively. These expressions are valid when the pump beam is focused at the center of the crystal and the state of the two photons is also measured there, that is, the center of the crystal is taken as the origin of coordinates $z = 0$.

As mentioned in the introduction, one could, in principle, make a change of the spatial basis, and express the plane waves in the OAM basis, but in this case a more elegant approach can be taken via the Schmidt decomposition [19]. The Schmidt decomposition of the two-photon state (Eq. 11.4) will give us the set of orthogonal spatial modes where the correlations are perfect; that is, only one of the modes in the Schmidt basis of the signal is correlated with one of the idler ones. The Schmidt bases for signal and idler do not have to be the same, but in this ideal case they are not only the same but they are also LG modes:

$$|\Psi\rangle_{s,i} = \sum_{\substack{m \in \{-\infty,\ldots,\infty\} \\ p \in \{0,\ldots,\infty\}}} (-1)^{|m|} (1 - z) z^{|m|/2+p} |LG_{m,p}\rangle_s |LG_{-m,p}\rangle_i \tag{11.6}$$

$$z = \frac{(A - B)^4}{(A^2 - B^2)^2} \tag{11.7}$$

Then, a few conclusions can be extracted from these equations. First, one can observe that there is perfect anticorrelation between the OAM modes. This is what we usually mean when we say that the collinear, nonbirefringent SPDC process conserves OAM. Also, it can be seen that the shape of the state can be easily changed with two parameters: the length of the crystal or width of the pump beam. Actually, the latter is the one which is easier to continuously change experimentally. The entanglement of OAM states represented as superposition of different product states can be quantified in different ways. For the case of pure states, the von Neumann entropy gives a good measure of entanglement, which in this case can be analytically calculated (Figure 11.3).

In Mair *et al.* [12], the correlation between OAM modes were measured and then the law of the conservation of OAM was derived (by using different pump beams). In Figure 11.4 we have sketched the measurement setup and some of their results that the authors used in their experiment. Also, found in this work is the first indication that the OAM state produced in the crystal is actually an entangled state, by measuring that there was also correlation between superposition of modes. The

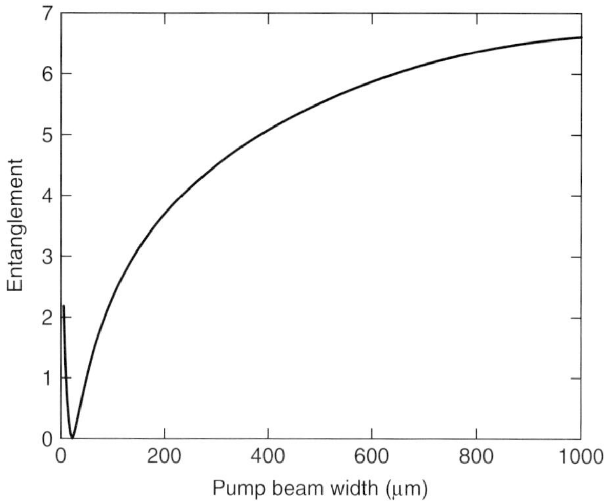

Figure 11.3 von Neumann entropy calculated for the state (11.4) with respect to the pump beam width (w_p). In this case $L = 30$ mm and $\lambda_p = 405$ nm. It is shown that the entanglement can be tuned from zero, that is, no entanglement, to a value close to seven, which will represent around 2^7 entangled modes. Note also that in the real case, when no approximations are made, the entanglement can never reach zero ([19]), although it is very close to that value.

experimental confirmation that the photons were entangled in their OAM was first performed via a Bell test [20] and then via a tomographic measurement [21]. These experiments were performed under an almost collinear situation, and the spatial walk-off was minimized by using a rather large pump width.

Finally, we would like to mention that the entangled state can be further controlled when other crystal geometries are used, which allow for noncollinear emission or for spatial walk-off. In these cases, the OAM modes do not have perfect correlations and new states can be generated. Nevertheless, under certain conditions (small noncollinear angle, large pump widths, etc.), the OAM conservation rule can be approximately fulfilled.

In the future, all these techniques will probably allow the generation of multiple entangled photons in OAM. One of the experimental possibilities could be the use of pulsed lasers and multipass schemes in the nonlinear crystal, simply mimicking the successful multiphoton generation schemes in polarization entanglement.

11.4 Applications in Quantum Information

The control of the OAM state of photons can be directly used in quantum information schemes; if nothing else, at least for proofs of principle of multidimensional schemes. It has already been proved that for some quantum communication schemes the use of higher dimensional states is beneficial. For example, the use

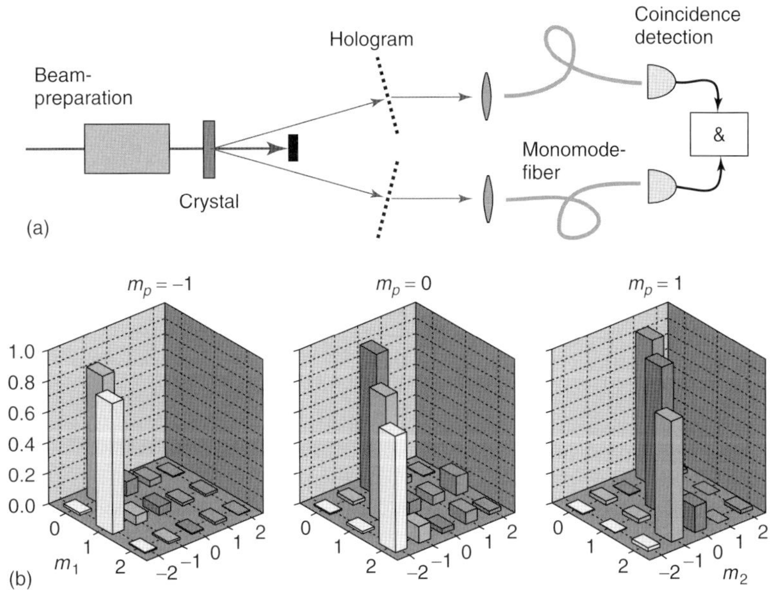

Figure 11.4 Conservation of OAM in the two-photon generation process : (a) Sketch of the experimental setup. (b) Results of the experiment, showing the relative number of coincident photons for different combinations of signal and idler modes. Every figure shows different pumping conditions. It can be noted how the OAM of the pump is transferred to the signal and idler and one only observes correlations in the outcomes when there is a conservation of the OAM between the three interacting modes. (Source: Reproduced from Mair *et al.* [12].) (Please find a color version of this figure on the color plates.)

of more dimensions in the codification of symbols allows for a higher information density coding and, surprisingly, a higher margin of security [22, 23]. In particular, in quantum cryptography, the use of N symbols and M codifying bases provides better security than that obtainable with qubits and two bases. Recently, a quantum cryptographic scheme for key distribution has been demonstrated, based on qutrits coded in OAM [24]. Another important quantum communication task is the distribution of entanglement. Again, theoretical investigations predict that physical systems with increasing dimensions can maintain nonclassical correlations in the presence of more hostile noise [25, 26].

We explain below in some detail, a quantum communication scheme that will allow a better understanding of the potential of higher dimensional quantum systems for practical applications. The communication task is called *quantum coin tossing* and it is the quantum version of a well-known classical cryptographic protocol, first stated by Blum [27]. In this protocol, two parties (say, the prototypical Alice and Bob) want to share a certain codified information (the result of a coin toss). The problem is that this "coin toss" should not be deciphered prior to a determined unveiling time, thus allowing the secrecy of the toss until the parties have bet on the result. After the bet, the result should be easily retrievable by any kind of manipulation.

This type of protocol is at the heart of other more complicated applications, where two partners are to realize transactions, but do not fully trust each other.

The classical solution for "coin tossing" is based on the computational security of one-way functions. The mathematical proof for such a solution does not exist and presently only relies on the technology limitations. The quantum version, on the other hand has been proved to be able to detect if one of the partners is cheating with a certain probability. A simplified quantum solution can be described in the following way: (i) Alice makes the coin toss and codifies the result in a quantum system, which is then sent to Bob. (ii) Bob bets on the result stored in the quantum system. (iii) The result is unveiled and the two parties can check whether the protocol was successful. This kind of protocol is based on the fact that it is impossible to retrieve an arbitrary state of a quantum system without *a priori* information. Then, when choosing the states to codify the "toss," two competing properties have to be taken into account. On one hand, the two states cannot be completely distinguishable, that is, Bob should not be able to retrieve all the information with a single measurement. On the other hand, the states should be properly differentiated once the result is announced, otherwise, Alice could always cheat and Bob would not have proper tools to discover her. These two seemingly opposite conditions find their trade-off in this kind of protocol when using mixed states in three dimensions [28, 29]. The use of pure states or two-dimensional states will result in a diminished security in the protocol, meaning that either Bob or Alice can cheat with a lower probability of being detected.

The need for three-dimensional mixed states in the optimal protocol makes the implementation with OAM-entangled photons very convenient. As we mentioned earlier, one possibility to codify a mixed state in one photon is to prepare an entangled two-photon source. Then, we would probabilistically perform different projective measurements on one of the photons, disregarding the result. This will effectively prepare the companion photon in a mixed state. This was the actual strategy used in the first implementation of the "quantum coin tossing" protocol in [30]. The authors used an SPDC source of OAM-entangled photons and prepared the measurements onto the three-dimensional OAM states with a series of beam-splitters and projective measurements, as explained in Section 11.2. In Figure 11.5 we show a sketch of the measurement setup and some of the results obtained.

11.5 Discussion

The above example of the "quantum coin tossing" experiment is very illuminating in the sense that it shows that the new physics of the Hilbert spaces of dimensions larger than two can be used in certain applications. The key element of the improved security in the "quantum coin tossing" protocol is the different properties of three-dimensional states as compared with two-dimensional ones. There are other properties of higher-dimensional spaces, which are rather intriguing, like the possibility of finding bound entanglement [31], optimization of certain quantum

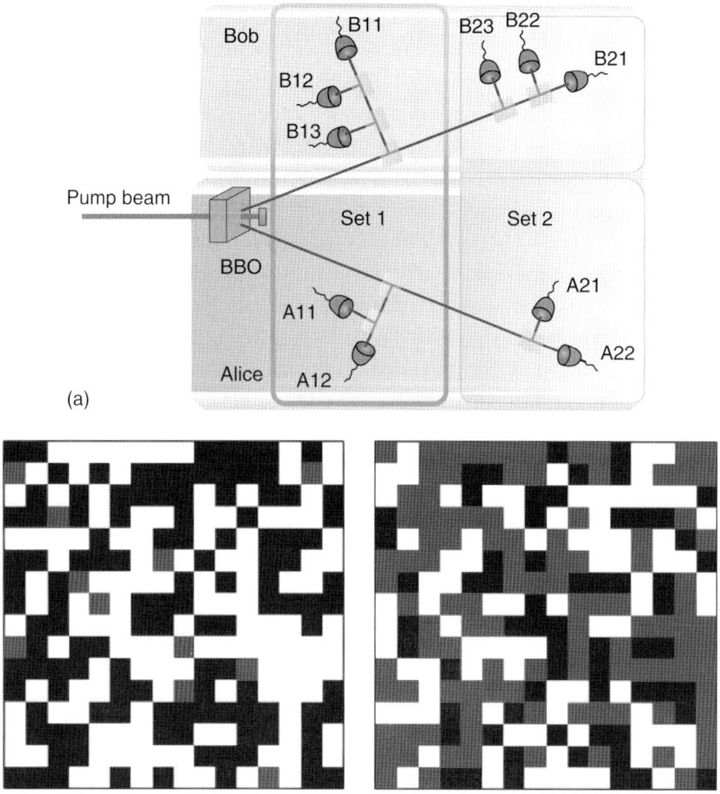

(a)

(b) (c)

Figure 11.5 Experimental demonstration of the quantum coin tossing. (a) Sketch of the experimental setup: the different labels from *A*11 to *A*22 indicate the different projective measurements that Alice carries out. Each projective measurement consists of a hologram which transforms the state into a superposition of two modes, and a single mode fiber. The combination of the two pairs of projective measurements is effectively preparing the photon sent to Bob in two different mixed states. At the final step of the protocol, Bob can measure the resulting state with a set of projective measurements, then implementing the measurement on two different bases. (b) Result of an honest protocol: each small square represents one photon successfully received and measured by Bob. The outcome of the protocol can be both parties agreeing on a "heads" result (white square), a "tails" one (black) or a "failure" (red) that is, not agreeing on the result. "Failures" can happen because of dishonest parties or due to experimental errors. (c) Result of a protocol where Alice was cheating: in this case, it can be seen how the number of "failures" critically increases, showing the presence of a dishonest party. (Please find a color version of this figure on the color plates.)

computation operations [32], or the relationship of the von Neumann entropy, and the degree of overlap between quantum states [33]. All these examples show that higher-dimensional spaces inherently have new properties that need to be explored and which could, in principle, be used for certain applications, following the example of the cryptographic protocol described earlier.

Encoding OAM information in photons has an important drawback in communication. This is that till date, a suitable way of reliably transmitting this information over large distances has not been found. The use of free space channels will only be possible over short distances or in the absence of atmospheric turbulence which tends to destroy the spatial information of light fields. In order to avoid this problem, important improvements in adaptative optics technology are required. Also, commercial optical fibers are not suitable for the transmission of spatial information. For short distances, one could, in principle, use specially designed fibers which allow for the transmission of a finite number of spatial modes.

On the other hand, owing to its ease of preparation and control, OAM states are very useful for testing the properties of higher-dimensional spaces. Also, other interesting possibilities of OAM states lie in the interface between quantum optics and nanophotonics, owing to their relation to the electromagnetic multipolar modes [7]. Controlling the OAM of photons could also lead to improved quantum metrology schemes.

11.6 Conclusion

In this chapter, we have tried to give a more detailed introduction to the experimental techniques of controlling the OAM of photons. We have reviewed the relation of the paraxial OAM modes with other sets of solutions of the electromagnetic field, showing the link between OAM and the total angular momentum of the light field. Then, we have described some of the techniques to control the OAM state of single and entangled photons. Finally, we have ended with the description of some applications and the possibilities of using the OAM of photons to study new physical phenomena.

References

1. Soskin, M.S. and Vasnetsov, M.V. (2001) Singular optics. *Prog. Opt.*, **42**, 219–276.
2. Loudon, R. (2000) *The Quantum Theory of Light*, Oxford University Press, USA.
3. Bouwmeester, D., Ekert, A.K., and Zeilinger, A. (eds) (2000) *The Physics of Quantum Information*, Springer Verlag, Berlin.
4. Allen, L., Padgett, M.J., and Babiker, M. (1999) The orbital angular momentum of light. *Prog. Opt.*, **39**, 291–372.
5. Molina-Terriza, G., Torres, J.P., and Torner, L. (2007) Twisted photons. *Nat. Phys.*, **3**, 305–310.
6. Cohen-Tannoudji, C., Dupont-Roc, J., and Grynberg, G. (1989) *Photons and Atoms: Introduction to Quantum Electrodynamics*, John Wiley & sons, New York.
7. Molina-Terriza, G. (2008) Determination of the total angular momentum of a paraxial beam. *Phys. Rev. A*, **78**, 053819.
8. Rajteri, M., Taralli, E., Portesi, Ch., Monticone, E., and Beyer, J. (2009) Photon-number discriminating superconducting transition-edge sensors. *Metrologia*, **46**, S283–S287.
9. Allen, L., Beijersbergen, M.W., Spreeuw, R.J.C., and Woerdman, J.P. (1992) Orbital angular momentum of light and the transformation of

Laguerre-Gaussian laser modes. *Phys. Rev. A*, **45**, 8185–8189.

10. Bazhenov, V.Y., Vasnetsov, M.V., and Soskin, M.S. (1990) Laser beams with screw dislocations in their wavefonts. *JETP Lett.*, **52**, 429–431.
11. Heckenberg, N.R., McDuff, R., Smith, C.P., and White, A.G. (1992) Generation of optical phase singularities by computer generated holograms. *Opt. Lett.*, **17**, 221–223.
12. Mair, A., Vaziri, A., Weihs, G., and Zeilinger, A. (2001) Entanglement of the orbital angular momentum states of photons. *Nature*, **412**, 313–316.
13. Vaziri, A., Weihs, G., and Zeilinger, A. (2002) Superpositions of the orbital angular momentum for applications in quantum experiments. *J. Opt. B: Quantum Semiclass. Opt.*, **4**, S47–S51.
14. Molina-Terriza, G., Torres, J.P., and Torner, L. (2001) Management of the orbital angular momentum of light: preparation of photons in multidimensional vector states of angular momentum. *Phys. Rev. Lett.*, **88**, 013601.
15. Molina-Terriza, G., Vaziri, A., Rehacek, J., Hradil, Z., and Zeilinger, A. (2004) Triggered qutrits for quantum communication protocols. *Phys. Rev. Lett.*, **92**, 167903.
16. Inoue, R., Kanai, N., Yonehara, T., Miyamoto, Y., Koashi, M., and Kozuma, M. (2006) Entanglement of orbital angular momentum states between an ensemble of cold atoms and a photon. *Phys. Rev. A*, **74**, 053809.
17. Walborn, S.P., de Oliveira, A.N., Padua, S., and Monken, C.H. (2003) Multimode Hong-Ou-Mandel interference. *Phys. Rev. Lett.*, **90**, 143601.
18. Chan, K.W., Torres, J.P., and Eberly, J.H. (2007) Transverse entanglement migration in Hilbert space. *Phys. Rev. A*, **75**(R), 050101.
19. Law, C.K., Walmsley, I.A., and Eberly, J.H. (2000) Continuous frequency entanglement: effective finite Hilbert space and entropy control. *Phys. Rev. Lett.*, **84**, 5304–5307.
20. Vaziri, A., Weihs, G., and Zeilinger, A. (2002) Experimental two-photon three-dimensional quantum entanglement. *Phys. Rev. Lett.*, **89**, 240401.
21. Langford, N.K., Dalton, R.B., Harvey, M.D., O'Brien, J.L., Pryde, G.J., Gilchrist, A., Bartlett, S.D., and White, A.G. (2004) Measuring entangled qutrits and their use for quantum bit commitment. *Phys. Rev. Lett.*, **93**, 053601.
22. Bruss, D. (1998) Optimal eavesdropping in quantum cryptography with six states. *Phys. Rev. Lett.*, **81**, 3018–3021.
23. Bechmann-Pasquinucci, H. and Peres, A. (2000) Quantum cryptography with 3-state systems. *Phys. Rev. Lett.*, **85**, 3313–3316.
24. Groblacher, S., Jennewein, T., Vaziri, A., Weihs, G., and Zeilinger, A. (2006) Experimental quantum crytography with qutrits. *New J. Phys.*, **8**, 75.
25. Kaszlikowski, D., Gnacinski, P., Zukowski, M., Miklaszewski, W., and Zeilinger, A. (2000) Violations of local realism by two entangled N-dimensional systems are stronger than for two qubits. *Phys. Rev. Lett.*, **85**, 4418–4421.
26. Collins, D., Gisin, N., Linden, N., Massar, S., and Popescu, S. (2002) Bell inequalities for arbitrarily high-dimensional systems. *Phys. Rev. Lett.*, **88**, 040404.
27. Blum, M. (1981) Coin flipping by phone. *CRYPTO*, **1981**, 11–15.
28. Ambainis, A. (2001) A new protocol and lower bounds for quantum coin flipping. *Proc. STOC*, **2001**, 134–142.
29. Spekkens, R.W. and Rudolph, T. (2001) Degrees of concealment and bindingness in quantum bit commitment protocols. *Phys. Rev. A*, **65**, 012310.
30. Molina-Terriza, G., Vaziri, A., Ursin, R., and Zeilinger, A. (2005) Experimental quantum coin tossing. *Phys. Rev. Lett.*, **90**, 040501.
31. Bennett, C.H., DiVincenzo, D.P., Mor, T., Shor, P.W., Smolin, J.A., and Terhal, B.M., (1999) Unextendible product bases and bound entanglement. *Phys. Rev. Lett.*, **82**, 5385–5388.
32. Greentree, A.D. et al. (2004) Maximizing the Hilbert space for a finite number of distinguishable quantum states. *Phys. Rev. Lett.*, **92**, 097901.
33. Jozsa, R. and Schlienz, J. (2000) Distinguishability of states and von Neumann entropy. *Phys. Rev. A*, **62**, 012301.

12
Rotating Atoms with Light

Kristian Helmerson and William D. Phillips

12.1
Introduction

An overarching theme of twentieth-century physics has been the control of matter at the atomic level. The arrival of quantum mechanics enabled a fundamental understanding of atomic structure and with it came the control of the internal states of an atom through the interaction with electromagnetic radiation. This understanding led to the development of the laser, which in turn afforded control over the translational states of atoms. The development of techniques, over the last decades of the twentieth century, to cool and trap atoms with lasers is testimony to the level of control that researchers now have over atoms. The reduction of the thermal motion of atoms has been exploited to build more accurate atomic clocks and realize new states of matter, such as Bose–Einstein condensates (BECs). While the "tools" to manipulate the linear momentum of atoms are now well established, such direct control over the center-of-mass rotational states of atoms (as opposed to their internal angular momentum states) has only recently been realized [1, 2].

In this chapter, we explain the techniques developed for manipulating and observing the rotational states of atoms using lasers beams that carry orbital angular momentum (OAM). We also describe applications of these techniques for generating and studying persistent currents in a superfluid atomic gas confined in a ring-shaped container. We do not cover the theoretical details of the interaction of atoms with light carrying OAM, except to provide equations, as necessary, to describe the underlying physics. Details about such interactions can be found in [3–5].

12.2
Orbital Angular Momentum of Light

Light, in addition to carrying linear momentum, also carries two kinds of angular momentum: Internal or spin angular momentum associated with its polarization and external or orbital angular momentum (OAM) associated with its spatial mode

Twisted Photons: Applications of Light with Orbital Angular Momentum.
Edited by Juan P. Torres and Lluis Torner

ISBN: 978-3-527-40907-5

[6, 7]. A convenient basis set for paraxial laser beams that carry OAM is the set of Laguerre–Gaussian (LG) modes (LG_p^l) [8, 9]. They are characterized by two indices l and p, where l is the winding number (the number of times the phase changes by 2π on a closed loop around the central singularity), and p is the number of radial nodes for radius $\rho > 0$. Each photon in the LG_p^lmode carries $l\hbar$ of OAM along its direction of propagation [6]. In contrast, spin angular momentum can only carry $\hbar$ of angular momentum per photon. The lowest order of LG mode that carries nonzero OAM is the LG_0^1 mode, where the electric field amplitude in polar coordinates (ρ, φ) at the beam waist varies as

$$LG_0^1(\rho, \varphi) = \frac{2}{\sqrt{\pi}} \frac{1}{w_0^2} \rho \exp\left(-\frac{\rho^2}{w_0^2}\right) \exp(i\varphi) \tag{12.1}$$

where φ is the azimuthal angle and the peak-to-peak diameter is $\sqrt{2}w_0$, where φ is the azimuthal angle. This mode carries $\hbar$ of angular momentum per photon and has a donut-like intensity profile since the amplitude goes to zero at $\rho = 0$. In general, the amplitude of an LG mode that carries nonzero OAM goes to zero at $\rho = 0$ because the phase is undefined on axis ($\rho = 0$). Light beams with OAM, however, have only recently been created [10, 11] and their effects on matter investigated.

12.3 The Mechanical Effects of Light

The interaction of light with matter inevitably involves the exchange of momentum. In the case of linear momentum, the mechanical effects of light range from comet tails to laser cooling of atoms. The coupling of optical spin angular momentum to atoms has been known for over a century [12] and has been verified in numerous experiments to control the internal states of atoms. The mechanical effect of the spin angular momentum of light on matter was first demonstrated 70 years ago in an experiment, where the spin angular momentum of circularly polarized light rotated a birefringent plate [13]. More recently, the circular polarization of light has been used to rotate birefringent, micron-sized particles held in optical tweezers [14, 15]. On the other hand, spin angular momentum of light cannot be used to change the rotational state of an atom, since it does not involve momentum exchange with the center-of-mass of the atom.

In contrast, the OAM of light does involve an azimuthal component of the wavevector or Poynting vector, which can couple to the center-of-mass of an object. Such coupling has been demonstrated by the rotation of micron-sized particles held in optical tweezers [16–18]. The coupling of OAM to atoms was first demonstrated in an experiment by Tabosa and Petrov [19]. In this and subsequent experiments by the same group [20], a diffraction grating corresponding to the interference of an LG mode with a Gaussian mode was written onto a thermal cloud of atoms. This atomic diffraction grating caused an incoming Gaussian beam to diffract into beams carrying OAM. Another experiment [21] used the OAM of light to create a rotating collective excitation of an atomic cloud. A subsequent Gaussian read-out pulse is converted to one with OAM by the reverse process. Although these experiments demonstrate the transfer of the spatial phase of the OAM beams to an ensemble of atoms, no direct measurement of the mechanical rotation of the atoms was observed. The thermal motion of the atoms, even for a laser cooled sample [19, 20], is so large compared to the coherent rotational motion of the atoms that it is impossible to observe the rotation directly. In order to definitively observe the mechanical rotation of atoms due to the quantized transfer of OAM from light, the atoms need to be prepared in a nearly pure motional state. Such a cold sample of atoms is available in an atomic Bose–Einstein condensate (BEC).

12.4 Rotating Bose–Einstein Condensates

The creation of BECs in dilute atomic vapors is one of the major triumphs of the quest to control atoms. In addition, the creation of BECs has renewed interest in the applications of coherent light to manipulate and control the states of atoms. Experiments to control the external states of atoms that used thermal atomic samples are analogous to early experiments in optics that used lamps. The creation of a BEC of a dilute atomic gas has provided a matter wave source analogous to the optical laser. The macroscopic occupation of the ground state of a trap by a BEC is similar to the occupation of a single mode of an optical cavity by photons. The atoms forming the condensate all occupy the same state in terms of their internal and external degrees of freedom. Hence the center-of-mass motion of the atoms in a BEC can be described by a single, macroscopic wavefunction, $\Psi = \sqrt{n}e^{i\phi}$, characterized by an amplitude and phase, ϕ. The square of the amplitude gives the atomic density n, and the velocity of the BEC is given by

$$\vec{v} = \frac{\hbar}{M}\vec{\nabla}\phi \qquad (12.2)$$

where M is the atomic mass of the atoms in the BEC.

BECs are nearly the ideal, monochromatic atomic source for manipulation by light, since the interaction of the atoms with an optical field typically involves the transfer of the photon momentum to the atoms. Because of the repulsive atom–atom interaction, which can be described by a mean field, the BEC swells to a size significantly larger than the ground state wavefunction of the harmonic trap confining the atoms [22]. The spatial extent of the resulting wavefunction can be several orders of magnitude larger than the optical wavelength. Hence the momentum width, given by the Heisenberg uncertainty principle, can be much less than the photon's momentum. Not all experiments will realize this reduced and intrinsic momentum width. The interaction energy may be converted to kinetic energy when the atoms are released from the trap. Nonetheless, the resulting additional momentum spread, due to the atom–atom interaction, can still be significantly less than the momentum of a single photon. Furthermore, the momentum spread due to the repulsive interactions can be, in a certain sense, coherent, like the divergence imparted to a coherent laser beam by a diverging lens. Such a divergence can be "recollimated" by an appropriate coherent manipulation.

The macroscopic wavefunction of the BEC implies that the atomic gas is irrotational, since the curl of the velocity (Eq. (12.2)) is zero. Such irrotational flow is characteristic of superfluids, where rigid body rotation is mimicked by an array of quantized vortices [22]. Early investigations in the superfluid character of atomic BECs prompted many proposals for generating and observing quantized vortices in BECs. The first generation of a vortex in a BEC involved a rapidly rotating Gaussian laser beam coupling the external motion to internal state Rabi oscillations [23, 24]. Later schemes included mechanically stirring the BEC with a focused laser beam [25] and "phase imprinting" by adiabatic passage [26]. Numerous papers proposed generating vortices in a BEC using stimulated Raman processes with LG optical fields that carry OAM [27–31]. This technique produces a single vortex with a chosen number of units of OAM (chosen charge). The process is fully coherent and can produce superposition of different vortex states with determined phase. Yet, despite the potential advantages and apparent simplicity of this approach, generating vortex states in a BEC by the transfer of optical OAM to atoms was not realized until 2006 [1].

12.4.1 Experiment to Transfer Orbital Angular Momentum to a BEC ($\Delta = 0$)

Our scheme for transferring OAM of photons is based on our technique of Bragg diffraction [32] for changing the momentum state of the BEC. It is similar in spirit to other proposed Raman schemes [27–31], but differs (at least in the original version of our technique [1]) in that we do not change the internal atomic states. Instead, we change the linear momentum states of the atoms, along with the OAM states. Our two-photon rotational Bragg scheme is shown in Figure 12.1a.

Figure 12.1b is a schematic representation of the experimental procedure used to transfer OAM from an LG beam to the atoms in a BEC. An atom of mass M, in the presence of counterpropagating LG_0^1 and Gaussian beams (Figure 12.1b, left and right beams, respectively) absorbs a LG_0^1 photon and stimulatedly emits a Gaussian photon, acquiring $2\hbar k$ of linear momentum ($k = 2\pi/\lambda$ with λ the photon wavelength). As with resonant Bragg diffraction with two Gaussian beams, the frequency difference between the two beams is $\delta\omega = 4E_r/h$, with $E_r = (\hbar k)^2/2M$ [32] the "recoil energy" or the kinetic energy change of an atom initially at rest after the absorption of photon momentum $\hbar k$. In addition to linear momentum the atoms pick up the OAM difference between the two photons. The additional energy due to the formation of a vortex is small and, for the pulse durations used in this experiment, does not affect the resonance condition. In our experiment, the rotational energy is in the order of 1 Hz, which is much smaller than the Fourier width of the Bragg pulse. Therefore, choosing the frequency difference between the two beams to satisfy the Bragg resonance condition of $\delta\omega$ ($2\pi \times 100$ kHz for sodium) [32] is sufficient to resonantly transfer both linear momentum and OAM to the atoms. The smallness of the vortex energy does necessitate changing either

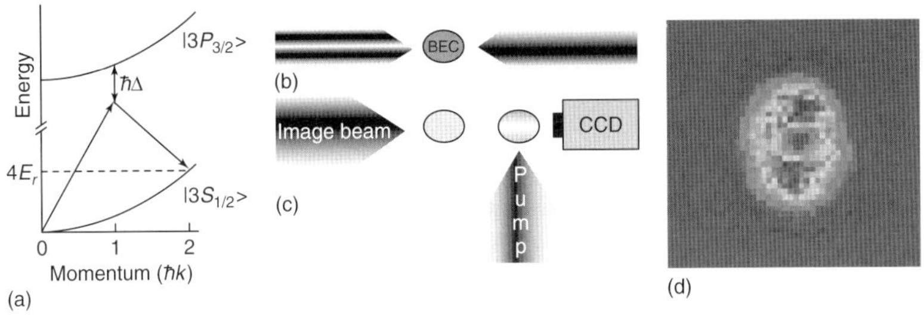

Figure 12.1 (a) Two-photon rotational Bragg scheme used to couple atoms between momentum states 0 and $2\hbar k$, transferring OAM in the process. (b) Schematic representation of the experiment. Counterpropagating LG_0^1 and Gaussian laser beams, with the same linear polarization and a variable frequency difference of $\delta\omega/2\pi$, are applied to a BEC. (c) The atoms that have undergone the Raman transitions (right cloud) have been separated from those that did not (left cloud). A spatially localized "pump" beam enables independent imaging of each cloud by absorption of a probe beam propagating along the direction of linear momentum transfer. (d) Absorption image of a cloud that has undergone the Raman transition, taken along the axis of the LG_0^1 beam. The vortex core is seen as a hole in the cloud. (Please find a color version of this figure on the color plates.)

the internal energy state of the atoms or, as we do, the linear momentum state in order to achieve good discrimination between initial and final states of the Raman process. If longer pulse lengths were used, it might be possible to directly induce a rotation of the condensate without changing the internal state or transferring linear momentum; however, such a process may be strongly suppressed since, in the Thomas–Fermi regime [22], the rotational energy is much less than the mean-field interaction energy. Such suppression was observed in an experiment to transfer small amounts of linear momentum to a condensate [33].

The linear momentum transferred by Bragg diffraction can be viewed as the result of the diffraction of atoms from a moving sinusoidal optical dipole potential generated by the interference of the counterpropagating Gaussian beams. The optical dipole potential generated by interference of the counterpropagating LG_0^1 and Gaussian beams is not sinusoidal, but due to the radial intensity profile and the helical phase of the LG_0^1 beam, the dipole potential generated is "corkscrew" like. Diffraction off this corkscrew potential produces a rotating state. This potential is the atom optics analog of a phase hologram, and one could generate any desired two-dimensional atomic state using a suitable hologram.

Our BEC consists of $1–2 \times 10^6$ sodium atoms in the $|3S_{1/2}, F = 1, m_F = -1\rangle$ state, confined in a magnetic time orbiting potential (TOP) trap such that the average characteristic size of the BEC (the Thomas–Fermi radius [22]) is $\approx$30 μm. The BEC exhibits some oscillation of the center-of-mass about the minimum of the trap due to the relaxation of the trap potential to its final value, but the momentum associated with this oscillation is less than $0.03\hbar k$. A Gaussian laser beam, detuned from the D_2 line ($\lambda = 589.0$ nm) by $\Delta = -1.5$ GHz ($\approx$150 linewidths, enough to prevent any significant spontaneous photon scattering) is split into two beams and passes through separate acousto-optic modulators (AOMs) in order to control their frequency difference, $\delta\omega$. One of the beams diffracts from a blazed transmission hologram [10, 11, 34] generating a LG_0^1 mode that propagates along x. The Gaussian beam propagates along $-x$. We apply these beams to the trapped atoms as a square pulse and then turn off the trap. After 6 ms time-of-flight (TOF), we image the released atoms by absorption of a probe beam resonant with the $|3S_{1/2}, F = 2\rangle$ to $|3P_{3/2}, F = 3\rangle$ transition. During imaging the atoms must be optically pumped from the $|3S_{1/2}, F = 1\rangle$ state into the $|3S_{1/2}, F = 2\rangle$ state by a pump beam resonant with the $|3S_{1/2}, F = 1\rangle$ to $|3P_{3/2}, F = 2\rangle$ transition. Atoms with momentum $2\hbar k$ from the Raman process will separate spatially during the TOF from atoms still at rest (see Figure 12.1c). We use a focused pump beam spatially localized along x to selectively image clouds of atoms in different momentum states with the probe beam propagating along x, the axis of rotation of the LG_0^1 beam. Figure 12.1d is an image of a cloud that has undergone the rotational Bragg process, where the vortex core is observed as a hole in the middle of the cloud.

12.4.2
Efficiency of the OAM Transfer Process

By measuring the relative number of atoms in the stationary cloud and moving vortex cloud, we find that the optimal transfer into the rotating state happens for $\delta\omega/2\pi = 97.5(5)$kHz.[1] The difference from the expected value $4E_r/h = 100$ kHz is attributed to initial motion of the BEC in the trap.

Figure 12.2a shows the transfer efficiency as a function of pulse duration for two different Bragg detunings, $\delta\omega/2\pi$. A maximal transfer of 53% was achieved for $\delta\omega/2\pi = 97.5$ kHz. The transfer is limited by the spatial mismatch between the LG_0^1 beam, where the intensity goes to zero in the center of the beam, and the (inverted parabolic shape) initial BEC, where the density is peaked in the center. A full three-dimensional, time-dependent simulation [35] based on the estimated spatial profile of the LG_0^1 beam and the BEC, shown in Figure 12.2b, predicts a maximum transfer efficiency of 60% (Figure 12.2a, red line) and 67% (Figure 12.2a, blue line) for $\delta\omega/2\pi$ of 97.5 and 100 kHz, respectively. The simulations confirm that because of the spatial mismatch, transfer of the entire BEC is impossible.

Other problems can arise because of the spatial mismatch between the LG modes and the initial density profile of the BEC. The spatial intensity profile of the LG mode, especially near the center, will result in a spatial variation of the Rabi frequency for coupling atoms in the initial BEC state to the final (rotating) state. The spatial variation of the Rabi frequency will result in a spatial dependence of the transfer efficiency, as shown in Figure 12.2c. This is observed in the experiment where maximum transfer of the atoms to the rotating state occurs at spatial locations where the Rabi frequency times the pulse duration corresponds to $(2n+1)\pi$, with $n = 0, 1, 2, \ldots$ It may be possible to eliminate the concentric

1) All uncertainties reported are one standard deviation combined statistical and systematic uncertainties.

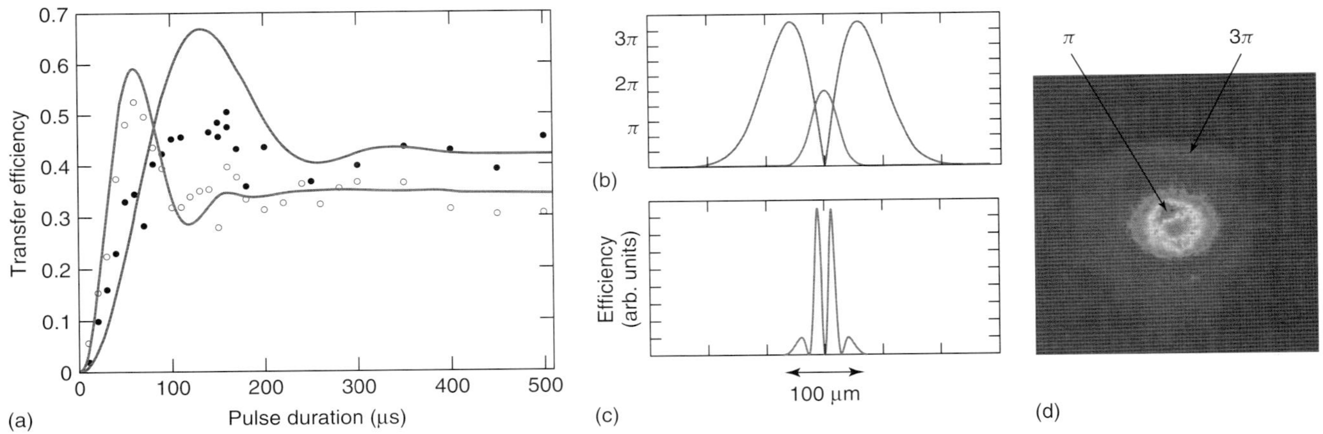

Figure 12.2 (a) Measured (dots) and calculated (lines) transfer efficiencies due to the mismatch of the spatial overlap of the Laguerre–Gaussian mode with the BEC. The experimental data points were obtained for Bragg detunings of 97.5 kHz (open dots) and 100 kHz (solid dots). The calculated curves are based on a full three-dimensional, time-dependent simulation of the BEC wavefunction [35], based on the spatial overlap depicted in (b). (b) Estimated spatial overlap of the LG beam with the BEC, approximated as a Gaussian. (c) Spatial dependence of the transfer efficiency (for a pulse duration of 130 μs) due to the spatial dependence of the Rabi frequency on the intensity profile of the LG mode shown in (b). (d) Observed spatial dependence of the transfer process of the BEC to the rotational state. The inner, intense ring results from an approximate π-pulse where the initial atom density is high, while the outer, weak ring corresponds to an approximate 3π pulse where the density is small. (Please find a color version of this figure on the color plates.)

rings observed in Figure 12.2d by using adiabatic rapid passage techniques [36, 37]; however, the spatial mismatch problem is best eliminated by having a ring-shaped BEC as the initial state.

12.5 Measuring the Rotational Motion of the Atoms

Figure 12.1d shows an image of a donut-shaped BEC corresponding to a vortex state, due to the transfer of OAM from a LG_0^1 beam to the atoms. It might be argued, quite justifiably, that the donut-shape occurs not because the atoms are in a vortex or rotational state, but because the coupling laser has a donut-shaped mode. That is, there are no atoms in the middle of the cloud, because the laser intensity is zero there and cannot transfer any atoms. Additional evidence is required to confirm that the atoms are in a rotational or vortex state. Three methods have been developed to measure the velocity field of a vortex state in a BEC. Two of these methods use interferometry, while the third method measures the Doppler shift of the moving atoms.

12.5.1 Interference of the Rotating State with a Nonrotating State

Since the velocity of the BEC is given by the gradient of the phase of the corresponding macroscopic wavefunction (Eq. (12.2)), interferometric techniques that measure the spatial variation of the phase of the BEC wavefunction can be used to determine the velocity field. Interferometry requires a reference phase that the measured phase is compared to. For most of the interferometry measurements, a spatially and temporally constant reference phase is desirable. The initial phase of the wavefunction of the BEC is typically spatially constant and temporally evolves on a timescale that is inversely proportional to the mean-field or interaction energy of the BEC [22]. Hence, the spatial phase variation of the vortex state can be determined directly by a comparison with the phase of the initial BEC state. This technique was used to map the phase profile of the first vortex produced in an atomic BEC [23]. In that experiment, a vortex state of atoms in one hyperfine state was created by a spatio-temporally varying, coherent coupling to a BEC of rubidium atoms in another hyperfine state. The atoms in the BEC that did not undergo transfer to the vortex state were subsequently interfered with the atoms in the vortex state using a spatially uniform, coherent coupling pulse.

In order to measure the spatial phase profile of our vortex state, we perform a similar interferometric measurement [23], but couple atoms in different linear momentum state rather than in different hyperfine states. A first pulse, consisting of a LG_0^1 beam at $\omega + \delta\omega$ and a counterpropagating Gaussian beam at ω (LG_0^1/G pulse) transfers some of the atoms to a state with linear momentum $2\hbar k$ and OAM $+\hbar$. Before the atoms with $2\hbar k$ of linear momentum have moved any significant distance, a second pulse, where the LG_0^1 beam is replaced by a Gaussian beam at $\omega + \delta\omega$, is applied to the atoms. This pair of Gaussian beams (G/G pulse) couples the similar two linear momenta states ($p = 0$ and $2\hbar k$), but without changing the OAM. Figure 12.2b is an image of the $2\hbar k$ cloud from the two-pulse sequence. The off-centered hole results from the interference between a state rotating with OAM $\hbar$ and a nonrotating state. The off-centered hole is expected for a $\hbar$ vortex state, which has a 2π phase winding, since at some location the phase of the vortex state is opposite to the phase of the reference BEC state and the two clouds of atoms will destructively interfere at that location. Thus the direction in which the hole is displaced is determined by the local phase difference between the two interfering states. (Such an interference pattern was also demonstrated in [23] and used to map the phase of a vortex state.)

Because stimulated Raman processes are coherent, we expect the relative quantum phase between the rotating and nonrotating states to be set by the relative phases of the laser beams used. We verified this with the same two-pulse experiment, but additionally interfered the LG_0^1 beam at $\omega + \delta\omega$ with the copropagating Gaussian beam, also at $\omega + \delta\omega$, on a CCD camera to measure the relative optical phase between the two beams, thereby determining the relative phase between the corkscrew and the sinusoidal diffractive structures generating the two interfering clouds. (Both pulses use the same counterpropagating Gaussian beam, so this common phase difference drops out.) In Figure 12.3d the measured phase of the atomic interference is plotted as a function of the measured optical phase, for 18 consecutive realizations of the experiment. They are correlated, as expected, even though the optical phase is measured 10 ms after the diffracting pulses. (The variation in the optical phase, mostly determined by mirror vibrations, appears to

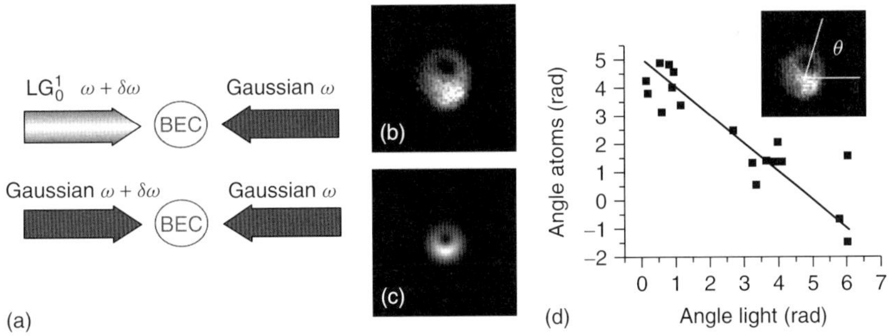

Figure 12.3 (a) Schematic representation of the laser pulse sequence used to generate and interfere the vortex state with the initial BEC. (b) Interference of a vortex state with $\hbar$ of angular momentum with a stationary state. The 2π phase winding associated with the rotating state results in a displacement of the hole, due to destructive interference. (c) Calculated interference pattern between a rotating and stationary state based on harmonic oscillator states. (d) Angle of the hole in the interference pattern between rotating and nonrotating atomic states as a function of the rotation angle of the optical interference pattern between the LG_0^1 and copropagating Gaussian beams. The straight line (to guide the eye) has slope −1. Inset: Image of the atomic interference between a rotating and nonrotating cloud. The hole is displaced from the center and its angular position θ depends on the relative phase between the interfering states. (Please find a color version of this figure on the color plates.)

be sufficiently small on this short timescale.) This demonstrates that, in principle, by controlling the relative phases of the Raman beams, atoms can be put into any desired coherent superposition of different rotational states.

12.5.2
Interference of the Rotating State with a Counterrotating State

A coherent superposition of a rotating and counterrotating state was generated and the resulting interference of the two states was observed (Figure 12.3b, c). To perform this, (Figure 12.4a) a first pulse consisting of the LG_0^1 beam at $\omega + \delta\omega$ and the counterpropagating Gaussian beam at ω (LG_0^1/G pulse) transferred about 20% of the atoms to a state with linear momentum $2\hbar k$ and OAM $+\hbar$. The same two beams were then used in a second pulse, but with the LG_0^1 beam at $\omega - \delta\omega$, to transfer about 40% of the remaining atoms to a state with linear momentum $-2\hbar k$ and OAM$-\hbar$. A third pulse, consisting of two counterpropagating Gaussian beams (G/G pulse) at the same frequency ω was then applied. The G/G pulse is resonant for a second order (four photon) Raman process between states with linear momenta $-2\hbar k$ and $+2\hbar k$ [32]. Again, there is essentially no delay between the pulses so that atoms with different linear momenta remain well overlapped spatially during the pulse sequence. Figure 12.4b is an image of one of the interfering clouds after the three pulses, and corresponds to the superposition of two clouds with OAM $\pm\hbar$ (Figure 12.4c). Since each of the diffracted atoms has absorbed or stimulatedly emitted one LG_0^1 photon, the interference pattern confirms that each LG_0^1 photon transfers $\hbar$ OAM to the atoms. Although interference has previously been used to observe vortex states [23, 38, 39], this is the first interference between independently generated, overlapping counterrotating vortex states. (A superposition of a left- and right-circulating vortex was created simultaneously to produce a soliton in a BEC [24, 40].)

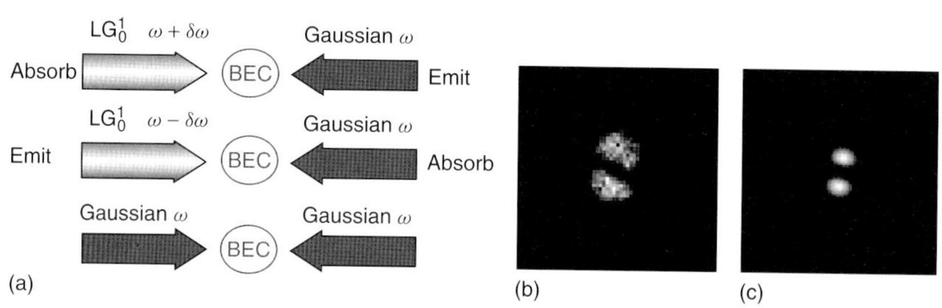

Figure 12.4 (a) Schematic representation of the laser pulse sequence used to generate and interfere the vortex state with an antivortex state. (b) Interference of a vortex state with $\hbar$ of angular momentum with a vortex state with $-\hbar$ of angular momentum. The interference of the rotating state with the counterrotating state results in a circular standing-wave. (c) Calculated interference pattern based on harmonic oscillator states between a rotating state (angular momentum $\hbar$) with a counterrotating state (angular momentum $-\hbar$). (Please find a color version of this figure on the color plates.)

12.5.3
Observation of Fork-Like Interference Structure

A related interferometry measurement involves interfering atomic clouds with a small linear momentum difference. For two otherwise identical clouds, the resulting interference pattern produces straight interference fringes or stripes, analogous to the interference of two laser beams with a slight angle between them. If one of the clouds contains a vortex state, then the resulting interference fringe pattern contains a fork-like dislocation. The resulting interference pattern looks similar to the diffraction gratings typically used to diffract Gaussian beams into LG modes [10, 11]. Such an interference pattern was observed in [25] to confirm the generation of a vortex state in a BEC by rotating the trapping potential. We also performed a similar interference experiment by using the pulse sequence described above for the interferometry experiment that generates the displaced hole (Figure 12.3a), except that the laser beams for the Gaussian/Gaussian pulse were not quite counterpropagating, as depicted in Figure 12.5a. The resulting interference pattern (Figure 12.5b, c) shows the characteristic fork-like structure owing to the presence of a vortex with $\hbar$ of angular momentum in one of the clouds.

12.5.4
Measurement of the Doppler Shift of the Rotating Atoms

A third method, which measures the Doppler shift of the moving atoms, was used to also confirm that the atoms are rotating after the transfer of OAM. This technique, often called *Bragg spectroscopy*, is also based on Bragg diffraction but uses longer pulses to achieve sufficient sensitivity to the Doppler shift to resolve small velocity changes [32, 41]. Two counterpropagating laser beams at frequencies $\nu_1 = \omega_1/2\pi + 100$ kHz and $\nu_2 = \omega_2/2\pi$, respectively, are used to Bragg diffract some of the atoms to the $2\hbar k$ linear momentum state. For sufficiently long

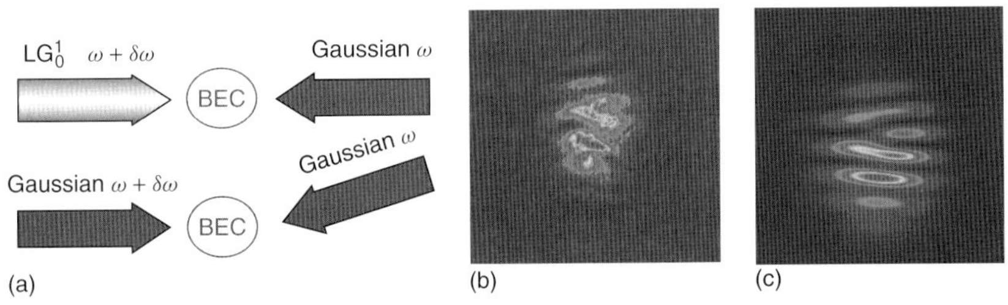

Figure 12.5 (a) Schematic representation of the laser pulse sequence used to generate and interfere the vortex state with the initial BEC. The second set of laser beams are not counterpropagating (in contrast to the first set), which results in the interfering clouds having a relative average velocity. (b) Interference of a vortex state with $\hbar$ of angular momentum with a nonrotating state. The fringes are a result of the relative velocity between the interfering clouds, while the singularity associated with the rotating state results in the fork-like structure. (c) Calculated interference pattern based on harmonic oscillator states. (Please find a color version of this figure on the color plates.)

pulses ($\approx$1ms) and taking 100 kHz to be the exact frequency offset for resonant Bragg diffraction at $v = 0$, the Doppler resonance condition $2kv = \omega_1 - \omega_2 = 2\pi(\nu_1 - \nu_2)$ will preferentially select out atoms at velocity v along the direction of the $\nu_1 + 100$ kHz laser beam to undergo the Bragg diffraction. For a rotating cloud (see Figure 12.6), the relative detuning between the lasers, $\nu_1 - \nu_2$, will select atoms at the location where they have the appropriate velocity component along the laser beams to be in resonance. For $\nu_1 > \nu_2$, atoms in the upper half of the rotating cloud will be more resonant for the two-photon process than those in the lower half and vice-versa (see Figure 12.6a). We measured the number of atoms in the upper half of the cloud N_{upper} minus the number in the lower half N_{lower} normalized to the total number of Bragg diffracted atoms ($N_{\text{upper}} + N_{\text{lower}}$) as a function of laser detuning $\nu_1 - \nu_2$ to obtain a Doppler profile of the rotating cloud. Such a profile is shown in Figure 12.6b for a cloud of atoms in a ring-shaped trap, rotating with $\hbar$ of angular momentum. The frequency difference between the two maxima of the profile (indicated by the arrows) can then be converted to the azimuthal velocity component of the rotating cloud using the Doppler shift equation given above. This technique has also been used to map out the velocity field of a vortex lattice in a rotating BEC [42].

12.6 Generating Other Rotational States of Atoms

The techniques described above are not limited to generating superpositions of rotational states with $\pm\hbar$ of angular momentum. In this section, we describe experiments generating vortex states of higher angular momentum and vortex states in spinor BECs. We also describe a matter wave amplification experiment on a vortex state.

12.6.1 Vortices of Higher Charge

The requirement that the phase of the wavefunction of the superfluid Bose gas be continuous gives rise to the circulation in the gas being quantized in units of $\hbar$ [22]. One manifestation of the higher circulation is the presence of vortices of higher charge. These are vortices with an associated phase winding of $\pm 2\pi n$, with $n > 1$. Vortices of 4π and 8π (charge 2 and 4, respectively) have been realized [26, 43] by "imprinting" a topological phase on a BEC. It is also possible to generate higher charge vortices through the transfer of OAM of photons to atoms.

We generated vortices of higher charge by transferring the angular momentum to each atom from several LG_0^1 photons (see Figure 12.7a). An initial LG_0^1/G pulse with $\delta\omega/2\pi \approx 100$ kHz transfers some of the atoms into the singly charged vortex state with linear momentum $2\hbar k$. A second LG_0^1/G pulse, with $\delta\omega/2\pi \approx 300$ kHz, transfers 80% of the atoms in the $2\hbar k$ linear momentum state into the $4\hbar k$ state and gives each of the atoms an additional $\hbar$ of OAM. The resulting state is a charged 2

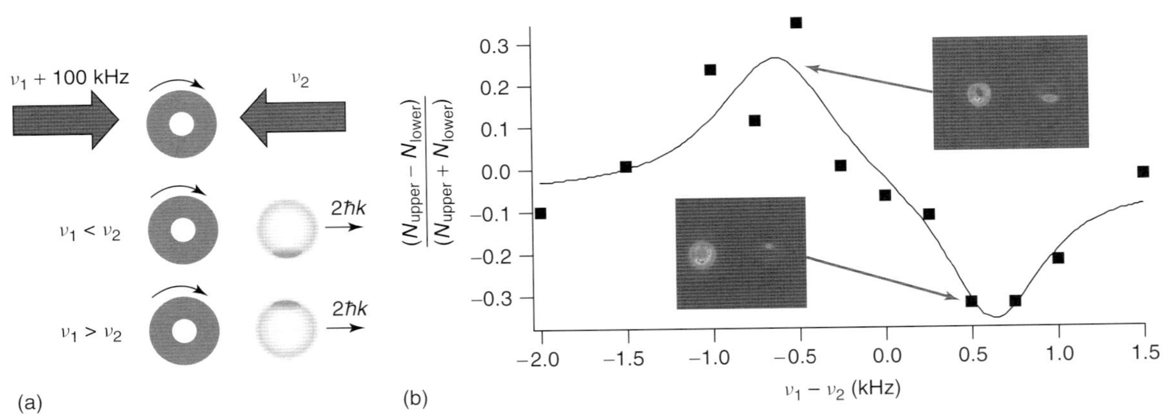

Figure 12.6 (a) Scheme for measuring rotational velocities using velocity-dependent Bragg diffraction. (b) The number of Bragg diffracted atoms in the upper half of the image minus the number of Bragg diffracted atoms in the lower half of an image, normalized to the total number of Bragg diffracted atoms, as a function of the relative detuning of the Bragg diffraction lasers. (Please find a color version of this figure on the color plates.)

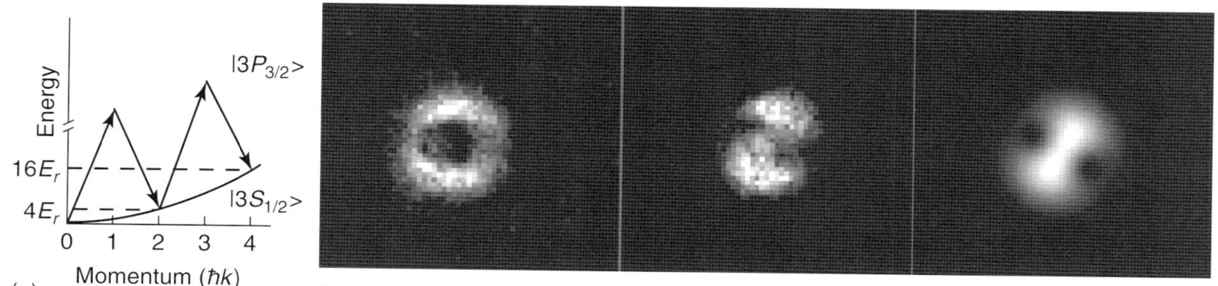

(a) (b) (c) (d)

Figure 12.7 (a) Two consecutive Raman processes transfer atoms into a doubly charged vortex state with linear momentum $4\hbar k$. (b) Absorption image of the doubly charged vortex cloud. The core is seen to be larger than for the single charged state of Figure 12.1d. (c) Absorption image of the cloud resulting from the interference between a doubly charged state and a nonrotating state. (d) Calculated interference pattern, based on harmonic oscillator states, between nonrotating and doubly charged rotating state.

vortex, moving with linear momentum $4\hbar k$. Figure 12.7b is an image of this state. The higher angular momentum of the charge 2 vortex state results in a larger core diameter compared to the singly charged state (see Figure 12.1d).

In order to verify that the state we generated is a charged 2 vortex, we first apply the procedure described in the previous paragraph, and then before the states with different linear momentum have a chance to separate, we apply a third, G/G pulse, with $\delta\omega/2\pi \approx 200$ kHz, which couples the 0 and $4\hbar k$ linear momentum states via a second order (four-photon) Raman process [32]. Figure 12.7c is an image of the $4\hbar k$ cloud generated by the three pulses, taken after 6 ms TOF. It corresponds to the interference between a nonrotating cloud and a cloud with angular momentum $2\hbar$. Because a charge 2 vortex has a phase winding of 4π, we expect two locations corresponding to destructive interference between the rotating state and the nonrotating state (see Figure 12.7d). Indeed, we observe two offset holes in the density distribution (Figure 12.7c).

Although we used a step-wise (multiphoton) process to transfer additional $\hbar$ of OAM per photon to the atoms, higher rotational states can also be generated by transferring higher amounts of OAM per photon. This can be accomplished using LG modes with higher l. This has the advantage of transferring larger amounts of OAM for a given linear momentum transfer.

12.6.2
Rotational States of Multilevel Atomic Condensates

BECs of weakly interacting gases have been realized with various atomic species and many of these species have multiple, ground-state sublevels. When the sublevels are magnetic spin states, such a multicomponent system is best described as a spinor BEC [44]. The multicomponent character of a spinor BEC can give rise to rich topological structures such as coreless vortices and spin textures. A coreless vortex has been generated in a spinor BEC using the topological phase "imprinting" technique [45] whereas spin textures have been observed in the rapid quench of a spinor system across the BEC phase transition [46].

The transfer of OAM from photons to atoms can also be used to generate vortices or rotational states of atoms in a spinor BEC. In an experiment with a ^{87}Rb BEC at the University of Rochester [47], a combination of LG_0^1, LG_0^{-1}, and Gaussian beams of appropriate circular polarizations were used to couple the ground state magnetic sublevels $|F = 1, m_F = -1\rangle$ and $|F = 1, m_F = +1\rangle$ via Raman transitions. The resulting state due to the OAM transfer from the LG beams was a coherent superposition of a vortex and antivortex in the $|F = 1, m_F = +1\rangle$ magnetic sublevel. The Rochester group also applied this technique to the ground state $F = 2$ manifold of ^{87}Rb BEC to create skyrmions and half-skyrmions in a spin-2 system [47].

An interesting aspect of the Rochester experiments is that the laser beams were typically copropagating, which results in no net linear momentum transfer from the Raman coupling. (This is in contrast to the experiments described in Section 12.4, where counterpropagating beams were used that resulted in

$2\hbar k$ of linear momentum transfer.) The selectivity of the transition between the degenerate magnetic sublevels within a particular hyperfine manifold ($F = 1$ in [48] and $F = 2$ in [47]) was achieved using appropriately polarized laser beams. OAM transfer to atoms with copropagating laser beams can also be achieved by going between internal states with sufficiently different energies. This was first demonstrated in an ^{87}Rb BEC at Rochester [2] using a Raman coupling between ground state $F = 2$ magnetic sublevels in the presence of a 1.7 mT bias field. The Zeeman shift due to the bias field separated adjacent sublevels by 12 MHz. The second-order (quadratic) Zeeman shift provided additional frequency discrimination to enable coupling only between two sublevels. More recently, we have demonstrated OAM transfer using copropagating laser beams in a sodium BEC. In this case, the Raman coupling can be either between the ground state $F = 1$ and 2 hyperfine levels, which are separated by $\approx$1.77 GHz, or magnetic substates within the $F = 1$ hyperfine manifold with the degeneracy lifted by the Zeeman shift in an applied magnetic field.

12.6.3
Matter wave Amplification of a Vortex State

The OAM transfer to atoms (rotational Bragg) described earlier in this chapter can be considered as the result of a four-wave mixing process involving two optical waves and two matter waves. The two optical waves (one with OAM) and a matter wave (the initial BEC) interact with each other to produce the fourth wave, which is the BEC with angular momentum. This four-wave mixing process can also be understood microscopically. An atom in the BEC absorbs a photon from one optical wave. Simultaneously, the second optical wave stimulates the emission of a photon from that atom, which then by conservation of linear and angular momentum recoils into a different translational and rotational state, corresponding to the second matter wave. An alternative four-wave mixing process for producing rotational states of atoms is for a matter wave to stimulate the emission of atoms into the desired final state. Four-wave mixing to produce linear momentum states has been demonstrated using only matter waves [49] and also two optical and two matter waves [50]. In the latter case, an atom in one matter wave absorbs a photon from an optical wave and then emits the photon. Simultaneously, as the atom is emitting the photon, the second (seed) matter wave stimulates the recoil of the atom to have the same momentum as the second matter wave. By conservation of momentum, the emitted photon has to have momentum equal in magnitude but opposite in direction as the recoiling atom. The net result of such processes is the amplification of a matter wave in a particular linear momentum state [50]. The same process can be applied to amplify a matter wave in a rotational state.

Figure 12.8a is a schematic representation of the laser beam arrangement and sequence for amplifying a rotational matter wave. A first pulse (Figure 12.8a, upper image), which consists of counterpropagating LG and Gaussian beams transfers OAM from the LG beam to a fraction of atoms in the BEC using the two-photon

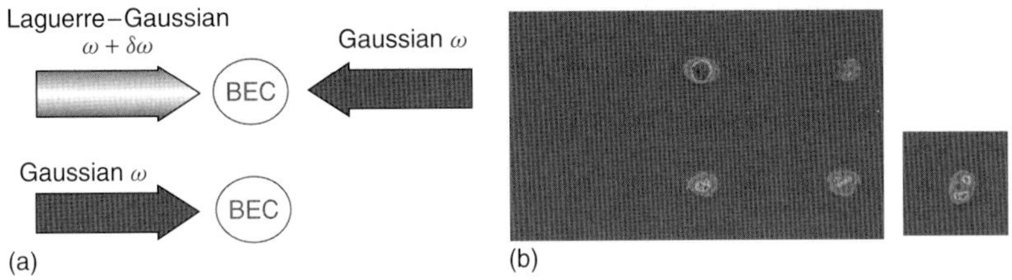

Figure 12.8 (a) Schematic representation of the laser pulse sequence used to amplify a vortex state generated in the initial BEC. (b) The upper image corresponds the generation of the "seed" vortex state from the initial BEC using only the LG and Gaussian beam pulse shown in the upper image of (a). The lower image shows the resulting amplification of the "seed" vortex from the two-pulse sequence of (a). The image on the far right is of the amplified vortex state taken along the direction of propagation. (Please find a color version of this figure on the color plates.)

Raman process described in Section 12.4.1. The upper image of Figure 12.8b shows images of the resulting clouds of atoms taken after several milliseconds of TOF following the application of the two-photon Raman pulse of light. The center cloud image is of the initial BEC with zero linear momentum. The cloud image on the right is of the atoms that have picked up the OAM of the LG beam (in this case, $\hbar$) along with the $2\hbar k$ of linear momentum from the two-photon Raman process.

In order to amplify the rotational state of atoms produced by the "seed" pulse, a second "pump" pulse is used (Figure 12.8a, lower image). This pulse, consisting of only a single Gaussian beam propagating in the same direction as the LG beam in the first pulse, is applied immediately after the first pulse, before the "seed" atoms have moved away from the location of the initial BEC. Atoms in the initial BEC simultaneously absorb photons from this beam and are stimulated to emit photons and recoil into the mode defined by the "seed" matter wave. The result of this two-pulse sequence can be seen in the lower images of Figure 12.8b. The cloud of atoms on the right, corresponding to the location of the "seed" matter wave has more atoms due to the amplification process. The cloud in the center (at the location of the initial BEC) shows a corresponding reduction in the number of atoms. The far right image in Figure 12.8b is of the amplified state, moving with $2\hbar k$ of linear momentum, taken along the direction of propagation. The characteristic hole of the vortex state is apparent.

In addition to amplifying the vortex state matter wave, the four-wave mixing process should produce an LG beam of light at frequency $\omega - \delta\omega$ propagating in the opposite direction with opposite OAM to the vortex state. Although we did not make an effort to detect this light, it has been detected in the experiments of Tabosa and collaborators [19, 20] with thermal atoms (which can also be interpreted as four-wave mixing experiments) as a signature of OAM transfer. Interestingly, there appears to be a faint cloud of atoms to the left of the initial BEC in the image of Figure 12.8b corresponding to the two-pulse sequence for matter wave amplification. This cloud may be the result of off-resonant Bragg diffraction of atoms from the combination of the LG beam (in this case with $-\hbar$ of OAM) at $\omega - \delta\omega$ produced from matter wave amplification and the Gaussian "pump" beam at ω.

12.7 Supercurrents

Quantized vortices are often considered a hallmark of superfluidity. Although vortices have been generated and observed in atomic quantum degenerate gases, the related phenomena of persistent or supercurrents have not been clearly observed in atomic BECs. Persistent currents in superconductors correspond to the flow of electrical current indefinitely. Similarly, persistent mass flow in either rotating bucket experiments or in a torus-shaped container has been observed with superfluid liquid helium. The generation of persistent currents in atomic BEC represents a striking manifestation of macroscopic quantum phenomena, opening the possibility for applications currently limited to superconductors, such as SQUIDS; and liquid helium [51], such as interferometers based on Josephson (weak-link) junctions.

12.7.1 Generation of a Supercurrent in a BEC

A vortex in a nonrotating, simply connected BEC is unstable [52]. In order to sustain a supercurrent in a BEC, we confine the condensate in a trap where the central region is excluded by a blue-detuned laser beam. A vortex (superfluid flow) would then be "pinned" by the blue detuned laser piercing its core, since the energy cost can be made quite high for the vortex to cross the region of high atomic density surrounding the core. A trap for Bose–Einstein condensation based on a magnetic quadrupole field with the zero field point "plugged" by a blue-detuned laser was first demonstrated at MIT [53], and more recently at Georgia Tech [54]. In our set-up [55], we focus the beam from a doubled YAG laser at 532 nm in the center of our TOP trap. The laser beam has a waist of approximately 5 μm and propagates along the weak (x) axis. Because the TOP trap is triaxial, the resulting "Mexican hat" potential produced is not rotationally symmetric. Hence any normal rotational flow generated around this potential would couple to other modes, since angular momentum is not conserved in our trap. A superfluid, however, is expected to flow indefinitely.

Figure 12.9a is an image of the BEC confined in the hybrid magnetic TOP and laser trap. The density of the BEC in the center goes to zero because this region is excluded from the blue-detuned laser beam. When the BEC is released from the hybrid trap, the zero density region in the middle fills in due to the mean-field expansion of the BEC [22]. This can be seen in Figure 12.9b, which is a TOF image of the released BEC after being held in the hybrid trap for 2 s. We can apply our Raman technique to the BEC in the hybrid trap to generate a state that is flowing

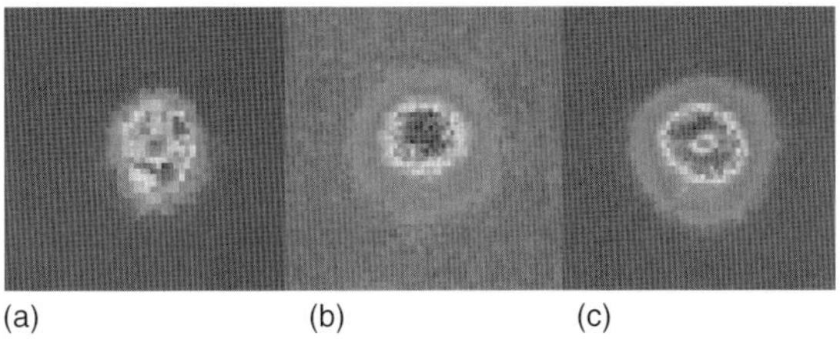

Figure 12.9 (a) Image of the condensate atoms in the hybrid magnetic TOP and blue-detuned laser trap. The absence of atoms in the central region is due to the laser. (b) TOF image of the condensate atoms released from the hybrid trap after being confined for 2 s. The interaction between the atoms causes the cloud to spread out and fill in the central region. (c) TOF image of the condensate atoms, which were in a vortex state about the blue-detuned laser beam. In this case the atoms in the vortex state were held in the hybrid trap for 2 s before being released. The presence of the core in the cloud of atoms after TOF indicates that the atoms were still in the vortex state when they were released from the hybrid trap. (Please find a color version of this figure on the color plates.)

around the blue-detuned plug. Figure 12.9c is a TOF image of the released BEC after being held in the hybrid trap for 2 s. In this case, the atoms were placed in a rotational state by transferring OAM via the stimulated Raman process with LG beams. In order to eliminate the linear momentum transferred in addition to the OAM, we use an initial Bragg diffraction pulse to put atoms in a nonzero linear momentum state from which they are subsequently transferred to a rotational state with zero linear momentum. (Alternatively we could use copropagating beams and drive transitions between different internal states in the atoms as proposed in [27–31].)

It is evident in Figure 12.9c that even after 2 s in the asymmetric trap, the atoms are still in a vortex state. That is, because of the superfluid nature of the condensate, the flow of atoms around the blue-detuned plug has persisted. We have observed persistent rotational flow of atoms around the blue-detuned laser plug for up to 13 s, a time comparable to the lifetime of the BEC at our vacuum pressures.

12.8 Conclusion

We have developed a new tool to generate arbitrary superpositions of atomic rotational states. This tool, together with the tools for controlling linear momentum and internal states, enables total control of an atom. We have shown that our technique can be used to generate a supercurrent in a quantum degenerate gas of atoms. Future applications of our technique range from generating supercurrents of atomic vapors in large diameter ring-shaped traps [56, 57] to superposition of macroscopic (Schrödinger cat) states [58]. Our experiments directly demonstrate that the OAM of a photon is transferred coherently to an atom in quantized units of $\hbar$. The exchange of OAM between light and atoms may also be useful in quantum information applications [31], for example, in quantum repeaters where the flying qubits are photons with OAM [34].

Acknowledgments

We gratefully thank the past and present members of the Laser Cooling and Trapping group sodium BEC effort, M.F. Andersen, G. Campbell, P. Cladé, A. Hansen, S. Muniz, V. Natarajan, E. Olimpio, A. Ramanathan, C. Ryu, A. Vaziri, K.C. Wright, and M. Zelan for the data presented in this chapter and for the work in progress on further developments with transfer of optical OAM to atoms. This work was partially supported by the ONR, NASA, and the NSF through the Physics Frontier Center at the Joint Quantum Institute.

References

1. Andersen, M.F., Ryu, C., Cladé, P., Natarajan, V., Vaziri, A., Helmerson, K., and Phillips, W.D. (2006) Quantized rotation of atoms from photons with orbital angular momentum. *Phys. Rev. Lett.*, **97**, 170406.
2. Wright, K.C., Leslie, L.S., and Bigelow, N.P. (2009) Optical control of the internal and external angular momentum of a Bose-Einstein condensate. *Phys. Rev. A*, **77**, 041601(R).
3. Babiker, M. and Andrews, D.L. (2008) Atomic and molecular manipulation using structured light, in *Structured Light and its Applications* (ed. D.L. Andrews), Elsevier Inc., Oxford, pp. 169–194.
4. Allen, L., Babiker, M., Lai, W.K., and Lembessis, V.E. (2003) Atom dynamics in multiple Laguerre-Gaussian beams, in *Optical Angular Momentum* (eds S. Allen, S.M. Barnett, and M.J. Padgett), IOP Publishing, Bristol, pp. 215–226.
5. van Enk, S.J. (2003) Selection rules and centre-of-mass motion of ultracold atoms, in *Optical Angular Momentum* (eds S. Allen, S.M. Barnett, and M.J. Padgett), IOP Publishing, Bristol, pp. 227–239.
6. Allen, L., Beijersbergen, M.W., Spreeuw, R.J.C., and Woerdman, J.P. (1992) Orbital angular momentum of light and the transformation of Laguerre-Gaussian laser modes. *Phys. Rev. A*, **45**, 8185–8189.
7. For a collection of papers, see: Allen, S., Barnett, S.M., and Padgett, M.J. (eds) (2003) *Optical Angular Momentum*, IOP Publishing, Bristol.
8. Siegman, A.E. (1986) *Lasers*, University Science Books, Mill Valley, CA.
9. Leach, J., Padgett, M.J., Barnett, S.M., Franke-Arnold, S., and Courtial, J. (2002) Measuring the orbital angular momentum of a single photon. *Phys. Rev. Lett.*, **88**, 257901.
10. Bazhenov, V.Y., Vasnetsov, M.V., and Soskin, M.S. (1990) Light beams with screw dislocations in their wavefronts. *Pis'ma Zh. Eksp. Teor. Fiz.* **52**, 1037–1039 [*JETP Lett.* **52**, 429–431 (1990)].
11. Heckenberg, N.R., McDuff, R., Smith, C.P., and White, A.G. (1992) Generation of optical phase singularities by computer-generated holograms. *Opt. Lett.*, **17**, 221–223.
12. Zeeman, P. (1897) The effect of magnetism on the nature of light emitted by a substance. *Nature*, **55**, 347.
13. Beth, R.A. (1936) Mechanical detection and measurement of the angular momentum of light. *Phys. Rev.*, **50**, 115–125.
14. Friese, M.E.J., Nieminen, T.A., Heckenberg, N.R., and Rubinsztein-Dunlop, H.A. (1998) Optical alignment and spinning of laser-trapped microscopic particles. *Nature*, **267**, 348.

15. Bishop, A.I., Nieminen, T.A., Heckenberg, N.R., and Rubinsztein-Dunlop, H.A. (2004) Optical microrheology using rotating laser-trapped particles. *Phys. Rev. Lett.*, **92**, 198104.
16. He, H., Friese, M.E.J., Heckenberg, N.R., and Rubinsztein-Dunlop, H. (1995) Direct observation of transfer of angular momentum to absorptive particles from a laser beam with a phase singularity. *Phys. Rev. Lett.*, **75**, 826–829.
17. Friese, M.E.J., Enger, J., Rubinsztein-Dunlop, H., and Heckenberg, N.R. (1996) Optical angular-momentum transfer to trapped absorbing particles. *Phys. Rev. A*, **54**, 1593–1596.
18. Simpson, N.B., Dholakia, K., Allen, L., and Padgett, M.J. (1997) Mechanical equivalence of spin and orbital angular momentum of light: an optical spanner. *Opt. Lett.*, **22**, 52–54.
19. Tabosa, J.W.R. and Petrov, D.V. (1999) Optical pumping of orbital angular momentum of light in cold cesium atoms. *Phys. Rev. Lett.*, **83**, 4967–4970.
20. Barreiro, S. and Tabosa, J.W.R. (2003) Generation of light carrying orbital angular momentum via induced coherence grating in cold atoms. *Phys. Rev. Lett.*, **90**, 133001.
21. Akamatsu, D. and Kozuma, M. (2003) Coherent transfer of orbital angular momentum from an atomic system to a light field. *Phys. Rev. A*, **67**, 023803.
22. Dalfovo, F., Giorgini, S., Pitaevskii, L.P., and Stringari, S. (1999) Theory of Bose-Einstein condensation in trapped gases. *Rev. Mod. Phys.*, **71**, 463.
23. Matthews, M.R., Anderson, B.P., Haljan, P.C., Hall, D.S., Weiman, C.E., and Cornell, E.A. (1999) Vortices in a Bose-Einstein condensate. *Phys. Rev. Lett.*, **83**, 2498–2501.
24. Williams, J.E. and Holland, M.J. (1999) Preparing topological states of a Bose-Einstein condensate. *Nature*, **401**, 568–572.
25. Madison, K.W., Chevy, F., Wohlleben, W., and Dalibard, J. (2000) Vortex formation in a stirred Bose-Einstein condensate. *Phys. Rev. Lett.*, **84**, 806–809.
26. Shin, Y., Saba, M., Vengalattore, M., Pasquini, T.A., Sanner, C., Leanhardt, E.A., Prentiss, M., Pritchard, D.E., and Ketterle, W. (2004) Dynamical instability of a doubly quantized vortex in a Bose-Einstein condensate. *Phys. Rev. Lett.*, **93**, 160406.
27. Marzlin, K.P., Zhang, W., and Wright, E.M. (1997) Vortex coupler for atomic Bose-Einstein condensates. *Phys. Rev. Lett.*, **79**, 4728–4731.
28. Bolda, E.L. and Walls, D.F. (1998) Creation of vortices in a Bose-Einstein condensate by a Raman technique. *Phys. Lett. A*, **245**, 32–35.
29. Dum, R., Cirac, J.I., Lewenstein, M., and Zoller, P. (1998) Creation of dark solitions and vortices in Bose-Einstein condensate. *Phys. Rev. Lett.*, **80**, 2972–2975.
30. Nandi, G., Walser, R., and Schleich, W.P. (2004) Vortex creation in a trapped Bose-Einstein condensate by stimulated Raman adiabatic passage. *Phys. Rev. A*, **69**, 063606.
31. Kapale, K.T. and Dowling, J.P. (2005) Vortex phase qubit: Generating arbitrary, counterrotating coherent superpositions in Bose-Einstein condensates via optical angular momentum beams. *Phys. Rev. Lett.*, **95**, 173601.
32. Kozuma, M., Deng, L., Hagley, E.W., Wen, J., Lutwak, R., Helmerson, K., Rolston, S.L., and Phillips, W.D. (1999) Coherent splitting of Bose-Einstein condensed atoms with optically induced Bragg diffraction. *Phys. Rev. Lett.*, **82**, 871–875.
33. Stamper-Kurn, D.M., Chikkatur, A.P., Görlitz, A., Inouye, S., Gupta, S., Pritchard, D.E., and Ketterle, W. (1999) Excitation of phonons in a Bose-Einstein condensate by light scattering. *Phys. Rev. Lett.*, **83**, 2876–2879.
34. Mair, A., Vaziri, A., Welhs, G., and Zeilinger, A. (2001) Entanglement of the orbital angular momentum states of photons. *Nature*, **412**, 313–316.
35. Simula, T.P., Nygaard, N., Hu, S.X., Collins, L.A., Schneider, B.I., and Mølmer, K. (2008) Angular momentum

exchange between coherent light and matter fields. *Phys. Rev. A*, **77**, 015401.

36. Unanyan, R., Fleicshhauer, M., Shore, B.W., and Bergmann, K. (1998) Robust creation and phase-sensitive probing of superposition states via stimulated Raman adiabatic passage (STIRAP) with degenerate dark states. *Opt. Commun.*, **155**, 144.
37. Weitz, M., Young, B.C., and Chu, S. (1994) Atomic interferometer based on adiabatic population transfer. *Phys. Rev. Lett.*, **73**, 2563.
38. Chevy, F., Madison, K.W., Bretin, V., and Dalibard, J. (2001) Interferometric detection of a single vortex in a dilute Bose-Einstein condensate. *Phys. Rev. A*, **64**, 031601R.
39. Inouye, S., Gupta, S., Rosenband, T., Chikkatur, A.P., Görlitz, A., Gustavson, T.L., Leanhardt, A.E., Pritchard, D.E., and Ketterle, W. (2001) Observation of vortex phase singularities in Bose-Einstein condensates. *Phys. Rev. Lett.*, **87**, 080402.
40. Anderson, B.P., Haljan, P.C., Regal, C.A., Feder, D.L., Collins, L.A., Clark, C.W., and Cornell, E.A. (2001) Watching dark solitons decay into vortex rings in a Bose-Einstein condensate. *Phys. Rev. Lett.*, **86**, 2926–2929.
41. Stenger, J., Inouye, S., Chikkatur, A.P., Stamper-Kurn, D.M., Pritchard, D.E., and Ketterle, W. (1999) Bragg spectroscopy of a Bose-Einstein condensate. *Phys. Rev. Lett.*, **82**, 4569.
42. Muniz, S.R., Naik, D.S., and Raman, C. (2006) Bragg spectroscopy of vortex lattices in Bose-Einstein condensates. *Phys. Rev. A*, **73**, 041605(R).
43. Leanhardt, A.E., Görlitz, A., Chikkatur, A.P., Kielpinski, D., Shin, Y., Pritchard, D.E., and Ketterle, W. (2002) Imprinting vortices in a Bose-Einstein condensate using topological phases. *Phys. Rev. Lett.*, **89**, 190403.
44. Ho, T.-L. (1998) Spinor Bose condensates in optical traps. *Phys. Rev. Lett.*, **81**, 742.
45. Leanhardt, A.E., Shin, Y., Kielpinski, D., Pritchard, D.E., and Ketterle, W. (2003) Coreless vortex formation in a spinor Bose-Einstein condensate. *Phys. Rev. Lett.*, **90**, 140403.
46. Sadler, L.E., Higbie, J.M., Leslie, S.R., Vengalattore, M., and Stamper-Kurn, D.M. (2006) Spontaneous symmetry breaking in a quenched ferromagnetic spinor Bose condensate. *Nature*, **443**, 312.
47. Wright, K.C., Leslie, L.S., Hansen, A., and Bigelow, N.P. (2009) Sculpting the vortex state of a spinor BEC. *Phys. Rev. Lett.*, **102**, 030405.
48. Leslie, L.S., Hansen, A., Wright, K.C., Deutsch, B.M., and Bigelow, N.P. (2009) Creation and detection of skyrmions in a Bose-Einstein condensate. *Phys. Rev. Lett.*, **103**, 250401.
49. Deng, L., Hagley, E.W., Wen, J., Trippenbach, M., Band, Y., Julienne, P.S., Simsarian, J.E., Helmerson, K., Rolston, S.L., and Phillips, W.D. (1999) Four-wave mixing with matter waves. *Nature*, **398**, 218.
50. Inouye, S., Pfau, T., Gupta, S., Chikkatur, A.P., Görlitz, A., Pritchard, D.E., and Ketterle, W. (1999) Phase-coherent amplification of atomic matter waves. *Nature*, **402**, 641.
51. Davis, J.C. and Packard, R.E. (2002) Superfluid He-3 Josephson weak links. *Rev. Mod. Phys.*, **74**, 741.
52. Fetter, A.L. and Svidzinsky, A.A. (2001) Vortices in a trapped dilute Bose-Einstein condensate. *J. Phys. Condens. Matter*, **13**, R135.
53. Davis, K.B., Mewes, M.O., Andrews, M.R., Van Druten, N.J., Durfee, D.S., Kurn, D.M., and Ketterle, W. (1995) Bose-Einstein condensation in a gas of sodium atoms. *Phys. Rev. Lett.*, **75**, 3969.
54. Naik, D.S. and Raman, C. (2005) Optically plugged quadrupole trap for Bose-Einstein condensates. *Phys. Rev. A*, **71**, 033617.
55. Ryu, C., Andersen, M.F., Cladé, P., Natarajan, V., Vaziri, A., Helmerson, K., and Phillips, W.D. (2007) Observation of persistent flow of a Bose-Einstein condensate in a toroidal trap. *Phys. Rev. Lett.*, **99**, 260401.
56. Gupta, S., Murch, K.W., Moore, K.L., Purdy, T.P., and Stamper-Kurn, D.M. (2005) Bose-Einstein condensation in a circular waveguide. *Phys. Rev. Lett.*, **95**, 143201.

57. Cozzini, M., Jackson, B., and Stringari, S. (2006) Vortex signatures in annular Bose-Einstein condensates. *Phys. Rev. A*, **73**, 013603.

58. Friedman, J.R., Patel, V., Chen, W., Tolpygo, S.K., and Lukens, J.E. (2000) Quantum superposition of distinct macroscopic states. *Nature*, **406**, 43.

Index

Twisted Photons: Applications of Light with Orbital Angular Momentum.
Edited by Juan P. Torres and Lluis Torner

ISBN: 978-3-527-40907-5